Praise for Virus X

"Frank Ryan is a compelling and engaging writer; *Virus X* is a wonderful read, a difficult book to put down. What elevates *Virus X* from other books is the unifying thesis he puts forward at the end. It will satisfy you even as it unsettles you."
— Abraham Verghese, Texas Tech University Health Center, Division of Infectious Diseases, author of *My Own Country: A Doctor's Story*

"Frank Ryan's thesis is provocative and disturbing. Ignore the notion of 'aggressive symbiosis' if you can. If, however, you agree with him, you will be compelled to face a frightening new prospect from the microbial world." — Laurie Garrett, author of *The Coming Plague*

"Extremely well written. . . . Frank Ryan has the page-turning and spine-chilling ability of a good novelist." — Matt Ridley, *Sunday Telegraph*

"Spellbinding. . . . *Virus X* is written like a detective thriller, emphasizing the human element in hunting down plagues." — Valerius Geist, *Nature*

"I was impressed and inspired by reading all the thoughts and data Frank Ryan has gathered from his direct experience, his encounters with scientists, and scientific magazines. . . . Each of the plagues — epidemic, pandemic, potential, real — demanded and will demand a specific global solution. This is what Frank Ryan's book is about: the ability to change perspective, from a retrovirus to the planet and back. Not many books — fiction or nonfiction — have this cosmic ability."
— Miroslav Holub, *Los Angeles Times*

"With gripping prose Frank Ryan shows that it *can* happen here. America has no magic immunity to contagious diseases that quickly kill both young and old. Dr. Ryan's imaginative theory of the role of killer viruses in nature's order may be the key to humanity's avoiding a catastrophic plague." — Michael J. Behe, author of *Darwin's Black Box*

virus X

ALSO BY FRANK RYAN

The Forgotten Plague

How the Battle Against Tuberculosis Was Won — and Lost

virus X

TRACKING THE NEW KILLER PLAGUES

FRANK RYAN, M.D.

BACK BAY BOOKS

LITTLE, BROWN AND COMPANY

BOSTON NEW YORK TORONTO LONDON

For my mother and my late father

Originally published in hardcover by Little, Brown and Company, 1997
First Back Bay paperback edition, 1998

Library of Congress Cataloging-in-Publication Data

Ryan, Frank.
Virus X : tracking the new killer plagues / Frank Ryan.
p. cm.
Includes index.
ISBN 0-316-76383-7 (hc) / 0-316-76306-3 (pb)
1. Communicable diseases — Epidemiology. 2. Communicable diseases — Popular works. I. Title.
RA643.R93 1997
614.4 — dc20 96-30495

10 9 8 7 6 5 4 3 2 1

MV-NY

Published simultaneously in Canada by Little, Brown & Company (Canada) Limited
Printed in the United States of America

I do not know what I may appear to the world;

but to myself I seem to have been only

like a boy playing on the seashore,

and diverting myself in now and then finding

a smoother pebble or a prettier shell than

ordinary, whilst the great ocean of truth lay

all undiscovered before me.

SIR ISAAC NEWTON

CONTENTS

Part II The Window on Life

PREFACE

The Question Why

Plagues frighten people. This is a very natural human reaction. Throughout history they have caused more death and terror than war or any other calamity. Plagues caused by new or "emerging" viruses are particularly frightening, since there is often no cure or preventive vaccine.

Threat of one kind or another is nothing new to life on earth. For half a century we have lived in the shadow of nuclear Armageddon. Today, with the thawing of the cold war, this fear, though omnipresent, seems to be subsiding, just as new worries are surfacing. Epidemic infections, once thought to be defeated, have surprised us all by making a worrying comeback. This has been paralleled by another very real worry, and one that may in some respects be linked with the threat of new epidemics: I am referring, of course, to the growing ecological disturbance of the world around us.

The most beautiful pictures I have seen are the views taken by the Apollo 8 astronauts as they performed the first lunar orbit. For the first time in our history, we saw the wonder of earth as it rose over the arid gray sand and stone of the moon. Life even from such a great distance was beautiful, in the ultramarine of our oceans, the mellowed golds and violets of our land and vegetation, the white swirl of wind-borne clouds. Yet today, with the possibility of global warming, the mass destruction of the rain forests, and the pollution and plundering of the oceans, that beautiful biosphere is under unrelenting threat.

Is there, as many people now wonder, a provable connection between the emergence of new plagues and the effects of human behavior, the vulnerable "monoculture" of twentieth-century population expansion, together with the associated effects of human exploitation and need on the delicate ecology of our world? This is as important a question as it is difficult to answer. Part of the problem lies in our emotional responses to such calamities. It has become fashionable to shout ecological rape. Yet without hard evidence of this link, the very people who might be in a position to change things simply will not listen.

A number of recent books have explored related territory, beginning some fifteen years or so ago with William H. McNeill's *Plagues and Peoples* and Richard M. Krause's *The Restless Tide,* complemented more recently by Laurie Garrett's encyclopedic *The Coming Plague,* Robin Marantz Henig's finely tuned *The Dancing Matrix,* and the wonderful *Emerging Viruses,* edited by Stephen S. Morse. Each has made an important contribution. So why produce another? There can be only one satisfactory answer: I believe that I have something new to say.

My purpose, though it necessitates some degree of overlap with each of these works, is to explore the scientific facts at a fundamental level and, through understanding, to arrive at the kind of synthesis that can only come from asking not only the question how but also the question why. The eminent science writer Stephen Jay Gould believes that difficulties in understanding result more from conceptual blocks than factual ignorance. To break through such difficulties, I have introduced some new and beguiling concepts.

These conclusions may prove a little controversial at first, but I hope the initial surprise will soon give way to new lines of biological research. To arrive at those conclusions I have traveled extensively and discussed these questions with many distinguished experts around the world.

Most of the experts I have spoken to believe that we are skating on ice. How thin that ice may be differs from one expert to another. New plagues, caused by "emerging" microbes, and new viruses in particular, increasingly threaten humanity: meanwhile there is a continuing, evolving menace from those that are already all too familiar.

But why do such plagues exist at all? At a time when we live under the threat of the lethal new epidemic of AIDS, where do new viruses, such as HIV, come from? Why do such viruses emerge? Is AIDS the worst that could happen or is our modern and highly technical world at risk of even more dangerous epidemics? The most important, and also the most frightening question of all: could a BSL-4 virus, such as Ebola, ever threaten us by aerosol spread, rather as a cold or influenza virus spreads from one person to another? This threat goes a long way beyond the responsibility of doctors. Indeed, this is a question for governments and even international health care organizations. For such a pandemic could be as catastrophic as a nuclear war. The question must be factually and responsibly aired. We need to know how real — or unreal — the possibility is. And if it is real, what precautions need be taken to help protect us from it.

ACKNOWLEDGMENTS

This book is a tribute to the assistance of colleagues who were prepared to interrupt busy schedules to accommodate my intrusive inquiries. In arriving at the conclusions outlined, I have of necessity learned vastly more from these colleagues than I could capture in these pages, and though not all of their courtesy and communication appears in the final narrative, it helped to formulate my ideas. Out of a very sincere respect and humility, I can only acknowledge the full depth of my debt to all of them.

These include Joshua Lederberg and Stephen S. Morse at the Rockefeller University, New York. In Gallup, Bruce Tempest and his colleagues at the IHS Hospital and Richard Malone, Deputy Medical Investigator at the McKinley County Courthouse. At the Navajo Nation in Window Rock, Duane Beyal, Press Officer to Peterson Zah, and Lydia Hubbard Pourier, Executive Director of the Navajo Division of Health. In Albuquerque, Frederick Koster, who so kindly arranged my itinerary, Howard Levy, Patricia McFeeley, Gregory Metz, Brian Hjelle, Steven Jennison, Tawnya Brown, her mother, La Donna, and her father, David, Lucille Long, Ralph T. Bryan, Tom Parmenter, and Terry and Nancy Yates, who so kindly introduced me not only to Terry's "Museum of Life" but also to the unforgettable Sevilletta; Steven Q. Simpson and Gustav Hallen at the VA Hospital, Jim and Roberta McLaughlin, Karen Hyde and Sandra Brantley; also the journalist Patricia S. Guthrie and Ben Muneta, who gave me a fascinating insight into Navajo life and philosophy. In Santa Fe, Gary L. Simpson, C. Mack Sewell, Ron Voorhoes and Edith Umland.

At the CDC, James M. Hughes, C. J. Peters, Thomas G. Ksiazek, Pierre Rollin, Jay Butler, Brian W. F. Mahy, Nancy Cox, Louisa E.

Chapman, James E. Childs, John W. Krebs, Stuart T. Nichol, Tom Folks, Rima F. Khabbaz, Patrick McConnon, Ethleen S. Lloyd, and Sherif Zaki. In San Francisco, Frederick A. Murphy and Don Francis. At Harvard, Paul Epstein, Richard Levins, and Max Essex. At Yale, Robert E. Shope. In Cambridge, Massachusetts, Thomas P. Monath. In Wyoming, Bill and Tine Close, and Bill's helpful assistant, Susie. Joel G. Breman in Bethesda, Maryland. Abraham Verghese in Texas. Christon J. Hurst in Ohio.

At the Pasteur Institute, Paris, Luc Montagnier, Charles Dauguet, Bernard Le Guenno, and Herlé Bourhy; also, the late Pierre Sureau, who kept such a lucid and vital diary of events in Yambuku. In Belgium, Jean François and Josiane Ruppol, Stefan R. Pattyn, Guido van der Groen, and Luc Eyckmans.

In Geneva, Jim LeDuc and Dr. T. Mertens. In Japan, the infinitely patient Mitsuaki Yoshida. In Germany, Werner Reisser. In South Africa, Margaretha Isaacson. In Pakistan, Joseph B. McCormick.

In the United Kingdom: David Simpson in Belfast; in London, Cedric Mims, Kevin De Cock, Cenydd Jones, and Dennis Mitchison. In Liverpool, David Smith and Barney Highton. Graham Lloyd at Porton Down and Patricia Webb in Scotland. At the John Innes Centre, Norwich, George Lomonossoff, Ian Bedford, and Mrs. Elizabeth Atchison. In Gloucestershire, Cliodna McNulty and Margaret Logan. Also, in Sheffield, a special thanks to Frank Needham, Christine Counsell, and, as always, to the ever-pressed Sue Nichol and her staff at the Northern General Hospital Library.

I would also like to take this opportunity to thank my teacher, Professor Mike McEntegart, who all those years ago encouraged my growing interest in microbiology. Mike, you little realized what you had set in motion!

Finally, my thanks to my wife, Barbara, and son, John, who read the typescript, my editors Catherine Crawford in New York and Michael Fishwick in London, and to my agent, Bill Hamilton, for his support from the very beginning of the project.

I take responsibility for those expressed concepts that might appear in any way controversial. Although I do not harbor any delusions of science as faith, in the sense of replacing that font of religion, I do regard its dedication to honesty and the intellectual search for the truth as the absolute guiding principle.

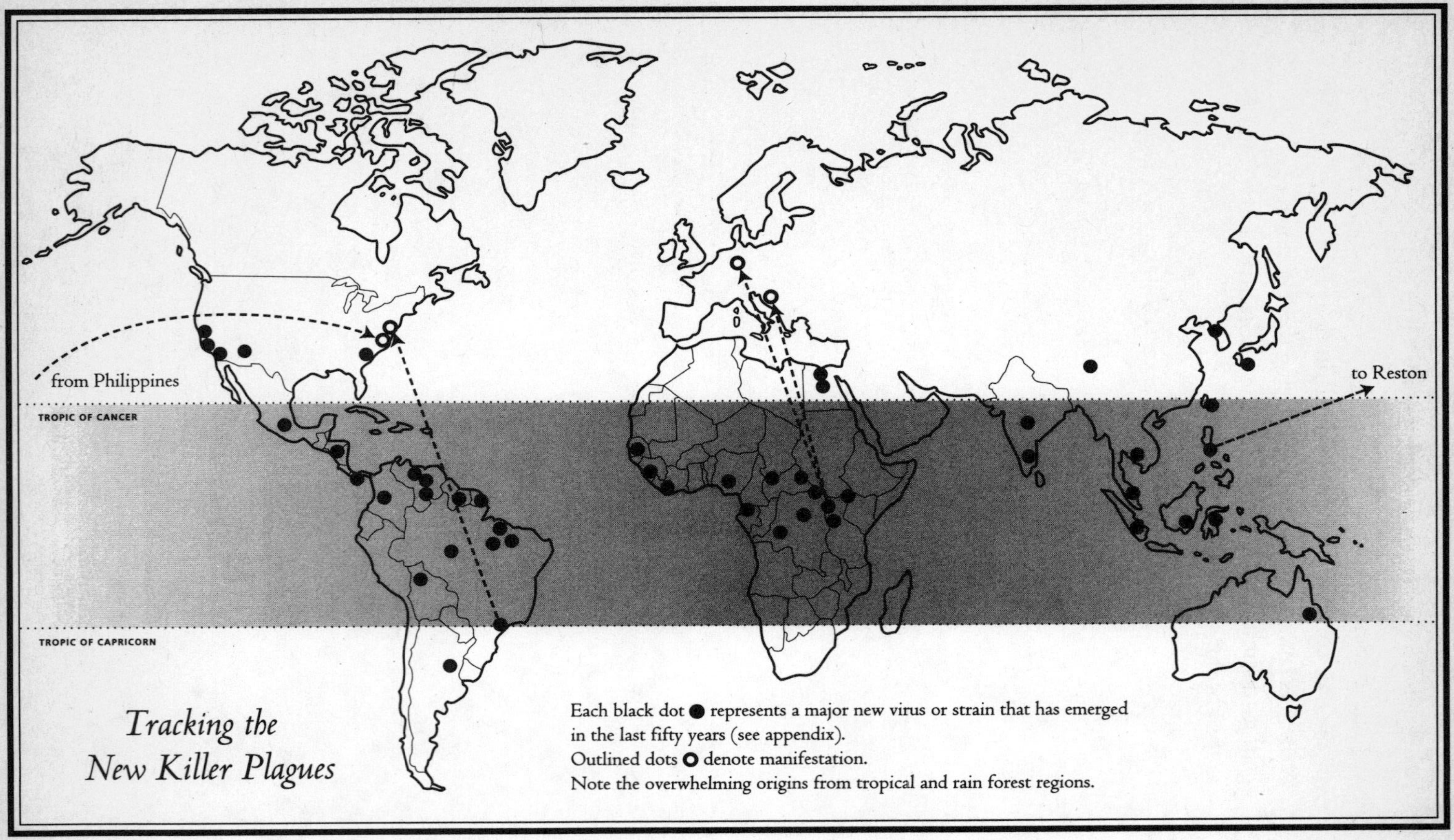

Tracking the
New Killer Plagues

Each black dot ● represents a major new virus or strain that has emerged in the last fifty years (see appendix).
Outlined dots ○ denote manifestation.
Note the overwhelming origins from tropical and rain forest regions.

PART ONE

The Emerging Menace

ONE

The Age of Delusion

Since the days of the cave man,
the earth has never been a Garden of Eden,
but a Valley of Decision
where resilience is essential to survival. . . .
To grow in the midst of dangers is the
fate of the human race.

RENÉ DUBOS
Mirage of Health

1

It is July 30, 1994, and outside a relief tent on the crown of a hill, a child's eyes glaze over and in a terrible moment, all the more disturbing for its seeming banality, he dies. Another child wanders among corpses wailing. His father, who is somewhere nearby, explains in a quiet voice to a reporter with outheld microphone that the child's mother died the day before. The father has another dead baby strapped to his back. The news photographer's camera pans a scene of pitiable desolation, a field of bodies, men and women, old and young, a scattering of colorful African waxes, exposed heads, naked limbs still clutching their pathetic bundle of transportable possessions, all recently dead, the crop of another morning at a Zairian camp for Rwandan refugees called Munigi.

On the evening news, day after day, the affluent world watched as exhausted men threw bodies off the backs of trucks into mass graves.

It was a scene, ominous in its portent, that could have been taken from the descriptions of the Black Death in the Middle Ages.

To most people these harrowing events must have evoked a sense of shock, of disbelief. Surely such ancient horrors, the great plagues of history — cholera, tuberculosis, typhoid, bubonic plague — were no more than nightmarish anachronisms. Those old-fashioned men of death might have been a romantic invention, etched in a gothic woodcut by Albrecht Dürer: if remotely recalled, it was with a breath of relief that assumed their place on the dusty shelves of history. Suddenly, in these vivid scenes, carried into people's living rooms by the inquisitive eye of television reporting, such perceptions were disturbingly shaken. Yet it was not so very long ago that the common perception was very different.

A few years ago, I visited the parsonage in the village of Haworth, in north Yorkshire, the home of Charlotte and Emily Brontë. I came away with two powerful impressions. The first was the small size of their gloves, which were so tiny they would barely fit the hands of a child today. The second impression was of a fearful mortality. Indeed, the first writer in the family was neither of the women but the unfairly maligned father, Patrick. When Charlotte and Emily were young children, their father was fighting a personal crusade against a mortality that carried away most of his parishioners before they were thirty. Open sewers coursed down the village streets. Half the children died in infancy.

By the turn of the nineteenth century, the mean life expectancy in Britain, as in the United States, Japan, Germany, and all developed countries, was still only forty-five years. Most of the human race still died from infections.

Sometimes this happened dramatically, as in the bubonic plague, which, in its most formidable manifestation, known as the Black Death, fell upon Europe from its origins in Southeast Asia, killing a third, sometimes even half the population of entire countries. Plagues seemed to attack out of the blue, or retreat in a similar aura of mystery, creating reigns of terror. Why did plagues happen? Where did they come from? Throughout most of our history, people did not know. But there were clues that might be recognized from the behavior of the plagues. Even during the Middle Ages, people looked hard at their deadly patterns, searching desperately for signs, extrapolating reasons. At the height of the Black Death, people recognized the infectious nature of "buboes" that disfigured the bodies of its victims. They also recognized that the hacking, bloodstained cough of plague victims could transmit the deadly contagion to others. The clothes and the bedcovers of the dead were burned and their bodies buried away from society in grotesque charnel pits quickened with lime.

Even the twentieth century has not been immune to Bunyan's so-called men of death. Fulminating epidemics of influenza and typhus scourged the weary population of Europe in the wake of the First World War, inflicting more casualties than all of the carnage. Though there are hieroglyphs from ancient Egypt depicting the typical wasting of limbs that results from poliomyelitis, the epidemic form of this paralyzing illness only emerged in the first half of this century. Even today, in the tragic dystopias of the developing world, the terror so vividly described by Daniel Defoe in his *Journal of the Plague Year* is apt to return, little altered from when he wrote it, almost three centuries ago.

Dramatically, between the 1940s and the 1960s, the medical and biological sciences discovered many new and effective answers. Thanks to the genius of Paul Ehrlich, the drug treatments of infection had started with salvarsan, used from the early 1900s to treat syphilis. Penicillin, discovered by Alexander Fleming in the late 1920s, was first manufactured for general use in the 1940s. Prontosil rubrum, the sulphonamide prototype discovered in Germany by Domagk, Klarer, and Mietzsch, had by then been generally available since 1935. Stimulated, ironically, by the military objectives of the Second World War, science had entered a new era of enlightenment that has been termed "continuous revolution." In 1943 the two antituberculosis drugs, streptomycin and para-aminosalicylic acid (PAS), were simultaneously discovered by, respectively, Waksman and Schatz in America and Lehmann and Rosdahl in Sweden.[1] These were quickly followed by a proliferation of further antibacterial discoveries, including Hubert Lechevalier's neomycin, the first antifungals, the cephalosporins, and the macrolides. Every known bacterial infection became treatable.

On all fronts, the biological and medical sciences, and the applications that derived from them, had entered a stage of acceleration more dramatic than had ever been seen before. And even those most refractory infections of all, those caused by viruses — formerly dismissed as untreatable because viruses disappeared into the inner labyrinths of the living cells, merging into the very genomes — were becoming amenable to early treatments: idoxyuridine, designed as an anti-DNA metabolite, proved useful for herpes simplex infections, as did virugon for influenza; methisazone, derived from Domagk's thiosemicarbazones, was showing potential in the treatment of early smallpox. A new front line of antiviral drugs — the interferons, based on the body's own antiviral mechanisms — seemed to harbor exceptional promise.

The global spirit of optimism that followed was aptly summed up in a speech made by U.S. President Richard Nixon before Congress in 1971. "I would ask for an appropriation of $100 million to launch an intensive campaign to find the cure for cancer. . . . The time has come in America when the same kind of concentrated effort that split the atom and took man to the moon should be turned towards conquering this dread disease."

Gone was the fear of infection. In declaring war on cancer, President Nixon was no more than iterating the zeitgeist of popular medical and lay opinion. On December 4, 1967, Dr. William H. Stewart, the U.S. Surgeon General, informed a meeting of state and territorial health officials that infectious diseases were now conquered. Under the umbrella of "A Mandate for State Action," he extolled the findings of the Centers for Disease Control a year earlier. Epidemic diseases such as smallpox, bubonic plague, and malaria were things of the past. Typhoid, polio, and diphtheria were heading in the same direction. While syphilis, gonorrhea, and tuberculosis were not quite so readily defeated, it was only a matter of time before every plague that had ever struck fear into the heart of decent Americans would be a distant memory. Cervical cancer was listed as one of the diseases that could be brought under effective control. Even cancer, it seemed, would be quickly solved. All that was needed was the money to lubricate the intellects of scientists.

In such a climate of optimism, Stewart urged the experts to focus the lion's share of health resources onto the "new dimensions" of ill health, the problems that would face the space age: chronic diseases.[2] Confidence was brimming over. It was as if in a single intoxicating summer, humanity had convinced itself that winter would never come.

Since 1960, when global war was declared against the disease, tuberculosis, which had killed roughly three quarters of a billion people in the century and a half up to 1960, was being beaten back on all fronts. Year by year, as a result of a massive and coordinated campaign, in America, in England, Europe in general, in Japan, and even Russia, its prevalence was declining. But it was never an easy victory. The battle still harnessed great expenditures of ingenuity, money, and manpower. Nevertheless, it seemed just a matter of time before those same methods that had worked for affluent countries would be put into effect to bring down the horrific death rate from tuberculosis throughout the developing world, where a staggering 3 million people were still dying from the disease each year.

Malaria, second in line for the distinction of greatest killer among the infections, was responding dramatically to antimalarial drugs, drug prophylaxis, and efforts at mosquito control. Plans were formulated, the banners aloft, for two of the greatest plagues that had afflicted humanity to be wiped off the face of the earth. Similar hopes existed for poliomyelitis, the plague that paralyzed children. Thanks to Albert B. Sabin's brave initiative with live virus coating sugar lumps, paralytic poliomyelitis was largely eradicated in the developed world. Triple vaccine was reducing the incidence of tetanus, whooping cough, and diphtheria to such low levels that young doctors might not see a case throughout the years of their training. Measles vaccine would soon be added.

These planners did not perceive themselves as overly optimistic. On the contrary, they were imbued with certainty, based on a sound understanding of the microbes they were battling and on the most up-to-date results of vaccine development, epidemiology, and antibacterial drugs. So, in a broad front of what amounted to a clandestine third world war, the men of death of the medieval imagination were being faced on the battlefield, beaten back, in some instances would soon be dead and buried.

When it came to plague viruses, none had ever caused such a fearful global mortality as smallpox. Believed to have originated in India in ancient times before first ravaging the Roman world as early as A.D. 165, since then it had scourged humanity in what amounted to a permanent pandemic, causing incalculable loss of life and misery through morbidity and disfigurement.[3] In its more virulent form caused by the virus *Variola Major*, it still caused up to 50 percent mortality in its victims. In 1958, when Russian doctors pressed for a concerted world campaign against it through the World Health Organization, 2 million people still died from its effects each year. The resultant global campaign against smallpox began in 1967, involved the vaccination of as many as 250 million people yearly, and was led by the tenacious American physician Donald A. Henderson. After ten years of backbreaking struggle, pressing to isolate cases and vaccinate populations through famine and war zones, success was finally announced in 1977. It was a remarkable achievement to match that of the defeat of tuberculosis, the realization of a hitherto impossible human dream. But it fueled a growing hubris among scientists. If we could eradicate smallpox, we could eradicate all of the viral plagues as well.

A certain vainglorious spirit of celebration was surely understandable. There was universal hope for the advance of human civilization. Other more subtle wars had been fought in real, philosophical, and sociological fronts against a historical legacy of tyranny, and those wars had, in part or in whole, been won. Outmoded despotic arrogance had been supplanted by a democratic spirit of emancipation, which, though led by Western enlightenment, was rapidly disseminating throughout the world. Civil rights and sexual liberation went hand in hand with the new expectation of health. In developed countries families condensed to the nuclear 2.2 average number of children: liberated from the fear of epidemics, women could make full use of the newly available systems of birth control to facilitate their own emancipation. The same generation that had created hydrogen bombs, jet air transport, and rockets to take men to the moon had every reason for optimism. While that spirit and courage might, in retrospect, appear naive, it was also laudable.

To them their aims were not unreasonable and they fought very hard to achieve them. It is tragic that their lofty ideals were not altogether realized. Perhaps it reflected, in part, a regrettable separation of clinicians from basic scientists.

In fact those people whose living depended upon a study of microbes, of their potential and durability, were never deluded. A prescient few, such as René Dubos, warned us openly that the optimism was unjustified.[4] But on the whole people were not inclined to listen. Most doctors, never mind members of the public, were infected with the prevailing overconfidence, hardly perceiving the growing threat of social changes inherent to the "global village." They seemed unable to grasp the new potential afforded to a very ancient peril arising from world travel. Diseases that once took months to cross the Atlantic with Columbus or the Pilgrim Fathers could now circumnavigate the globe in a single day.

Today, as one after another of the dismissed plagues returns to haunt us, as new plagues every bit as deadly as anything seen in previous history threaten our species, it is obvious that the postwar years were an age of delusion. It was comforting, a very understandable delusion, but a delusion nevertheless.

2

At first glance, there is something very curious about this ability to delude ourselves. Until the discovery of the antibiotics most of the human race, even in developed countries, died from infection, often the so-called "common" infections. Common these might have been, but that did not imply insignificant. Deaths from these infections were often very unpleasant. How was it possible, therefore, that we had forgotten this? It is every bit as strange as if humanity had within a generation lost its fear of war.

Perhaps, in the words of Matthew Arnold, we forget because we must and not because we will. There is an understandable human need to forget misery and embrace happiness. The average member of the public has little real conception of microbes, their ubiquity, and the vital role they play in the cycles of life on earth. Television advertisements in the 1960s had tidy housewives banishing germs from the home with a flourish of a spray or bottle of bleach, as if humanity were at last embarked upon a world of inner cleanliness, a world in which bacteria and other microbes were banished into the sewers of history. In science also, a dramatic metamorphosis of attitude, born out of understanding, was also taking shape. The discovery of DNA had triggered the change. Virology, at its intellectual cusp, would move closer to the developing world of molecular biology and its scope would expand far beyond the domain of doctors.

Gone was any residual simplistic notion of a static chemistry of life — in its place, a mysterious and much more complex tapestry, a hybrid picture not static at all but constantly changing, with the most subtle implications. The real landscape inhabited by viruses was perceived to be the landscape of the genome. And the genomes, or books of life, were far from the stable monoliths once envisaged.

Viruses, it now seemed, were much more than parasites. As perceived by Joshua Lederberg, "Our view of a virus as a parasite is complicated by that of a virus as a genetic element, a two way channel."[5] It was the business of viruses to weave in and out of the genomes of every form of life on earth. As a result, terrestrial life had become a dense web of genetic interactions. The relatively simplistic notion of host and parasite was eclipsed by this more complex and far more interesting world. Yet it was nevertheless a world in which viruses were as dangerous as ever, perhaps all the more dangerous for their being less predictable.

But one should not allow oneself to be seduced by the reductive rationalizations of molecular biology. To understand plague viruses, it is equally important to remember the other landscape: this wonderful diversity of life on earth, in which the human expansion and colonization must be just as intimately linked with viral behavior.

While people might, in their naïveté, dream of a perfect world in which there were no microbes, in reality, a world so sanitized as this is more than just an impossible dream, it is a nightmare. Early in this century, the Russian scientist Vladimir Vernadsky invented the "biosphere" as a holistic concept that embraces all of life on earth, together with its interactive atmosphere, soil, and oceans. Microbial life is an integral part of this living whole. A world devoid of microbes would mean death to all life on earth. The cycles of life would stop, plants would cease to grow, and with the death of plants, oxygen would slowly disappear from our atmosphere and the animals that depended directly or ultimately on these great symbiotic cycles of nature would become extinct.

Several years ago, a scientist called George Poinar teased out germs from a 30 million-year-old bee preserved in amber. The world was astonished at this miracle of survival, a miracle that was all the more profound for the fact the germs were almost identical to those found in bees today.[6] In September 1994, scientists working on ocean sediments off the coast of Japan reported the finding of hitherto unknown species of bacteria living in sediments at depths of 520 meters under the ocean bed, in a world devoid of oxygen and at pressures that would seem to preclude life. The bacteria were similar to primitive types already known to inhabit mud, but the extent of the seas and the depths at which the germs were found meant that they significantly increased the diversity of life on earth.[7] In another extreme example of fecundity, others have found bacteria two miles beneath the earth's surface, eking out a precarious living in the Stygian dark between the very grains of the rocks.[8]

Three and a half billion years ago, microbes were the earliest evolution of life on earth. For aeons they colonized and ruled every habitat, from the sea to the sulphur-laden air, from the beds of the oceans to the primeval mud of shore and lake, where the earliest forms of life began. Unlike the dinosaurs, those smaller and simpler elements never became extinct, always prepared to modify their genetic book of life, through natural selection, to hold their occupation of every ecosphere on earth.

We live in a world that teems with microbial life. It swarms in unimaginable numbers and variety in the soil, in manure heaps, in hedgerows and the floors of forests, in the waters of river, lake, and sea. Perhaps even more important, it swarms over and through every known form of life. Two thirds of the bulk of human feces is made up of enteric microbes, living within us in an age-old symbiosis. These are not harmful. Quite the contrary, they appear to play a benign role in human health.

Infection, in the sense of an interaction between our lives and a small minority of microscopic life forms, is a fact of life on earth as solid as the mountains and as predictable as the seasons and tides. Not only is nature not benign, nature cannot be benign. You cannot watch a television program on life in the wild without becoming rapidly aware of the tenuous nature of existence for very many species on earth. The governing relationship is often that of predator and prey. A big cat pounces on a gazelle or a wildebeest, sinking its canines into the vulnerable neck. Within seconds it seems the prey's eyes glaze and, as if anesthetized, it sinks from life, choked in a manner that is a daily routine in the wild. We live in a predatorial world. No species is exempt from this fear of predation, just as no species, including our own, is exempt from the threat of extinction.

The microbes that kill people, particularly those that kill huge numbers in sweeping epidemics, follow, in many ways, the same universal law of predator and prey. It is part of this complex gestalt that the balance is shaped by the behavior of the prey. If the prey moves — if it changes, if its numbers increase or decrease, if its ecology alters — the predator must move with it. To understand what is now threatening us, we must look to ourselves. In what way have we, the prey, changed?

Throughout the developing world, assisted by the beneficent efforts of entities such as the World Health Organization and the United Nations, and the social revolutions that have resulted from advances in agriculture, speed of international travel, and remarkable achievements in engineering, such as the Aswân Dam, the result has been a massive burgeoning of population. At present close to 6 billion people inhabit a world where there were no more than 1.5 billion a century ago. The resultant human needs, the exploitation that necessarily derives from them, have put the world's ecology under threat. Plagues spread a good deal more easily in overcrowded surroundings. And the greatest expansions of population are in the poorest countries, where the medical infrastructure and money spent on health is at its lowest.

There are signs that can only appear ominous. The present generation has seen a proliferation of new plagues and a frightening recrudescence of old ones. The burgeoning human population, the ease of spread through international travel, the changing patterns of human behavior, including the sexual liberation that began in the 1960s, have all put great strains on our ability to cope with these new and renewed burdens. In the face of such massive new demands, how justified is the optimism in science's ability to cope with the threat of epidemics? A question, unpalatable, perhaps even frightening, as it might seem, must nevertheless be asked.

Could the genuine miracles of science, the very real benefits of the social revolution, the mass production of food, the miracle drugs such as antibiotics and vaccination, fail us?

TWO

Unexplained Deaths

The harmony within the universe is such that everything lives within the balance of the whole. Man is not at the top of the pyramid but is a part of it. When a Western doctor says an illness is caused by some infectious agent, the Navajo will say there is a disturbance to the hozhon.

BEN MUNETA
conversation with the author

1

Shy and inward-looking, with intense dark eyes, Michael was more comfortable speaking Navajo than English.[1] His family home was the small town of Torreon on the 25,000 square miles of Navajo reservation, the Navajo Nation, that straddles the Four Corners states of New Mexico, Arizona, Utah, and Colorado. Like many young Navajo he left the reservation and completed high school in 1992 as a boarder at the Indian school in Santa Fe. But after graduation, he returned home to his friends and people and the old ways.

Friends would describe him as pleasant and sensible. They would also describe him as athletic, a very good runner. In photographs, Michael is usually shown running, a handsome young man, his face filled with concentration and his thick black hair swept back by the wind. He liked to test himself out in the dry sunny hills of home rather than on the artificial surface of field or track. Just twelve months after he had finished school, he was three times runner-up in the state marathon championship.

Michael had first met Rosina when they attended the same junior high school. She was two years older, always a class or two ahead. Rosina was pretty. She was a little more outgoing than Michael, a happy-go-lucky sort of person, full of energy and given to joking in Navajo and English. Later they met again when she was manager of the high school track club. Michael and Rosina had fallen in love and planned to marry. In the tradition of the Navajo, Michael had gone to live with Rosina's family in their compound in Littlewater. Relatives and friends would describe theirs as a very happy relationship. In January 1993, when Michael was nineteen years old and Rosina twenty-one, she had their first baby, delighting Michael with a healthy son.

In time, it would be this aspect that would shake the community around Littlewater: they seemed so young and happy, only recently blessed with a baby, a couple who had a good and fruitful life ahead of them.

New Mexico, where Michael and Rosina lived, is the fifth largest state in America. It is sparsely populated, with a statewide total of just 1.6 million. There is one large city, Albuquerque, accommodating a little more than half a million people. This population is rapidly expanding, drawing immigrants from Mexico, young Navajo from the giant reservation, and Anglos from California fleeing the threat of earthquakes and the after-effects of the Los Angeles riots. To get to Gallup, the "Indian Capital of America," you take Interstate 40 westward from Albuquerque, a journey of 135 miles. Travel on a little farther and you arrive in Arizona at the eastern cliff face of the Grand Canyon.

Michael and Rosina traveled this road every time they left home for Santa Fe. The landscape is hauntingly beautiful. It varies from arid flatland to mountains and seasonal river valleys, or arroyos, that are parchedly beautiful and deceptively rich in a variety of wildlife. Buttes of rock, of every gorgeous shade of red and gold, soar out of the desert, worn and fissured in the furnace of the sun. Towering sandstone cliffs change color with the dawn and sunset, and wooded canyons reveal blue-green foliage bathed in heat haze. To the Navajo, whose 25,000 square miles of reservations encompass Gallup, this land is sacred, their ancestral homeland. For them it is the fifth world, the world of *hozhon*, or harmony, into which First Man and First Woman brought soil from the disharmonious third world that would be forged into the four sacred mountains. You pass directly by one of these landmarks at Grants on the interstate — Mount Taylor, to the Navajo, *Dzil Dotlizi*, or the Turquoise Mountain. Here it is easy to imagine that you are close to that creative enchantment.

On May 4, 1993, Rosina felt ill. Her symptoms resembled flu, but there was more stomach upset than usual. A few days after her illness began, she deteriorated and was admitted as an emergency to the Indian Health Service hospital in Crown Point. Michael could not believe it when she died there, only a day or so after her admission. Even the doctors were mystified.

He was so upset he felt sick himself, with a vaguely flulike illness, starting with headache, then progressing to a fever with aching in his muscles. He lost his appetite. His bowels turned loose. By May 12, he felt so much worse he drove to the clinic in Crown Point, where the examining doctor diagnosed flu combined with grief over the loss of his fiancée. When the doctor offered to admit him to the hospital for observation, Michael refused. With no regular job, he had little money, and he needed to visit relatives and friends to raise the cost of Rosina's funeral.

On May 14, at about nine in the morning and still feeling ill, he climbed into the backseat of Rosina's father's car, to be taken into Gallup for her funeral. Starting out from their home in Littlewater, near to Crown Point, their journey took them southward along the broad and dust-blown Route 371 that would join up with the interstate at Thoreau. Along the way, Michael became progressively more distressed. What had appeared a minor if distracting illness was rapidly worsening. The muscle aches evolved to a constant torment, he felt lancinating pains in his chest, and he was becoming more and more breathless.

The deterioration was frighteningly rapid. Within ten or fifteen minutes, his hands and lips were blue and he was gasping for breath. He could not sit still, he would straighten up, then hunch over, try to stretch out through the window of the car. It made no difference. Nothing seemed to ease him. Rosina's father, alarmed at his condition, pulled over the car in the town of Thoreau, veering onto the dirt parking lot at the side of the road.

In the lazy heat of mid-morning, the B.J. Kountry Store perfectly represented Thoreau: its broad, low gable, friendly and unpretentious, with a simple mural of jousting stags, was framed by the coral-red cliffs of the Mesa de los Lobos. The matching sun-bleached boards of its neighbor carried the Navajo title for Thoreau, *Dio-Ay Azhi,* which means prairie dog town. As the helpful shopkeeper dialed the emergency ambulance on the old-fashioned black telephone, Michael was rapidly deteriorating. Confused and struggling to breathe, he staggered and paced restlessly around the parking lot before collapsing on the sun-baked dirt outside the store.

In the ambulance screaming westward, Michael continued to deteriorate. Throughout the thirty-mile journey, the crew performed cardiopulmonary resuscitation.

Half an hour later, with emergency lights still flashing, the ambulance reversed into the admission bay of the Indian Health Service hospital in Gallup. The ambulance crew had already called the hospital switchboard on their radio, and the emergency staff were waiting by the entrance as the rear doors of the vehicle burst open and Michael was rushed through the double swing doors, then left into the emergency room, where he was transferred by metal gurney into the resuscitation area. Curtains were hurriedly drawn round him as all the while the cardiopulmonary resuscitation measures continued. The emergency beeper sounded on the breast pocket of the medical internist on call.

There are six primary care internists working at the hospital, each with his or her own specialty. That morning the senior internist, Dr. Bruce Tempest, was conducting his round when the intern on duty was called away. As soon as the round finished, Dr. Tempest hurried down to see what was going on.

Resuscitation measures were tragically unsuccessful. Michael was pronounced dead at 11:53 A.M.

2

Gallup is a small city of twenty thousand people and the IHS hospital caters exclusively to Native Americans. It is small, as city hospitals go, but well equipped and friendly, with its brightly painted corridors decorated with Native American patterns. Most of the staff of the hospital are recruited locally, so there is a powerful sense of community. Losing somebody under such circumstances is distressing, even for the most experienced of doctors and nurses. It was into this emotionally charged atmosphere that Bruce Tempest arrived after the ward round.

After some discussion with the emergency room doctor and the admission nurses, who were understandably shaken and upset, Tempest walked over to an illuminated wall viewer to inspect Michael's chest X rays. What he saw startled him. Instead of the feathery translucency of healthy lungs, all he could see was a solid opaque white. The delicate air sacs, or alveoli, of Michael's lungs had been flooded, leaving no room at all for air to get in. Michael had literally drowned in his own body secretions.

With a pang of apprehension, Tempest remembered a similar case at the hospital only a month earlier.

At that time a thirty-year-old woman, also a Navajo, had been rushed to the emergency room unable to breathe. She too had suffered from an illness that vaguely resembled flu in the few days before her admission. She had also died soon after admission, with a "whiteout" on her chest X rays. At autopsy, her lungs had been so laden with fluid they weighed more than twice the normal. The pathologist had been so baffled by the strange findings she had been unable to issue a death certificate. Like Michael, the young woman had drowned in her own secretions. At the time, her death had been attributed to "adult respiratory distress syndrome," familiar to doctors under the acronym ARDS. This is far from a precise diagnosis. It is a blanket designation that covers many possibilities, including heart failure, overwhelming infection in the lungs, shock from inhaling heat or fumes or from massive lung injury, for example as a result of being close to an explosion. No such explanation had been found.

In New Mexico all unexplained deaths are reported to the Office of the Medical Investigator, which is situated in Albuquerque. The investigator works with a group of forensic pathologists, who perform autopsies on accidental or criminal deaths and who also investigate non-criminal deaths where a doctor cannot give an adequate explanation. In the more outlying rural areas, the investigator employs local policemen or detectives who are deputized to investigate their own territories. The deputy examiner for McKinley County is Richard Malone.

Malone is a young-looking man with neatly trimmed dark hair, a moustache, and calm brown eyes. He has a gentle voice, an expressive face, a tendency to wrinkle his forehead and narrow his eyes while pausing a moment for thought before replying to questions. His office is in the McKinley Courthouse, a striking brown-pink pueblo on Hill Street, which stands like an island in a sea of dusty cars and pickup trucks. At 12:35 P.M. that Friday morning, he received a call from a physician at the hospital asking if he could take jurisdiction over the young man's death. Needing an autopsy, the doctor anticipated legal technicalities in getting permission. The doctor told him that he had absolutely no idea why this young man had died.

Malone's office is about two miles from the Indian Health Service hospital. When he arrived in the emergency room, he listened to the disturbing story of the young man's presentation and the emergency room doctor showed him the dramatic chest X rays with their whited-out lungs. Then Malone went to look at Michael's body.

A graduate in criminology in the campuses of New Mexico and Indiana, Richard Malone was an experienced officer who had worked as a police investigator before taking his current post with the medical investigator's office. In fourteen years, his work had never lost its fascination for him. It did not take him long to inspect Michael's body. Sometimes you get clues from the skin color, the color of the whites of the eyes, obvious bruises, wounds, lacerations. But there was nothing to be deduced from this superficial examination. There was no jaundice of the skin and eyes, as might be seen in a death from alcoholism, no extreme bruising, none of the puncture tattoos about the veins that expose the secret horrors of intravenous drug abuse, nothing at all, apart from the common findings that result from the emergency resuscitation measures themselves. Malone could only shake his head.

He was no wiser than the doctor as to the cause of death, and that presented him with a sensitive and rather difficult problem. Michael's family members were now gathered in a hallway outside the emergency room. He needed to formulate his thoughts for a few moments before going out to talk to them.

Michael's parents were so stricken by the suddenness of events, they could not believe that their son was really dead. How could this happen, when Michael was a good enough track athlete to have represented his high school as a marathon runner. Malone sat down with them and patiently inquired about their circumstances, where Michael lived, the necessary probing of occupation, hobbies, social life, and habits. "Could you just try to carefully retrace the last few days of Michael's illness prior to the collapse?" They could add little to what he had heard from the doctor.

Malone was born in Thoreau. He has had plenty of opportunity to get to know the Navajo and understood their culture. During his interview with Michael's parents, he expressed his condolences and then, out of courtesy, asked them where Michael was headed today on his journey. They told him he was coming into town to attend the funeral of his fiancée, Rosina, who had died five days earlier.

"I'm real sorry to hear that," Malone commiserated. "What happened to Rosina?"

To his amazement, the family described identical symptoms to those of Michael's illness.

Malone felt a sense of growing shock as he continued to ask questions, eliciting more and more worrying information. The details of Rosina's illness matched Michael's, symptom for symptom. "At that moment, I realized we had a major problem. I didn't know what was going on, but whatever it was, we were dealing with something very serious."

Two young healthy people, living in the same household, had suffered identical symptoms. Rosina had lived on the reservation, a sovereign and self-governing body, which was not required by law to report deaths on Indian land. In practice, the Navajo Nation were very good about reporting homicides, suicides, and deaths by accident. But this had been regarded as a death from natural causes.

On the wall of his office, Malone has two certificates awarded to him for outstanding service. There is a large street map of Gallup with pins fixed wherever somebody has died of exposure. His mind is by nature methodical. He is a man who studies patterns, the algorithms of human behavior, the geography of death. Now he thought back to a month earlier when he had been puzzled by the death of another young woman who had died from an illness similar to that of this young man and his fiancée — the same case Dr. Tempest had recalled upon inspecting Michael's chest X rays. At the time, her case seemed to be an isolated one. It was also still unresolved. As he reflected on her detailed symptoms and findings, he realized that she too appeared to fit the strange pattern, symptom for symptom, of what he was seeing here.

"When is the funeral?" he asked the family.

"It starts in fifteen minutes."

Malone excused himself and hurried to a telephone within the hospital, then called Patricia McFeeley, the pathologist working in the medical investigator's office in Albuquerque. It was Dr. McFeeley who had performed the autopsy on the young woman in April. Malone gave her a summary of what he had just discovered. He asked her if she remembered the other woman's case from a month ago. She remembered it very well. She agreed that they needed to stop the funeral. They had urgent need of those two more autopsies. In Navajo culture, great importance is placed on privacy and religious tradition. During the customary four-day mourning period, when the *chindi,* or spirit, of the dead is still wandering the earth, the deceased is not even spoken of. Gazing on the body of the dead is frowned upon. Any irreverence might cause the spirit to linger, leading to disharmony, tragedy, or even worse. Autopsies are particularly unwelcome.

"You can't imagine the thoughts that were going through my mind as I realized I would have to go across the street and tell her family, only fifteen minutes away from the service, that they couldn't have a funeral — that they couldn't have a burial."

Malone called the mortuary to give them notice of his intentions, so it wouldn't appear to be an ambush when he arrived. Perhaps they could gather together the essential family relatives in the arrangement room? Walking across the parking lot of the hospital to the mortuary, he met with Rosina's family in the silent room. He introduced himself, then explained that he had just come from the hospital, where Michael had just died. Expressing his sympathy, he allowed some moments for them to come to terms with this additional grief. Then he said: "Unfortunately, we have a problem here because, as far as I can see, all of Michael's symptoms match the symptoms of Rosina's illness."

With tact and sympathy, Malone explained the possibility of a contagious illness. "We need to do an autopsy because there is a very real danger that the rest of your family could be in jeopardy from whatever has happened to these two young people."

The surviving relatives lived together as an extended family on the Navajo reservation, Rosina's mother and father, her siblings, perhaps even grandparents. Malone outlined how, if this illness proved to be contagious, everybody close to them, including the parents themselves, the infant, and their neighbors, were at risk. In spite of their shock and grief, the family listened to him and were courteous in reply. His words frightened and concerned them. They consented to Rosina's autopsy. Malone suggested that they should go ahead with the funeral service but then instead of going on to the cemetery for burial, the family should go home, leaving the body at the mortuary.

Malone returned to the hospital. By now Dr. McFeeley and her team were standing by in Albuquerque. He made arrangements for the two bodies to travel down Interstate 40.

3

Bruce Tempest had first joined the Indian Health Service in 1967, starting work on the western end of the reservation, where the people were more traditional. "When I came out here I realized that I could put to use here all the things that seemed important in my life up to that time. I really felt needed." He found himself fighting life-and-death battles against the old-time infectious diseases, rampaging tuberculosis, epidemics of diphtheria, pneumonia and dysentery. Not only had he come to know and like the Navajo, he knew their illnesses. Michael's sudden death, those bizarre X-ray findings, troubled him greatly. Back in his office, Tempest began a parallel investigation, talking it over with his colleague Larry Crook, compiling his own litany of disturbing similarities, evolving to further inquiries. He got on the phone to Crown Point about Michael's May 12 investigation and treatment. The recorded history confirmed that muscle pains had been a prominent symptom. The doctor had wondered about an obscure infection called a mycoplasma, caused by a microbe, halfway between a bacterium and a virus. He had prescribed a suitable antibiotic.

Tempest now thought mycoplasma unlikely. He already knew about the death of the young man's fiancée and that of the thirty-year-old woman in April. Now he remembered being consulted about a strikingly similar case by a colleague in Arizona the previous November.

There are eight small rural hospitals scattered throughout the vast area of the Navajo Nation, each responsible for a different geographic region, and these hospitals work closely together. It is not uncommon for other doctors to approach the experienced Tempest for advice. He now thought back to the details when a colleague at Crown Point had called him after the death of the young man's fiancée. It had only been a few days ago, and he remembered his puzzled colleague asking his opinion. He knew he had not been of much help: he just could not make any real sense of it. Today, with the death of the young man, he picked up the phone and called his colleague for more detailed information. With totally whited-out lungs, a particular diagnosis played on Tempest's mind. He was considering the possibility of plague.

Plague is of course the most infamous epidemic disease in history, the cause of the Black Death that swept through Asia and Europe in the Middle Ages. Such a diagnosis might seem exotic, but the disease is endemic in the New Mexico area and a handful of cases occur each year, particularly in the spring and early summer. There are two clinical patterns of plague, bubonic or septicemic, both caused by a germ *Yersinia pestis,* which is transmitted in the bite of an infected flea. The two forms vary in their clinical evolution. In one, there is a big swelling of lymph glands in the region draining the bite. This swelling, called a bubo, localizes in a matted collection of infectious pus, giving the disease its name, "bubonic plague." In the other form, it doesn't localize at all but spreads through the bloodstream with a clinical picture like septicemia or blood poisoning. If caught early enough, the disease is still potentially curable using modern antibiotics. But speed of diagnosis is essential, particularly in the much more dangerous septicemic form, where if not diagnosed early, it progresses to an overwhelming blood-borne spread. The septicemic form can seed the lungs, giving rise to a whiteout on the chest X ray. In this "pneumonic" form, the plague can spread from person to person through coughing, making it extremely contagious.

Tempest called his colleague at Crown Point to ask him if they had screened the patient for plague. In fact they had looked into this possibility and seemingly ruled it out.

The other young woman had come into the Gallup Medical Center six weeks before. She had arrived in the hospital about midnight, was intubated and put onto a ventilator, yet she died just four hours later. In his mind, Tempest located her geographically. Unlike Michael and Rosina, who had lived northeast of Gallup, she had lived a good distance south of the town. If it was a contagious infection, it was extending over a wide territory. He made some more phone calls, confirming what Malone had also recalled, that they had managed to obtain an autopsy on this woman. But the pathologist, Patricia McFeeley, had come up with no new information from examination of the internal organs. Nor had the bacterial or viral cultures subsequently grown anything. The implications were becoming more and more sinister.

While Tempest was requesting her records, he remembered yet another occasion, where he had been asked to review the case notes and observation charts of a patient who had died about six months before. Another young and previously fit woman, she had come into a reservation hospital fifty miles west of Gallup. Her presentation was a carbon copy of the others, with fever and severe muscle pains. She hadn't seemed very sick at all on admission. Four days later, and despite every measure to save her, she was dead. Tempest remembered studying the details of her hospitalization, which had lasted several days. Her physicians had treated her with broad-spectrum antibiotics, so they were convinced they were dealing with an infection. From the sound of it, they had done all the right things, yet she had deteriorated and died.

Sitting at his desk, Tempest knew he had the case histories of four people, all of whom sounded very much alike. He picked up the phone again and called the hospital that had looked after this woman. Doctors often request tests, such as viral cultures, which take a long time to produce results. These might not have been included in the charts they had sent him. His colleague could offer no new information on the woman, but during the conversation he suddenly volunteered a new statistic to add to Dr. Tempest's growing compilation. "Well listen! There was this guy that died here last week who sounded just like her."

Hearing this description of yet another case, Dr. Tempest was alarmed.

In just a couple of hours on that Friday, May 14, 1993, he had identified five people who lived within the Navajo reservation, all of whom had contracted a mysterious illness. They sounded so very much alike they had to be suffering from the same illness. He had no idea what that illness was. All five people had been young and previously fit. All five had rapidly deteriorated and, in some cases despite hospitalization and urgent medical treatment, all five had died.

4

The Office of the Medical Investigator (OMI) stands in a discreet corner of the campus of the University of New Mexico hospital in Albuquerque. The entrance doors are of smoked glass and the front elevation is garlanded with bougainvillea.

On Friday, May 14, Dr. Patricia McFeeley was the pathologist on call for the weekend. At three in the afternoon, she was sitting in her office when Richard Malone called. McFeeley has a deep respect for Malone: in her opinion he is one of the most experienced of the medical investigators attached to the department. This was why, when Malone gave her brief details of the two unexplained deaths, she sanctioned his request for the autopsies.

A few minutes later, worried they might be dealing with something contagious, McFeeley called her friend and colleague Dr. Edith Umland. Umland was chief of the state public health laboratory, two floors above in the same building. She was also a good pathologist, somebody McFeeley had worked with closely in the past and whose opinion she trusted. Now Patty drew Edith's attention to the similarities between the puzzling autopsy a month earlier and the deaths of the young couple, similarities that had been shrewdly spotted by Malone. In their discussion, they probed those similarities and what they might mean, the differing ages of the young couple and the thirty-year-old woman, where these people lived. The thirty-year-old woman lived more than a hundred miles from Littlewater. It seemed unlikely they had ever come into contact with each other.

While waiting for Umland to come down to her office, McFeeley had called the state department of epidemiology, which is based in Santa Fe. The call was logged at 3:35. The message was essentially anticipatory: "We may have something of a problem."

Patricia McFeeley is a tall, slim woman in her forties, with fair hair and light blue eyes. She enjoys her work in the medical investigator's department. "What you do makes a difference to people. It really matters to the family of the dead person, certainly to a person who may be on trial and in whose case I am giving evidence." She gets a particular satisfaction in counseling the parents of children who have suffered violent or accidental deaths. On that Friday afternoon, when Patty McFeeley called up the Department of Public Health in Santa Fe, the officer she spoke to was the medical epidemiologist on call. At this time McFeeley and Umland were convinced they were dealing with plague. There were other possibilities, some more or less improbable, but they weren't really considering anything dramatically out of the ordinary. The epidemiologist was interested but could add nothing to their investigations. He would wait for the result of the autopsies. He also gave McFeeley the name of his colleague who would be covering over the weekend. "If you find something, give us a call."

The bodies arrived late that night. Patty was enjoying a night out with her husband and daughter, having dinner at a friend's home. After taking the call, she drove back into the department, arriving at about nine o'clock. She was joined in the morgue by Edith.

Given the late hour and anticipating a readily demonstrable diagnosis, the two doctors only performed limited dissections. McFeeley opened both chests just enough to get lung biopsies that could be tested for infection and examined under the microscope. In Michael's case, she also took blood that could be used for serology. They worked quickly, extracting samples that were as fresh as possible. Then, back in their two laboratories, they began an extensive series of tests. They examined blood, mucoid smears from the lungs and from throat and nasal swabs. They spread the smears over slides and stained them for bacteria. Edith arranged fluorescent antibody tests that would detect the presence of *Yersinia pestis,* the bacterial cause of bubonic and pneumonic plague.

The test is technically finicky and takes about two hours to perform. Essentially you smear lung tissue onto a slide, then apply specific antiserum to the *Yersinia,* which has been labeled with a fluorescent dye. The slides are then washed to remove unattached antibody before being allowed to dry. You inspect the results using a fluorescence microscope, where, if present, the plague bacilli light up a bright apple green.

When, with her senior technician, Edith inspected the slides, no yellow-green fluorescence showed.

She performed Gram stains on the smears — these would show up most of the common pathogens — and she set up routine bacterial cultures for plague and a host of other germs. She included another bacterium that occasionally causes a fatal pulmonary disease, *Legionella.* This causes Legionnaire's disease, a novel infection that had been first diagnosed in America in 1976 among army veterans. It seemed prudent to inoculate tissue cultures for likely viruses, particularly influenza.

The tests had taken them a good deal longer than they had anticipated. All of the quick readings had proved negative, though the cultures might yet prove revelatory. About midnight, Umland called the epidemiologist in Santa Fe, who had been anticipating her call. The people in the public health offices had become a little more concerned. They too had been convinced it would turn out to be plague. It was close to one in the morning when the two exhausted women headed for home.

The following morning they returned to read the overnight cultures. Edith brought the news downstairs to Patty: they had all drawn a blank. McFeeley knew now that she would have to perform full dissections on the two bodies. On her arrival, she had found a large number of other autopsies waiting, so she asked some residents to take over the routines. Summoning up a renewed vitality, she started work again on the bodies of Michael and Rosina. It was now 8:30 A.M. as Umland stood by, watching the operations and accepting specimens. McFeeley began the crude but formal incision, from the throat to the pubic bone, prizing open the entire chest and abdominal cavities. It is normal practice at autopsy for the pathologist to cut out each individual internal organ and weigh it. Michael's lungs felt heavy. When she put them on the scales, they weighed more than a kilogram each, more than three times the normal.

Where somebody has died of an overwhelming bacterial pneumonia, from plague, for example, the lungs would be full of pus. Michael's lungs were not full of pus. They felt and looked rubbery. They had the consistency of a good-quality sponge, allowed to soak until it was turgid with water. In medical language, this excess fluid accumulation is termed "edema." McFeeley considered these lungs as "very edematous." Cutting into their firm rubbery mass with her long-bladed knife, a watery fluid seeped out, frothy, bubbling up from the sliced surface and lacing the stainless steel blade with fine pinkish bubbles.

Patty McFeeley felt a powerful sense of déjà vu. Michael's findings were exactly the same as those of the thirty-year-old woman a month previously. His lungs were waterlogged. The very air sacs, through which oxygen crossed into the capillaries of the blood and carbon dioxide diffused out, were completely flooded. In medicine, pathology is said to be the final recourse to the truth. She was gazing at that truth, the pathological confirmation of the whiteouts on the X rays. Michael had literally drowned in his own secretions.

Drug overdose flickered across McFeeley's mind: you certainly saw edematous lungs like these in some cases of death from drug overdose. In Michael's case, there was also some free fluid in the pleural cavity, the potential space between lungs and the chest wall. Again it seemed to fit with the edematous lungs.

In Rosina's case, the embalming process ruled out any meaningful bacterial or viral cultures. Any likely tissue effusions that might have given useful information had been removed by the undertakers. It was still useful, however, to perform the autopsy since tissue samples could still be examined under the microscope. At the very least, that histological examination might show up commonalities between her disease and that of Michael's, a clue to the cause of their deaths. In the gross pathology McFeeley did discover a feature the young couple had in common. They had both suffered bleeding from the lining membrane of the stomach. To the pathologists, it seemed to confirm the possibility of a drug- or poison-related cause of both deaths, what medically would be included under the umbrella term "toxins."

Certain drugs, taken as part of a lovers' pact, will cause bleeding from the lining of the stomach. Aspirin is notorious for this, and it is a common self-prescription in such tragic circumstances.

McFeeley saw little else that would point to any particular disease. Nothing even remotely helpful. In particular, the spleens were not enlarged, as would be found in an overwhelming bacterial infection, such as blood-borne plague. She cut out chunky samples from various organs, more than enough to search for microscopic evidence of every possible obscure disease and for chemical analysis for potential drugs and poisons. The diagnosis was as elusive as ever.

By Friday night, Patty McFeeley and Edith Umland had already gone a long way toward excluding plague as the cause of death. By Saturday morning, they thought a virus pneumonia due to influenza unlikely. They were already wondering if they were dealing with a much rarer infection, such as inhalational anthrax. McFeeley called the OMI's histology technicians at their homes. She explained it was an emergency and asked them to come in and run the specimens through so they could get slides out the next day. By Sunday McFeeley and Umland had micron thin sections of the tissue samples mounted on slides and stained to look for any unusual pathological change, for the odd, the curious, and the rare among bacteria, protozoa, and fungi.

By Sunday they had found no organisms of any description on any of their tests. Routine cultures had ruled out the commonly invasive bacteria, the aerobes, like the abscess-causing staphylococcus and the beta hemolytic streptococcus, which would subsequently become notorious in the 1994 flesh-eating bug scare in Britain. They had ruled out chlamydia and Legionnaire's disease. They had dry runs on anaerobes, the bacillus that causes anthrax, the bacterium that causes the milk-associated brucellosis, the epidemic germs of whooping cough, diphtheria, the intestinal pathogen *Campylobacter,* even the fungus-like *Nocardia,* mostly found in the immunocompromised.

That Sunday McFeeley spent hours just looking down the microscope. Umland came downstairs to join her. McFeeley would subsequently remember her colleague sitting there, willing her on to find something, watching and waiting. By this time they were surer than ever that they were dealing not with two but three cases. McFeeley had pulled out the slides on that previous autopsy about a month before, looking for commonalities in all three cases. Patty turned to Edith in frustration and said, "Go ahead, use my microscope. Sit here and look yourself."

Patricia McFeeley and Edith Umland were more than colleagues. They were also friends. Patty was the taller of the two, Edith with slightly darker hair, also blue-eyed. Where Patty tended to wear contacts, Edith wore wire-rimmed spectacles. They were of much the same age, having first met when they had attended the UNM medical school. They had shared the same pathology rotation in their postgraduate training. In their working relationship, their mutual sense of trust, they were comfortable with one another. At this moment, Patty was exhausted from looking down the microscope. Edith had just arrived and her eyes were fresh. Patty had some family business to see to — she needed to pick up the kids and take them somewhere — so she left her office, with Edith rummaging through the slides all over again.

When McFeeley returned, later that same day, Umland had gone home. But McFeeley found a hastily scribbled note, a jotting of her friend's thoughts, dated May 16. "To my eye, all three have the same thing, more or less, whatever it is." Umland then summarized most of what would ever be demonstrable under light microscopy of the pathology in the mystery illness. There was mild damage to the liver and slightly more damage to the kidneys. Her final line was a prediction: "I sure hope we get a virus out of the lung tissue."

THREE

The Panic Spreads

People would hear about somebody they knew
dying with symptoms that appeared not much
different from an ordinary cold.
At the first sign of a headache or mild fever,
they jumped to the terrifying conclusion:
"That's me — I have all of those symptoms."

PATRICIA GUTHRIE
conversation with the author

1

The University Hospital, in Albuquerque, is an attractive cluster of modernist buildings linked through an ironwork bridge with the tree-lined main campus across Lomas Boulevard. A colorful ethnic mix of Spanish, Anglo, and Native Americans jostle cheerfully about its wards and corridors, which are shaded by tinted glass from the full glare of the New Mexican sun. Dr. Fred Koster is the Professor in the Department of Medicine. On Monday, May 24, Gary Simpson, Director of Infectious Diseases for the state, came down from Santa Fe to speak at the weekly departmental case review conference. Simpson had been alerted by Bruce Tempest and now he told the UNM clinicians about the strange cluster of five cases. Fred Koster is six feet tall, gray-haired, with a moustache of the same color. He wears metal-framed spectacles and is filled with a restless impatient energy that seems to marry well with his enthusiasm for research into the immunological defenses against infectious diseases. Koster would subsequently remember the reaction to Simpson's presentation: "We all thought that something very strange was going on."

A couple of days later, Koster took a call from a doctor at Crown Point asking him to take over the care of a young man who had come down with a fever. Koster listened to the symptoms of muscle aching, shortness of breath, and a cough. Although the patient was not seriously ill, the IHS doctor suspected the new mystery illness.

"Why," Koster asked him, "do you suspect the mystery illness?"

"This is the brother-in-law of the marathon runner. His sister also died from it."

Rosina's brother Frank normally lived in Washington State but had come to stay in the family compound to attend her funeral. Hearing that Frank was now sick, Koster felt it prudent to transfer him to the intensive care unit.

When Frank made his reluctant journey to the UNM hospital, his wife, Dolores, traveled with him. Frank was very frightened, so Dolores refused to leave his side, even when she needed to sleep. Concerned relatives of hospital patients as a rule are offered accommodation on campus. Occasionally, when matters are critical, they will stretch out on a couch in the waiting room of the intensive care unit. But Dolores, who was thirty-four weeks pregnant, refused even this separation. She took some blankets and made her bed on the floor of her husband's room in the intensive care unit.

Frank was not as desperately sick as Rosina and Michael. Nevertheless, he had a chest X ray performed every twelve hours. Fleeting shadows would appear and as quickly disappear again. It seemed likely that he had a milder version of the illness that had killed his sister and brother-in-law. He never became breathless and there was no question of putting him onto a ventilator. There was another curious finding. His blood film showed abnormal lymphocytes.

Lymphocytes are white blood cells, circular in outline and with a large nucleus. They do not ingest foreign matter in the amoeboid fashion of the polymorphonuclears. Their role is more subtle, recognizing foreign antigens, helping in the production of antibodies. They also contribute to "cellular immunity" — part of the body's defenses against foreign antigens, such as organ transplants, and long-term infections like tuberculosis. An increase in lymphocytes in a patient's blood film is suggestive of a viral illness.

Each morning at 7:00 A.M. the doctors would arrive to conduct their round. Masked and gowned as a precaution against a possible aerosol-spread infection, they would enter the room to talk with Frank. They would listen to his symptoms, examine his chest, look at the charts. Then they would have a few words with Dolores.

"How do you feel?"

"I feel fine."

Three or four days later, Frank's fleeting exudates had cleared and he was feeling so improved the doctors stopped his oxygen. They were thinking about letting him go home. A medical student who had been chatting to Dolores drew the doctor's attention to the fact that she looked sick.

"Well," she told the doctors on today's round, "I've been getting all these backaches and muscle aches but I thought it was just from lying on the floor. I think maybe you're right. Maybe I am sick."

When they took her temperature, it was markedly elevated. They arranged a chest X ray, and to their consternation the pictures of her lungs resembled a snowstorm. It was apparent that Dolores had the illness. Her lungs were already filling up with the strange infiltrates.

Her illness advanced with a fearsome acceleration. She was admitted into a room further along the corridor and hour by hour the doctors followed her progress. Within eight hours of first noticing she was sick, they watched her crash. She became extremely breathless and her chest X ray showed a whiteout identical to that of her late brother-in-law, Michael.

Horrified by what was happening to this young woman, Fred Koster was now certain they were dealing with a very unusual condition. He talked it over with Howard Levy, the director of the intensive care unit, who was taking immediate care of Dolores. Earlier in the week, they had read through the pathology reports of Michael and Rosina. What baffled them was the fact that so much pointed toward an overwhelming infection yet no infectious agent had been found. Meanwhile, with the close monitoring of Dolores, they noticed another unusual feature.

Her platelet count was tumbling. Platelets are tiny particles that circulate in blood and are vital in forming blood clots. Apart from their function in wounds and accidental injuries, they play an important role in sealing off small leaks and damage to the walls of tiny blood vessels. This is happening all the time through normal wear and tear within the body. So the fall in Dolores's count was significant.

Frank had also shown a plummeting platelet count. Now they noticed that Dolores was also showing atypical lymphocytes in her blood. These were bigger than normal and stained differently, a royal purplish blue. They were immature precursor cells, of a kind normally found only in the spleen, lymph nodes, and bone marrow: doctors called them "immunoblasts" or "blasts." In Koster's mind a syndrome, inchoate as yet, was struggling into creative recognition: and those atypical lymphocytes and the low platelet count were an integral part of it.

A "Dear Doctor" letter had been circulated statewide by the concerned Public Health Offices in Santa Fe a few days earlier, and it was already generating a response. While Frank and Dolores were being treated in the intensive care unit, more cases were being referred into the UNM hospital. By the time of Dolores's decline there were at least six cases, all thought to have the mystery illness, on the same intensive care unit. At this stage there was a call from Crown Point. Dolores's five-year-old son had a temperature. Crown Point received permission to send him straight to the UNM hospital for observation.

On Thursday, May 27, Fred Koster picked up the phone to have an urgent conversation with Gary Simpson in the Public Health Department in Santa Fe.

Dr. Koster's call arrived just as Simpson was attending a crisis meeting with senior colleagues, including C. Mack Sewell, the state epidemiologist. Everybody was extremely tense with the worry they might be dealing with an aerosol spread of a highly contagious and lethal microbe. One of the technicians in the medical investigator's office, who had assisted in the original autopsies on the couple, had just been admitted to the UNM hospital with fever and a dropping platelet count. That same afternoon came news that two other health workers who had cared for the original couple were also sick with febrile illnesses of unknown cause.

Howard Levy was considering closing the intensive care unit at UNM hospital to all cases other than the mystery illness so they could triage suspected cases into there. "We don't know what we are dealing with," Koster told Simpson. "We had better get together and think this through."

Events were moving at such an alarming pace that Simpson welcomed Koster's suggestion. At eight in the evening of Thursday, May 27, an urgent gathering of seven or eight senior doctors took place at Simpson's home in Placitas, about twenty miles north of Albuquerque. Those present included Norton Kalishman, the Chief Medical Officer for the State of New Mexico, Edith Umland, Fred Koster, Mack Sewell, together with his deputy, Ron Voorhoes, and Jim Cheek, head of epidemiology for the Indian Health Service.

With two decades of experience in public health, Sewell is not easily ruffled. But in the last forty-eight hours, he had received word of many more suspected cases. Earlier in the afternoon, he had called his colleagues in Arizona, Utah, and Colorado. A week ago they had not seen a single case. But now the first reports of suspected cases were arriving into their public offices of health. Said Sewell, "In my position you find yourself sitting there and you have to weigh and to judge what is really going on, how much resources you need to put into an investigation." Since Tuesday, May 18, there had been a series of communications between Maggie Gallaher in Sewell's department and the Centers for Disease Control (CDC) in Atlanta.

The CDC has the most experienced disease investigators in the world. But its federal charter does not permit it to interfere in state health affairs unless invited to do so by the state authorities. That afternoon everybody had agreed they should call them in.

Sewell knew Ed Kilbourne, a senior officer at the CDC, working in the Center for Environmental Health. They had worked closely together during the investigation of an outbreak known as the eosinophilic-myalgia syndrome. Later in the afternoon, Sewell had spoken to Kilbourne on the phone and had formally requested the CDC's assistance.

Sewell's presence at the brainstorming meeting at Gary Simpson's home would be constantly interrupted by telephone calls and requests for updates from colleagues in other states and from senior officers at the CDC as the formal request for assistance moved through the divisions. While Sewell was fielding the inquiries, his deputy, Ron Voorhoes, briefed the small gathering on a trip he had made to the site of the outbreak. On the Thursday morning, Voorhoes and Jim Cheek had made a special trip to talk with twenty or so members of Rosina's extended family in Littlewater. Aerosol spread was still a worrying possibility and Voorhoes described the atmosphere of alarm now pervading the community around Crown Point.

Fred Koster startled everybody with his description of Dolores's fulminating decline. The small gathering talked about that. Perhaps they should subject her to a lung biopsy? Sometimes this is the only way in which doctors can diagnose a baffling pulmonary illness. But it would involve some degree of risk, and on the whole his colleagues were opposed to it. Again and again the conversation veered back to the question that was on everybody's mind: what exactly were they dealing with?

Was this really something new? Or was it something familiar they were failing to recognize? There were a number of diseases that could behave in just as deadly a fashion if they remained undiagnosed and, consequently, untreated. They spent time thrashing out a list of possible diagnoses. It was a very long list. Fred Koster felt that he now had the rudiments of a clinical and laboratory picture to go on, enough perhaps to give other doctors in the southwestern states a clearer idea of how to recognize the illness.

Since the circulation of the "Dear Doctor" letter, the evolution of alarm had greatly accelerated. How could they even hope to keep colleagues updated on circumstances that altered alarmingly from hour to hour? The closer circles within the state were no longer adequate. With cases now suspected in neighboring states, there was a cogent need to alert colleagues wider afield — but just how wide did the net extend? Certainly to the Four Corners states, and perhaps, Simpson pensively wondered, even nationwide? There was a momentary silence as they pondered the ramifications of that. Were they all going to look very stupid when a common explanation eventually turned up?

But Koster's awe was contagious. The illness seemed a virtual death sentence. The eclectic choice of victim over the reservation was frankly terrifying. The pattern was too unusual, too dangerous, not to take the risk. There was no option but to set bigger wheels in motion. The public offices of health would take on the arduous labor of keeping colleagues statewide — nationwide if it came down to it — apprised of the day-to-day developments.

The problems were so daunting it was the unanimous conclusion that they needed to hold a major conference, thrown open to every conceivable medical expert in the state. The conference would take place just two days later, on the Saturday of the Memorial Day holiday weekend.

2

On Thursday, May 27, the same day the small group met at Simpson's home in Placitas, the story broke in one of the state newspapers. The *Albuquerque Journal* carried a front-page headline: MYSTERY FLU KILLS 6 IN TRIBAL AREA. The report was written by staff writer Leslie Linthicum.

A mystery flu-like illness has killed six people on or near the Navajo reservations in the past six weeks and has sent four others to the hospital for treatment.... Moving on to describe three cases in the index family, the piece was ominously complete with photograph of Michael, his real name, the name of his fiancée, the name of their baby, and their address in Littlewater.

The state health department would subsequently praise the local reporters, who showed considerable restraint in their handling of sensitive local issues. Yet this very first report gave fair warning. There would be likely conflicts of priorities between journalism and medicine in the coming weeks. This problem aside, the article was remarkably accurate. Two out of the ten suspected victims of the mystery disease were identified as "Anglos"; it should have been clear from the very beginning that this could not be specifically a Navajo disease.

The following day, Patricia Guthrie, a staff reporter on the *Albuquerque Tribune*, was out interviewing people on the Navajo reservation. She was in the position to write two separate reports, highlighting the mysterious nature of the illness together with the fact that yet another possible death had been confirmed. The latest victim was a Navajo male in his early twenties who had died on Tuesday at the Zuni Pueblo hospital after being transferred from the neighboring Pine Hill health clinic, in Ramah. His death, together with another at an as yet undisclosed location, brought the total fatalities to eight. In New Mexico as a whole, the suspected total of infected had now risen to fourteen. The only glimmer of relief was the indication that Rosina's baby and Dolores's son had both been excluded from infection.

By now the story was hitting the U.S. national media. Reporters and film crews from around the country descended upon the reservation, where, in the words of Duane Beyal, public relations assistant to President Peterson Zah, "they trampled the rights and the privacy of these poor families who were already stricken with this scary mystery disease." While the local journalists understood the traditional reticence of the Navajo culture and their need for a four-day period of quiet mourning, out-of-state reporters had little conception of the culture shock their arrival into the introverted and highly conservative local community would bring. They would go out and swarm around families, photographing funerals, printing victims' names, disturbing customs, and invading people's privacy.

In a typical incident, a television crew arrived at the home of a recently bereaved family, pulled up outside the hogan in one of their big trucks, set up their link-up vans with satellite on top, and then knocked on the door. In more extreme examples, reporters sneaked into funeral parlors and attempted to photograph or to interview relatives in the very intimacy of the church or the service. One went so far as to trick his way into a hospital, where he attempted to interview a woman who was being ventilated in the intensive care unit. The following day the woman died. To the grieving Navajo, it seemed that "even on your death bed, you could find no peace."

The medical authorities in Santa Fe were frantically trying to warn physicians and the public at large that what appeared to be the flu might not in fact be the flu at all. Inevitably, as the symptoms were spelled out in the local newspapers, as the staggering lethality of the disease became more widely known, people jumped to the worst conclusions. By the Memorial Day weekend, the emergency waiting areas of both main hospitals in Gallup were besieged by people, anxious they had the plague.

The symptoms were so nonspecific it was impossible at this stage to placate this mass anxiety. Terror of a malignant unknown, the bafflement of the doctors themselves, fueled the public panic. In every newspaper article, the symptoms were described as vaguely flulike. And flulike symptoms could mean just about anything.

By Monday, May 31, the tally had risen to twenty-five victims. Most of these were young and previously fit, including eighteen Navajo, five Anglos, one Hispanic, and a single Hopi Indian. Though Dolores would survive, the death rate remained desperately high in spite of the efforts of both the doctors and the press to persuade people to report in early. One patient who had been sitting up in bed in the morning eating breakfast was on a respirator by afternoon and dead that same night.

News reports branded the epidemic a "Navajo disease." Before the medical experts had been able to demonstrate that the disease was not spread from person to person, fear of contagion led to the resurrection of old prejudices.

A school bus of Navajo children traveling to Los Angeles to see their pen pals was turned back at the California border. Restaurant meals were presented to people on paper plates, or the serving waitresses wore rubber gloves. A car with New Jersey license plates was seen driving across the reservation with its passengers wearing surgical masks. A Navajo couple were detained at a large airport and kept for a time in quarantine conditions, because the airport authorities feared they might be carrying plague.

Day after day there would be some new horror story, some petty episode of discrimination, some indignity imposed on the frightened people. When Ben Muneta, a Navajo doctor with the Indian Health Service, arrived in Gallup to conduct investigations, people avoided him in the street. "In the shops in Gallup, they wouldn't touch me. I'd give them money and they would throw it back at me." The shopkeepers in the small town, highly dependent upon the tourist trade, were blaming the Navajo for the loss of business. In one of the most unpleasant resurrections of prejudice, gangs of thugs roamed some of the border towns, beating up Navajo.

Little wonder that the residents on the reservation began putting up NO MEDIA signs on the approach roads. A more tragic consequence of this circus-like intrusion, coupled with the fear that the medical authorities were openly cooperating with the media, releasing names and addresses, was the refusal of local people, sometimes even the families and neighbors of victims, to cooperate with the ongoing medical investigations.

The paranoia now gripping the reservation was excusable. The arrival of a plague is one of the most terrifying experiences any community will ever experience. When Dr. Muneta drove out to reassure his mother, who lived in a hogan at the top of Coyote Canyon, she told him about a neighbor who had suddenly taken sick and had been carried off by helicopter to Albuquerque. The slow swelling ripple of death continued to widen, following that same mysterious pattern that had been established that fateful day of May 14 and claiming the youngest and fittest in the local population.

Panic also bred internecine tensions. People in Littlewater were apprehensive about one another. The Navajo community at large was looking at Littlewater somewhat apprehensively. In turn the Navajo people were being regarded with apprehension by the people in the border town areas who suffered the same suspicions from the people in Albuquerque; and with the arrival of CNN and the national media, these ever-widening ripples extended throughout the country.

In the midst of this fear and mistrust, some Navajo asked themselves if it had been Anglo tourists who had brought the disease into the reservation in the first place. Unlikely as this might seem, history for the Native Americans carried some terrible precedents.

In 1763, Sir Jeffrey Amherst, commander in chief of the army in North America, ordered that blankets laden with smallpox should be distributed to the Pontiac Indians, who had been fighting a successful and courageous battle for their lands and culture against the encroachment of the European settlers. This ruthless use of germ warfare destroyed the Pontiac and, in the explosive contagion that typifies plagues in a virgin population, straddled the Rockies to involve the Sioux and Plains Indians in a wholesale slaughter. It is believed that smallpox, measles, and a host of other plagues carried with the European settlers, more often introduced accidentally than deliberately, may have wiped out as many as 56 million of the native inhabitants of continental America.[1]

Over the tribal lands, people remembered the odd climatic conditions of the last year. They talked about a greenish-yellow mist that overhung the Southwest. Ben Muneta would remember how the winds would come in from the dry desert bringing a similar haze over Albuquerque. The changes in the weather and local ecology had been so remarkable that some of the Navajo elders, who spent their lives studying such changes, had predicted, six months before the sickness appeared, that people would die.

If the elders had a long and venerable tradition of wisdom that gave reasonable grounds to their fears, elsewhere the scientific vacuum threw up some rather more arcane and crazy theories. Every oddball psychic and psychotic took the opportunity to ring in with an offbeat theory. Somebody from California called the medical authorities to say that the disease was an effect of earthquakes, which caused microfissures in the ground from which coccidioidomycoses spores were escaping from quartz crystals. Another faxed the reservation to say that Russians were going around in UFOs and dropping the illness onto the reservation.

Wild speculation now swept through the community that an Andromeda strain of virus had escaped from some supposed biological weapons bunker at Fort Wingate. Some people even wondered whether the Navajo were the victims of some ruthless germ warfare experiment.

3

The medical intensive care unit would bear the responsibility for the urgent care of all suspected cases of the illness arriving at the University of New Mexico hospital. Howard Levy is a youthful man with swept-back dark hair, sensitive features, and intelligent green eyes. He wears a white coat belted at the back and decorated with a Navajo cross on the breast pocket.

It had been part of Levy's duties when newly appointed to plan the design of the ICU. He was proud of his modern facility of fourteen rooms, each fitted with multiple-channel video monitors linked to a master network housed in a control room. In this master center, a hive of space-age technology, with banks of monitors, each scrolling four to eight differently colored traces, followed the vital signs of every patient. Here they could observe fourteen continuous EKGs on a single screen, fourteen respiratory rates, arterial gases, or the pressure waves from Swan-Ganz catheters that had been inserted through a subclavian vein and wound through the right-sided chambers of the heart into the capillary bed of the lungs. At a glance you could pick up the moment when somebody was breathless, if they turned over in their sleep, or their life was suddenly threatened with a cardiac arrhythmia.

Howard Levy graduated in medicine at the University of Witwatersrand in Johannesburg, where he took a special interest in pulmonary critical care. He remembers how his colleagues had to go into quarantine when a nurse was infected with a devastating hemorrhagic fever called Marburg disease, which she had contracted from a young Australian backpacking across Africa. The horror and drama of that imbued him with a healthy respect for plague viruses. Now, with the increasing alarm surrounding Dolores, he was very glad that he had included high-grade barrier treatment at a time when those precautions might have seemed extravagant. "One day," he had predicted, "a formidable infection will hit us and we will be prepared for it."

Levy is gifted with an engaging wit that peppers his conversation. On the afternoon of Friday, May 28, the senior staff at the hospital called a meeting to review the situation, but Levy knew nothing about it. "They called the crisis meeting and forgot to invite me." At 3:00 P.M. his pager sounded and he was instructed to hurry along immediately. He entered a cramped room through a door in the back of the main canteen. "Everybody who is anything in the hospital was there," he recalled, "sitting and waiting. It was kind of terrifying to realize they were expecting me to explain what was happening." He painted a graphic picture of the cluster of deaths and explained the findings at autopsy. In response to their anxious questioning he assured them that he and his staff were ready and willing to look after these patients.

By this stage two suspected cases had already died here in his unit. The staff were clearly frightened. Since the cause of the disease was unknown, it was impossible to predict just how it was spread. As soon as he returned from the meeting Levy gathered the unit nursing and medical staff together so he could explain what he had in mind. The intensive care unit had been designed with two rooms exhausted to the outside air. With more patients already in transit, this would clearly be inadequate. As soon as he had accepted the fourth and fifth cases, Levy closed the unit to all other cases. All of the beds except four, which needed to be allocated for the more usual purposes of the intensive care unit, were now standing by for the epidemic. From that moment everybody working there was instructed to wear masks incorporating particulate filters, which were designed to trap a virus. Now his nurses gave him reason to be proud of them: not a single member of staff, from senior to junior, asked for leave or resigned.

The interview room was converted to a dressing area, where people would gown up before entering the quarantine zone. On the entrance they placed notices, warning people to gown up and to wear gloves and masks. "We had no need to place a guard on this door — nobody wanted to come in."

In the two hours of relative peace beginning each morning with the 7:00 A.M. ward round, Koster, Levy, and another intensivist called Steve Simpson would rack their brains in a determined attempt to understand the physiological disturbance that caused people to die. What could be happening in the tissues and organs of the sick patients; what might be the cause of the whiteouts, the sudden crashes that overwhelmed the defenses of even the fittest victims?

For Fred Koster, there was no one moment when everything clicked. It was a gradual accumulation of small details, beginning with Dolores's astonishing progression right there in front of them, to recognizing that the curious blood patterns were repeating themselves with every victim, to studying very carefully the clinical and laboratory findings and titrating those against the tragic outcome in the fatal cases, when the tissues were subjected to minute examination at the autopsies now being continued by Kurt Nolte.

They had already demonstrated that when fluid was sucked out from the waterlogged lungs, it looked to the naked eye like straw-colored water. They had sent this to the lab for analysis and the report had confirmed their suspicion. It was hardly what was to be expected from lungs inflamed with a lethal infection. When the pathologists examined it for pus cells, they found virtually no cells at all. If you took a sample of blood and allowed it to clot, then filtered away all the red and white blood cells, all the solid components of the blood clot itself, this is what you would be left with. This, by definition, was serum. It was both a clue and a puzzle.

It was increasingly apparent that there was some unexplained anomaly in the way the patients were dying. "Once we tumbled to the fact that this deterioration was extremely rapid, we could aggressively oxygenate them. Yet even though they were very well oxygenated, still they died." This suggested that although the pulmonary disease was an important part of the crash, there was some other, as yet unknown, component. By degrees the doctors became more convinced that there really was some vital yet mysterious aspect to the shock that killed these people.

On Saturday, May 29, a task force of forty specialists parked their cars on the sun-scorched campus of the UNM hospital to join the meeting now taking place in a large conference room on the ground floor of the building that housed both the OMI's office and the state public health laboratories. They were joined by Jay Butler, an enthusiastic young epidemiologist newly arrived from the CDC and head of a small team including a fellow EIS officer, Jeff Duchin, and a toxicologist called Ron Moolinaar. As one by one the arriving experts signed the attendance sheet, also de facto joining what was now the working group, people in the university hospital a block away were dying from the disease.

At first the possibilities were so confusingly numerous, they had to be tabulated on flip charts. But the meeting participants moved through this encyclopedic list very quickly. Many of the possibilities had already been ruled out by the hard work of McFeeley and Umland or during the Thursday meeting at Placitas. Nobody seriously believed they had missed bubonic and pneumonic plague. No more did they believe they had missed any of the other uncommon yet lethal germs that might have caused the picture. The pattern, when you took everything into consideration, just did not fit with anything they were familiar with.

They drew up a remaining shortlist of three broad diagnostic categories. One possibility that still worried many of them was a virulent form of influenza. Influenza A, as the virologists reminded their colleagues, was still causing serious infection in the community. In 1918 a pandemic strain had killed 25 million people. The second possibility was poisoning. For example, some local farmers used phosgene, a First World War poison gas, to put down rodents in their burrows. If pockets collected underground and subsequently leaked out, it could cause a similar pattern of pulmonary disease. The majority did not believe that this was the explanation. That focused people's minds onto the third diagnostic possibility. People began openly to consider a rather shocking scenario. They were not only dealing with something new, a new plague caused by a microbe nobody had ever seen before, but those negative findings pointed to the most worrying possibility of all — what the scientists now call an "emerging virus."

FOUR

Emerging Viruses

A virus is a piece of bad news
wrapped in a protein.

SIR PETER MEDAWAR[1]

1

The fundamental principle — the quest, if you like — behind this book is understanding. It serves little purpose merely to be scared by viruses. But it serves a good deal of purpose to understand them.

What do scientists mean when they talk of a virus? This is not quite so elementary as some people might believe. In *The Shorter Oxford English Dictionary,* a virus is defined as "a morbid principle, or a poisonous venom, especially one capable of being introduced into another person or animal." The dictionary takes its cue from the Latin *virus,* which denotes a slimy liquid, a poison, an offensive odor or taste. It is a colorful definition, redolent of medieval notions of disease origins in evil emanations, but it offers little by way of scientific understanding.[2]

A common confusion is to equate viruses with bacteria. For example, a recent article in a London broadsheet carried a headline about a virus when the offending microbe was clearly a bacterium. Although the confusion is understandable, it is also misleading.

Both bacteria and viruses are very tiny life forms. They are both classed as microbes. Both may cause human disease. Nevertheless, to equate them is rather like lumping together birds with flying insects on the basis that they both have wings. Bacteria are free-living life forms. They imbibe their own nourishment and multiply with a profligate independence. Most bacteria are not parasitic at all but inhabit the soil, making a vital contribution to the cycles of life on earth. Viruses are altogether different.

In the seventeenth century, Antonie van Leeuwenhoek was the first to observe bacteria and protozoa, through his invention of the first simple microscope. He did not, however, see viruses, because viruses are invisible to the light microscope. The first hint of their existence was the observation by Adolf Mayer in 1885 that the mosaic, or leaf-spot, disease of the tobacco plant was caused by a strangely invisible contagious agent. A little later Dimitri Ivanowski, contemporaneously with Martinus Willem Beijerinck, found that the tobacco mosaic disease agent was not only invisible but would pass through the finest pores of porcelain filters. Beijerinck proclaimed it a "contagious living fluid."[3]

The infective agent was not, however, a fluid but a particulate living entity so small it would readily pass through a porcelain filter. We now call it the tobacco mosaic virus. Twenty-five years would pass before any further advances were made in the study of this extraordinary life form.

Although some viruses may be a fifth the size of small bacteria, on the whole they tend to be orders of magnitude smaller, a hundredth or even a thousandth the size of bacteria. The flu virus, for example, is 80 to 100 nanometers (thousandths of a millionth of a meter) in diameter, whereas the bacterial cause of typhoid measures 2 to 4 microns (millionths of a meter). This means that they are not usually visible through the light microscope. It was not until 1933, with the invention of the electron microscope by Ernst Ruska, that their true anatomy could be seen for the first time. For those early virologists, it was an exciting introduction to an entirely new dimension of life.

The ancient Greeks defined beauty as fitness for function: nothing could more adequately describe the austere beauty of viruses, their tendency to symmetry of form, which can give rise to wonderfully aesthetic if geometric realms of perfection.

Even today, there remains a sense of mystery about viruses that seems not to apply to other life forms. If I were to ask a series of different experts what they meant by a bacterium, there would be a pretty close general agreement. Not so with viruses! As I traveled about in many countries and spoke at length with various experts, I would routinely ask them to define what a virus meant to them. Of all the questions I asked, this provoked the longest pauses: it seemed that the more the expert knew about viruses, the less easy the answer. How revealing it was that each expert gave me a somewhat different definition.

As you might anticipate, each had evolved a perspective that tallied with the nature of his or her work. Molecular biologists tended not to perceive viruses as living entities. What they envisaged were nanomachines with their own inbuilt system of replication, controlled by a chemical coding analogous to a computer program.

There is a good deal of truth to this. Outside its host, a virus does appear to become inert, to behave as a complex organic chemical. Viruses can aggregate to form structures that resemble crystals.

Medical virologists, on the other hand, see viruses solely as the harbingers of disease. To them, they appear the ultimate in predators, minuscule time bombs of nascent malevolence, invariably parasitic, contributing nothing positive to life on earth. Yet other experts, in particular the Nobel laureate Joshua Lederberg, have a totally different perception: for them viruses are vehicles for genetic exchange between the disparate species that make up the matrix of life on earth. Seen from a particular perspective, every definition contains its own implicit truth. If viruses were human, they would appear to have multiple personalities.

In my opinion, a virus is very definitely a form of life. It has the same physiology and biological chemistry as all other forms of life on earth. DNA, with its counterpart RNA, is of course the template of life and heredity. It is also the environment in which a virus comes to the full realization of its nascent potential. This landscape, the landscape of the genome, is the world a virus, in human terms, might call home. No other form of life inhabits this extraordinary ecological niche. And in the way of all life, the virus changes the landscape just as the landscape molds the virus. What power such a primal association would seem to confer!

Viruses are certainly very small. They are so minuscule that billions could fit into a drop of water — or a drop of human blood. They are life honed to the absolute minimum. In the opinion of John Holland, an expert at the University of California, they probably go back to the first origins of life. Though they possess no nucleus, they contain the nuclear template of DNA and RNA, no cytoplasm, but some viruses do contain the absolute minimum of enzymes — chemicals that control and facilitate the everyday processes of life.

Viruses have no processes of locomotion, yet in the words of Stephen Morse, "they move around the world." Viruses do not breathe, they do not taste, hear, or see. They do, however, have a single representative of the five senses: they have a kind of sensation that could be classed as intermediate between a rudimentary smell or touch. However, they do not feel or smell through a sensitive skin, with nerve endings that conduct electrical stimulation to a perceiving brain. Viruses have no skin, no nerves, and for that matter no brain. But they have a way of detecting the chemical composition of cell surfaces. Every virus has a chosen host cell, whether it is the leaf of a tobacco plant in mosaic disease or the CD4 subset of T lymphocytes in a human sufferer from AIDS. The virus has the most exquisite ability to sense the right cell surfaces. It recognizes them through a perception in three-dimensional surface chemistry. But then so too does each of our human senses derive ultimately from a similar complex chemistry.

Though devoid of a mind, viruses undoubtedly possess an alternative means of control, what I shall subsequently refer to as a "genomic executive intelligence." Consequently they do not think in a human sense at all. They are amoral, in the true sense — a complete absence of morality. This is a point that needs emphasis because people so readily project along these lines in relation to plagues and disease.

Of course there is a "purpose," speaking from an evolutionary point of view, to this extreme frugality of anatomy. It means that viruses do not need a physiology of movement, of breathing, not even of growth. Their plan of life has made all such processes redundant. Everything is honed down, with incredible single-mindedness, to the pursuit of the only function that matters from an evolutionary standpoint: the need to create more viruses like themselves.

2

When compared to the complexity of the cells that make up a human being, viruses do appear structurally simple. But the comparison is misleading. In their behavior, viruses are far from simple.

The late Dr. Bernard Fields, who originated and edited the masterly text known as *Fields Virology,* marveled at their ability to survive. In their invasion of human victims, he compared them to spaceships voyaging into the most dangerous and hostile of alien worlds. Devoid of even the simplest of vision and hearing, devoid even of their own means of propulsion, viruses must devise ways of spreading from host to host, of battling their way through powerful defense barriers to gain entry into the body. And once inside, these tiny life forms must withstand the unrelenting attack of the cells and antibodies of the human immune system.

Virology is a relatively new field. It is also a rapidly changing field, as important new steps are made both in our understanding of viruses and their ecological consequences throughout the animal and plant kingdoms. Viruses don't just cause diseases in people, they infect every form of life on earth. There are viruses that infect chimpanzees, whales, molluscs, and the bacteria that swarm in the soil or lie on the sunken floors of the seas.

Fred Murphy, Dean of the University of California at Davis, has spent many years of his life classifying them into genera and species. He wrote the taxonomic chapter of *Fields Virology*, in which he describes approximately 4,000 classified species, of which about 150 are known to infect people.[4] On the other hand, the ecologists estimate that there are 30 million species of life on earth, each of which probably has at least one virus species that infects it. In the yawning chasm between the known and the estimate, we glimpse the boundaries of a great unknown, an extraordinary submicroscopic universe that science has hardly begun to explore.

Viruses do not grow, like other forms of life. They are created in an identical assembly, every member at once a replica of the others of its species. When, as part of his Cambridge doctoral work, the botanical molecular biologist George Lomonossoff mixed the separate chemical structure of the tobacco mosaic virus in a test tube, the component structures self-assembled to form the virus. They achieved this without the aid of any outside sources of power, without the need of enzymes: an awesome vision of alien potential, scary and thrilling in equal parts.[5]

There are aspects to virus infection that demand explanation if we are to attain any fundamental level of understanding. For example, very many, perhaps the majority of viruses, infect people, animals, and plants without causing any symptoms of distress or disease. Yet other, admittedly rarer, species of virus are devastatingly destructive in their infections, with a very high degree of lethality. People were aware of the harmful effects of such viruses long before they realized their nature or existence. Viruses have caused some but not all of the frightening epidemics that project like tombstones through the fabric of human history. People were more likely to notice plagues that struck panic and terror in their own communities, perhaps afflicting their own families. Farmers could not fail to notice blights that fouled their crops, or husbandmen the epidemics that swept through their herds of cattle, sheep, pigs, and poultry. Long before the first virus was seen, the elders and doctors of afflicted communities had recognized the contagious nature of such pestilences. Rituals were handed down from shaman to shaman, from doctor to doctor, and from mother to daughter, serendipitous observations on coping with such miasmas, rituals of avoidance, of assisting the victims, of purification of the land and community after the evil's passing.

Plagues were named after the animal or plant that was afflicted, or in human terms after the patterns of disease they caused or the place where the tragedy first appeared. The viruses that caused some of these plagues have inherited the same appellations. The yellow fever virus, for example, causes infection in the liver, which gives rise to yellow discoloration of the skin and eyes. The smallpox virus causes holes, or pocks, in skin, smaller than the holes caused by the "great pox," now believed to have been the epidemic manifestation of syphilis. The common cold tends to infect people in cold weather. Such traditions of naming plagues continue even today. O'nyong-nyong is the Ugandan name for a mosquito-borne virus that was first diagnosed in 1959, causing explosive outbreaks of fever and severe joint pain — the African name means "joint-breaker."

But as our knowledge and understanding of viruses increased, these simple designations and classifications were found to be inadequate until finally, the International Committee for the Taxonomy of Viruses, an international committee meeting in Moscow in 1966, began the first orderly taxonomic classification. Viruses are "polyphyletic" in their origins; in other words, different viruses evolved from different origins, so they cannot be grouped under any single kingdom or phylum of life. The notion of a species, in the sense of compatibility of reproduction, has much less meaning for viruses, but the term "strain" and "substrain" carries a kind of equivalence. Viruses are classed in a number of definitive ways: by size and shape, on the presence or absence of an enclosing capsule, on the properties of their genomic makeup — whether DNA or RNA based, single or double strand, negative or positive sense — on their serological antigenicity, their component proteins, their manner of replication, and other physical and biological properties.[6] This breaks directly into families, their component genera and strains and substrains (species).

For example, the hantaviruses are a genus within the family of *Bunyaviridae,* or Bunyaviruses, and the herpes simplex virus that causes cold sores, *Herpes simplex labialis,* is a member of the genus of simplex viruses within the family *Herpesviridae.* With every discovery of a new virus it is classified on the basis of its similarities to or differences from the large number of known viruses.

And new viruses, indeed new microbes of every description, are still being discovered at an astonishing rate. While the majority do not cause much in the way of disease, a significant minority come to our attention because of the diseases they cause, whether in people, animals, or plants. This whole group of new diseases, whether caused by a virus, a bacterium, a protozoan, or possibly even a non-living particle called a "prion," are linked together under the heading "emerging infections." Emerging infections are one of the greatest dangers facing our world today.

Viruses pose a particular threat, one that is qualitatively different from that of bacteria and protozoa. Few virus infections are amenable to drug therapy. Even vaccines, which were first thought to be the answer to viruses, have controlled only a few. This means that emerging viruses, though far less important than bacteria and protozoa in the day-to-day fight against disease in hospitals and family practices throughout the world, nevertheless pose a potentially more dangerous global threat than any other form of infection. Even in the case of viruses, most of the new emergers do not cause epidemics of fatal disease. Roseola, for example, is a ubiquitous infection of children, which manifests with a pinkish rash and a mild febrile illness. Although it has been recognized since at least 1910, the virus that causes it, a herpesvirus known as HHV-6, was only discovered in 1986.[7]

Some emerging viruses are a good deal more serious. Common examples are the hepatitis viruses, B and C. The hepatitis B pattern of illness was recognized at the end of the nineteenth century, yet the virus itself was not isolated until 1963.[8] The surface antigen of this one virus is being carried by 176 million people globally, and the infection causes a vast amount of illness and death, including most of the fatal cases of liver cell cancer. More exotic examples of emerging viruses include Crimean-Congo hemorrhagic fever virus, a member of the family of *Bunyaviridae*, which emerged independently in Africa and Asia, spreads by tick bites, and causes 10 percent mortality in its wide distribution across Africa, the Middle East, and Asia. The previously mentioned O'nyong-nyong, a member of the family of *Alphaviridae*, emerged in 1959 to cause a series of epidemics that would eventually infect 2 million people in East Africa. And the various Ebola viruses will be discussed in detail later in this book.

Another way in which a disease may be regarded as emerging is when one that is already long established changes its pattern and begins to spread more rapidly so as to become a new danger. Dengue fever is a typical example.

The first epidemic of a disease resembling dengue was reported in Philadelphia as long ago as 1780. Today we recognize four different strains of dengue virus that can cause it, all belonging to the family *Flaviviridae*, which also includes many of the most dangerous arboviruses known, such as yellow fever and the insect-borne viruses that cause brain infection, such as Japanese, Murray Valley, and St. Louis encephalitis. Today dengue fever is one of the most rapidly emerging diseases in tropical parts of the world, with millions of cases occurring each year. Puerto Rico, which suffered five dengue epidemics in the first seventy-five years of this century, has had six epidemics in the last twelve years.[9] At the same time, there has been a massive increase in the numbers of cases throughout many South American countries, extending to Cuba and the Caribbean. The lethal form of dengue, dengue hemorrhagic fever, is also increasing. And with the arrival of the tiger mosquito, *Aedes albopictus*, into America in shipments of used tires from Asia, the circumstances favor invasion into the United States.

All of these viral diseases are regarded as newly emergent or representing a growing threat as reemergents, and the number of new diseases caused by emerging viruses appears to be increasing. In the words of the authoritative Murphy and Nathanson, "The list of newly emergent viruses of humans and animals is impressive, indeed, and is seemingly prophetic of more to come in the future."[10]

Viruses have a phenomenal capacity for mutation: RNA viruses, which have no proof-reading mechanism during replication, mutate at thousands of times the rate of human cells. In previous decades scientists were convinced that such viruses appeared largely as a result of mutation. Today, however, we think that this is much less commonly the case. Emerging viruses do not arise from spontaneous creation any more than other forms of life; rather, they are life forms that have been around a long time but have never been identified with human disease before.

In the words of Stephen Morse, a leading virologist working at the Rockefeller University in New York, "Newly evolved viruses will usually descend from a parent that already exists in nature."[11] While evolution of a novel strain of virus through genetic reassortment or mutation can be important in emergence, for example with new "Asian" strains of flu, a more common route to human infection is through a virus that normally infects a very different species. For such species hopping, Morse has coined the term "viral traffic." Whether as a result of such cross-species trafficking or mutation, when a virus emerges as a totally new entity or as a renewed assault from a new strain of a previously recognized species, the virus always emerges, as it were, into human consciousness. And often human behavioral factors play a leading role in inviting the virus to traffic across species.

There appears to be an inexhaustible supply of such new viruses, with a new addition appearing at least once a year and often a good deal more frequently than that. They emerge from every ecological niche of nature, from rain forests to deserts, from the seven oceans to the decorative plants you buy in a garden center. The vast majority of these are uninfectious to people, but the list of human viruses is gradually increasing. Some years ago, the Rockefeller Foundation was the fulcrum of eight satellite studies spread about the world, including Africa and South America, all actively looking for new viruses that had the potential to infect humans. Over the twenty-year period between 1951 and 1971 they discovered no less than sixty new viruses.[12]

It is fortunate that few such viruses infect us: if they did, the human species would be ravaged by an unrelenting series of epidemics. It is actually quite rare for an emerging virus to cause a global epidemic, what scientists call a pandemic. But it does happen from time to time, and some of the familiar viral plagues of history began as emerging infections caused by new viruses in this way. Smallpox and measles, for example, are thought to have emerged as new diseases at the time of the Roman Empire. Both these viruses spread by respiratory aerosol, the most contagious route for any infection, and when they first emerged, they caused catastrophic epidemics.

Among the more serious new viruses seen in the twentieth century, we can list the pandemic strains of influenza, dengue fever, and HIV. Another important group of emerging virus infections are known as the hemorrhagic fevers.[13] These have certain clinical manifestations in common, a tendency to high fever and prostration, with bleeding from the body orifices. The hemorrhagic fevers include some of the most lethal viruses known to man. They are caused by a miscellany of different viruses, some new, others all too familiar. A familiar hemorrhagic fever virus, and one that has caused an appalling catalog of death and suffering for centuries, is the mosquito-borne yellow fever. Many other hemorrhagic fevers are caused by emerging viruses that have erupted into human awareness through lethal epidemics in many different countries over the past half century, from South America to Africa and Asia.[14]

One such genus of hemorrhagic fever viruses is the hantaviruses. These take their name from the prototype, a virus called Hantaan, which was named after the river that flows through the epidemic area in Korea.

3

Hantaan virus first came to the attention of U.S. military doctors during the Korean War, when more than three thousand United Nations troops were afflicted by a mysterious plague. It caused high fever, weakness, and prostration, many cases showing signs of hemorrhage, often no more than small bruises under the skin or in the lining of the mouth. But in some cases the hemorrhage could be more prominent. People would bleed into the whites of the eyes, they would develop large bruises under the skin or bleed profusely from the bowels. The epidemic was called Korean hemorrhagic fever. Roughly one in ten of those infected died.

Although the disease was entirely new to the Western doctors caring for the troops, it had long been familiar to the Russians and Japanese. They strongly suspected a virus, but in spite of every effort nobody could find it. It was twenty years later that a Korean doctor, Huang Lee, and his colleagues working at the Korean University Medical School in Seoul discovered the Hantaan virus.[15]

After two years of searching, they found the reservoir of the virus in nature. It was a small rodent, *Apodemus agrarius,* commonly known as the striped field mouse. The mouse, it seemed, was the natural host to the virus, contracting the infection from other mice when young. After infection, the virus seemed to do the mouse little harm, living in some curious accommodation with the animal for the remainder of its life, being constantly shed in the saliva, urine, and feces. Epidemics of the virus were now explained. They broke out at the time of the rice harvest, when people disturbed the mouse's burrows, raising clouds of dust contaminated with the dried urine and feces. That dust, thrown up by the pounding of feet and flails, was a teeming effluvium of virus that coated the skin and was inhaled and swallowed by the harvesters.

The striped field mouse extends across most of eastern Asia, throughout all of southern Russia, and into the eastern countries of Europe. Since the discovery of Hantaan virus, scientists have recognized similar viruses as the cause of other hemorrhagic fever epidemics in many countries throughout Asia and Europe. The patterns of illness have been similar to Korean hemorrhagic fever, with fever, hemorrhages and, in a minority of sufferers, death from kidney failure. The syndrome is called Hemorrhagic Fever with Renal Syndrome, or HFRS.

As early as the 1930s a similar illness was being reported in Scandinavia, where, almost half a century later, researchers would discover a different strain of hantavirus, named, after a small Finnish town, the Puumala virus. In 1986 another variant was diagnosed in Greece, Albania, and parts of what was then Yugoslavia. So it went, new strains of Hantaan and Puumala viruses being diagnosed in localized epidemics afflicting one country after another, including Korea, China, Japan, the USSR, Scandinavia, France, Italy, Belgium, Hungary — even Scotland within the British Isles. Another hantavirus, the Seoul virus, was discovered in a fatal human case in Korea. This was a different strain of virus again from Hantaan. It did not infect rice harvesters. Instead, it caused illness and kidney failure in urban residents. The natural host was again a rodent, not a mouse but a rat — or to be more specific, different species of rats, including *Rattus norwegicus*, the brown rat, and *Rattus rattus*, the black rat.[16] A wave of disquiet began to ripple through the biological circles working in the great university conurbations. Rats were common city denizens. One or other of the two rats is to be found everywhere in the world with the exception of Antarctica. Tentatively, in a number of different cities at once, biologists started to screen the rats in their own backyards.

In 1982, Huang Lee joined forces with a Nobel laureate, Carleton D. Gajdusek, to look for hantaviruses in America. Gajdusek recruited his ten adopted New Guinea children to scour his own property on Prospect Hill in Frederick, Maryland, for likely rodents. They quickly discovered a new hantavirus, the Prospect Hill virus, which infected two local species of voles, *Microtus pennsylvanicus* and *Microtus californicus.*[17] Gajdusek and his colleagues at the National Institutes of Health looked for a related disease in humans but could find no convincing evidence for it.

4

In October 1987 the U.S. military staged a field exercise, close to the demilitarized zone in Korea, that involved a simulated attack by thousands of assault troops including large numbers of amphibious marines. Following the exercise, one of the marines reported sick; he died soon after being flown out to a hospital in Seoul. Within days several other troops fell sick with identical symptoms.

The engineering corp seemed afflicted with a relatively high sickness rate, with three cases among a group of around eighty. These were advance teams, responsible for the pre-exercise constructions. Their job was to make sure the following marines could get through the natural obstacles. They would lay out platforms for headquarters tents, build bridges, grade small roads, lay down tracks for tanks. During these activities they lived outdoors, sleeping in tents. They also disturbed a lot of soil. By the time most of the troops had returned to base in Okinawa, a total of thirteen additional marines had reported sick and three more had died.

In 1983, a tall, bespectacled man arrived to work at the U.S. Army Medical Research establishment at Fort Detrick in Frederick, Maryland — familiarly known as USAMRIID. Lieutenant Colonel Tom Ksiazek was a veterinary microbiologist in the epidemiology branch of the disease assessment division. At the time, few medical or scientific centers in the world were interested in funding the kind of research that was taking place at USAMRIID.

Plague microbes were a familiar threat to armies long before their deliberate use as weapons. It costs very little to develop extremely dangerous microbes, which could be all too readily turned against an enemy. As a result of this, biological warfare offers poorer nations a means of terror every bit as dangerous as nuclear weapons. In the 1960s, overwhelming public revulsion caused these centers to be abandoned in Europe and America. But it seemed prudent to continue the defensive aspects of this research. One such defensive strategy at

USAMRIID was the search for vaccines against dangerous infections, whether of natural origin or the brainchild of ruthless biological experimentation. Tom Ksiazek's role was to discover and improve upon the methods of diagnosis. He had a particular interest in the hemorrhagic fevers, including those caused by hantaviruses.

In the fall of 1987, when Ksiazek was attending a meeting for overseas research laboratory staff at the Walter Reed Hospital in Washington, a doctor from Okinawa gave him some specimens of serum from the war exercise outbreak. Tom took the specimens back to USAMRIID for testing.

At this time, it was still difficult to diagnose a hantavirus infection. The usual test used an indirect fluorescent antibody assay, the same test that had been developed by Dr. Huang Lee, who had discovered the Hantaan virus. This was reasonably accurate, but it was subjective, and laboratory staff had to maintain a supply of infected field mice, passing live virus from one generation of mice to another. For scientists working with dangerous pathogens, the greatest risk comes from handling infected animals. They squirm, kick, scratch, bite, introducing the unpredictability of life into an already dangerous situation. German colleagues had devised a safer way of testing for certain virus antibodies, but these had not included hantaviruses. At USAMRIID, Tom Ksiazek modified the test for hantaviruses.

His test was called an immunoglobulin capture assay. Since the test would work just as well with dead virus, scientists could kill the virus first, removing the hazard of handling live virus or infected animals. If the person's blood contained antibodies to Hantaan, the dead virus would attach to the captured antibody layer. A dye, attached to a virus-detecting system, would show up green in the well. Tom Ksiazek tested the serum from the first dead marine at USAMRIID and the wells turned green. Now he and a colleague, A. J. Williams, flew out to Okinawa, where they set up a laboratory within the hospital to test the troops.

Three and a half thousand marines were now screened for antibodies. Although the assay was hard work for the laboratory staff, testing of a new case could be completed in four hours. Ksiazek's test system proved dramatically accurate in the field setting. Doctors were able to diagnose cases the same day they arrived at the hospital. But two years later, such optimism for the test would be confounded by experience in Central Europe.

In 1989, there was an outbreak of hemorrhagic fever in Sarajevo, in what was then a united Yugoslavia. To make the situation more difficult, it was known that there were at least two types of hantaviruses infecting local rodents. One was closely related to Hantaan, while a second was a variant of Puumala. When Tom Ksiazek arrived, his immunoglobulin capture test was used to diagnose suspect cases coming into the local hospital. But here in Sarajevo he made an important observation. As anticipated, some patients were confirmed as suffering from Korean hemorrhagic fever. But he also found that a high proportion of these people were not infected with Hantaan virus at all but with Puumala. This created serious diagnostic problems but also an interesting dilemma.

With two viruses that are of the same genus but different strains, there is some degree of cross-reactivity on antibody tests. In other words, a patient infected with virus A will have a weak positive reaction to virus B, and vice versa. Tom, who had hoped his test would screen for all hantaviruses, found that the cross-reaction between Puumala and Hantaan was deceptively weak. There was no problem if he obtained a strong antibody reaction to the Hantaan virus, when he could confidently diagnose Korean hemorrhagic fever, but in those with a weak positive to Hantaan he could not be sure if it meant Puumala or nothing at all.

A test system designed only for the Hantaan virus was not going to cover the other hantavirus. For the army scientist it had proved a valuable lesson.

But there was an additional, more subtle, implication to these hantavirus researches and discoveries. The genus of the virus was distributed widely around the globe, where new strains of the virus — the equivalent of species — were increasingly being discovered. Hantaviruses, it seemed, were distributed in a vast archipelago of species throughout whole continents. There was some curious relationship between the virus and its host that must signify an ecological stability, an intercalation with nature, that ran counter to the classical thinking of viruses as simple predators.

In 1991, Tom Ksiazek left USAMRIID to take up a post as chief of the Diagnostic Section in the Special Pathogens Branch of the Centers for Disease Control in Atlanta.

FIVE

The Special Pathogens Branch

"We have discovered the secret of life,"
Crick told everyone within earshot over drinks
that noon at the Eagle.
It was not the entire secret of life,
yet truly for the first time
at the ultimate biological level structure
had become one with function.

HORACE FREELAND JUDSON[1]

1

The Centers for Disease Control is situated in Clifton, a pleasant suburb of Atlanta, attractively wooded with hickories, oaks, gums, and loblolly pines. There is a small island where cabs can pull up to the entrance, with the Stars and Stripes fluttering around a lofty flagpole. Inside, the Virus and Rickettsial Division is approached along a catwalk of plain concrete with thick tubular steel guards, fenced with a steel mesh. There is the absurd feeling you are striding the clanking corridors of a high-security prison. The catwalk ends abruptly with a door set into the wall. A notice reads: WARNING. RESTRICTED ENTRY. AUTHORIZED CARD KEY ACCESS ONLY. Here the visitor's pass will take you no farther. A security card, restricted to essential staff, must be inserted into a slot, when a green light shows and the door opens.

Inside the building, you descend three floors by elevator before emerging onto a wide atrium, floored with six-inch-square tiles of red terra-cotta. The atrium divides the Virus and Rickettsial Division into two separate but interconnected buildings. A constant throbbing and humming fills the visitor's ears. For a moment the floor appears uncertain, as if you have clambered aboard the swaying deck of a ship. Your eyes are drawn to the walls that surround you. Those on the left have windows: essentially five floors of offices with some laboratories behind them. But on the right is a sheer cliff face of dead-wall concrete, with a single row of inscrutably dark windows halfway up its towering reach. This is the maximum security laboratory: in the language of virology, it is a biosafety level four facility, a BSL-4.

There is only one other facility like this in the United States and that is at USAMRIID. Russia has two. Canada has just built its own. Europe has no identical facility, though Britain, at Porton Down, has an upgraded BSL-3, which is designated to work with BSL-4 agents.

The Infectious Diseases Center of the CDC, and particularly the Special Pathogens Branch, has a protective role that extends far beyond the United States. In the words of Jim LeDuc, who heads emerging viruses for the World Health Organization, this is the public health service laboratory for the world. By the very nature of their vocations, the people who work here spend their lives in close proximity to danger. The department is on permanent standby. If the virus of our worst nightmares were to appear somewhere on earth — a pandemic strain that would threaten all of humanity — the front line, with its terrible dangers and heavy burden of responsibility, would be led by the handful of people who work in this building.

The tissue specimens and the blood from five or six suspected victims of the New Mexican mystery plague first arrived on the desk of chief pathologist Sherif Zaki, who works a block away from the Virus and Rickettsial Division. On the Friday before the Memorial Day weekend, he examined the formalin-fixed specimens of lung, liver, kidney, spleen, and lymph glands, including those taken by Patricia McFeeley at the autopsies of Michael and Rosina and forwarded from New Mexico. Zaki's initial findings corroborated those of Umland and McFeeley. There were nonspecific changes in various internal organs. The lungs showed inflammation in the tissues between the alveoli, an appearance termed "interstitial pneumonitis."

Zaki had, however, noticed something that might be important. In three of seven cases whose autopsy samples he obtained, the changes were different from the others: where the others had shown differing patterns, suggesting a variety of unrelated diseases — for example, one patient who had inhaled food after a fit — these three exhibited a pattern in common. This suggested they were suffering from the same if unknown cause: and they included Michael and Rosina. It seemed to Zaki that the microscopic appearance followed the pattern he would expect of a virus.

Sherif Zaki's formal title is Chief of Molecular Pathology and Ultrastructure. He has techniques that will probe for the elusive traces of viruses within the labyrinthine complexity of the ultramicroscopic cellular landscape. On Monday, he began a second line of approach, subjecting the formalin-fixed tissues to a range of immunohistochemical stains that would target unusual viruses. He used probes of exquisite delicacy generated to single viral antigens, called monoclonal antibodies.

While Zaki hunted with these refined tracking techniques, his suggestion of a virus attracted the interest of the Special Pathogens Branch.

This disease was lethal. If they were dealing with a virus, then the frozen sera and fresh tissue specimens were highly contagious. On Monday, May 31, all the fresh sera and tissue biopsies were passed to Tom Ksiazek in Special Pathogens. The handling of hazardous specimens of this nature is performed under BSL-3 conditions in a laboratory directly behind the offices. To gain entry, you need a restricted access security card. These laboratories are under negative pressure with regard to the corridor outside. When the door is opened, air is sucked into the laboratory, rather than the other way around.

Viruses classed as BSL-2 include measles, mumps, influenza, and many of the common respiratory viruses, such as colds, parainfluenza, and flu. An average hospital laboratory would feel comfortable dealing with these. Many are highly infectious, but they are rarely lethal. The human immunodeficiency viruses, HIV-1, and HIV-2, are often handled in BSL-2 because, though shockingly lethal, in air they are incongruously fragile and hardly contagious. The BSL-3 laboratory provides a further degree of security against lethality and danger.

The air here is not recirculated. You progress through two more doors, each a zone of decreasing pressure, so the airflow follows your progress. The next door reads BSL-3. CAUTION: GLOVES REQUIRED IN THIS AREA. DISPOSABLE GOWN REQUIRED IN THIS AREA. CAUTION: MASK REQUIRED IN THIS AREA. DANGER: EYE PROTECTION REQUIRED HERE. The international biosafety hazard logo stands out in startling carmine: the central seed with its spreading, intersecting ripples.

It was here, on Tuesday, June 1, that Tom Ksiazek and a French scientist, Pierre Rollin, began the virological examination of the specimens from New Mexico. They wore portable full-face pressure masks that covered their eyes like a windshield, the air they breathed coming in through a small blower fitted to their belts. The task was tricky, dangerous, and laborious. They worked in laminar flow cabinets, or "hoods," their protected hands and arms inserted through an eight-inch slit in a Perspex wall, all the air within the cabinets drawn in and taken away through HEPA filters.

With a slow deliberation, they chopped up the tissue specimens with scissors, then homogenized them, creating a potentially lethal spray of aerosolized tissue that was whisked away by the airflow. The laminar flow cabinet is fitted with a warning. If things go wrong a light flashes or an alarm sounds, or both. And if this happens, the probability is that lives are now in danger. As the experienced Ksiazek and Rollin toiled away, no bell sounded and no light flashed. The homogenized fresh tissues were now resuspended in saline. Some samples were put into cell culture lines in a hood fitted with an ultraviolet light that sterilized everything after use. Others were put aside for the moment.

Using the same precautions as were employed with the tissue specimens, Ksiazek and Rollin divided the sera into a series of smaller samples. Some of these were again inoculated into tissue cultures in BSL-3. But some of the most dangerous viruses are extremely difficult to grow in tissue cultures. It is standard practice to inoculate suspect material into animals, to observe the development of illness and to search in the sacrificed bodies of the sick animals for evidence of the mystery pathogen.

2

Still wearing BSL-3 protective clothing, Ksiazek and Rollin carried those samples up two levels, crossed a bridge, and emerged outside a door that would permit entry into the BSL-4 laboratory. The key-cards at the CDC are all individually coded, so they allow each individual specific access through designated doors. Very few cards are coded to allow entry here. If any unauthorized person tries to open this, even if they are in possession of the right card, the lock will not open. A light in the mechanism displays red. A second security code must be dialed.

As you stand outside the entrance to BSL-4, you cannot ignore the sound of the huge pumps, driven by engines like those of jet aircraft, powering the movement of air and exhaust from every corridor and room within this building. The walls are two feet of concrete, designed to withstand a nuclear attack. Everything is built in duplicate — air supply, water supply, cooling system, decontamination systems — part of the omnipresent spirit of caution. The laboratory has its own electrical generators in case of need, its own water supply, separate from the citizens of Atlanta.

Ksiazek and Rollin went through a careful ballet of maneuvers to pass through this series of doors. For here is where the most dangerous viruses on earth are stored and handled. There are no treatments for these viruses and no vaccines. Fortunately for humanity, none of these viruses has ever spread from human to human by aerosol. Every now and then the death of a laboratory scientist will hit the headlines: some unfortunate man or woman who, in spite of a lifetime of toil with appropriate caution, has suffered an accidental exposure through working with a virus such as these. BSL-4 viruses are mercifully few, at about a dozen, but the list is slowly increasing. They include the hantaviruses, such as Hantaan and Puumala, and various strains of Ebola.

Most of the people whose lives depend upon Tom Ksiazek will never meet him, let alone even know that he exists. Yet a very great deal depends on people like him, the quiet, hard-working people whose lives are spent in the laboratory background. When it comes to the identification of new or emerging viruses, Tom is one of the most experienced diagnosticians in the world.

In an anteroom, the two scientists removed their office clothes, then passed through an air-lock shower, closing the vaultlike door behind them. From the shower they entered a changing room where they donned green surgical scrubs, socks, and surgical gloves. From here, they progressed to another room containing heavy rubberized suits, a striking cerulean blue in color, that were hanging from a rack on the wall. As they picked their individually labeled suits off the rack, they first inspected them for damage, the thick overgloves getting particularly close attention, then they climbed into the thick blue rubber, with its inbuilt bootees. Before pulling the hoods over their heads, they put on hearing protection. From now on the world receded in terms of sound and touch. To the suits they attached the heavy metal couplings of their air regulators. The final step was a literal one, placing their feet into Wellington boots before attaching the air regulators to dangling coils of ceiling hoses, then listening to the sound of the air rushing into the stiffly swelling suits.

Each man, for the time he worked in BSL-4, was weighed down with an extra twenty-five pounds of pressurized suit: his eyes looked out on the world through the Plexiglas visor, his hearing, considerably dampened by the ear protectors, was further impaired by the incoming air. Without the ear protectors, the rush of air sounds like a gale force wind. Even wearing the compulsory protectors, there is a tendency over long periods for diagnosticians who work in these conditions to suffer permanent hearing loss. From the time of uncoupling, they had thirty seconds to reconnect with another of the lifelines dangling at intervals from the ceiling. They had entered the normal working environment of the BSL-4 laboratory.

An hour working in here is equivalent to two or three under high tension on the outside. Very few human beings can withstand this mental pressure. "It is very common for people who start off well to burn out after a number of years. They begin to worry after a near accident, or they hear about somebody else who has had an accident and they get to thinking a little harder about what they are doing. Little by little the consequences start to prey on their minds in a way that didn't happen when they first started. They eventually opt to do something else."

In the animal room, they began the painstaking task of inoculating mice, gerbils, and guinea pigs. Every day, from now on, one or the other of the men would return to inspect the animals for signs of illness. But their work in BSL-4 was not yet complete.

There was an even more restricted area, the most restricted of all: they entered, through a door marked with the logos of both the biological and radiation hazard warnings, into a windowless room, a sanctum within a sanctum. This is a killing ground for dangerous viruses. It contains a radioactive source, the GAMMACEL, based around a core of cobalt 60 so hot it has to be sealed within a ten-ton lead vessel. Some viruses are extremely difficult to kill. Where, for example, a lethal dose for humans would be 600 or 800 rads, the dose for viruses can be as high as 5 million.

To enter this room, Tom Ksiazek used a third code-key, the highest security precaution of all. Here he irradiated a batch of serum samples to kill any presumed virus they might contain. These dead viruses, which still contained all of the viral antigens, would provide the samples that would be distributed widely throughout the CDC, to laboratories that did not have the containment facilities of the Special Pathogens Branch. Some of these samples would also go forward into the third phase of investigation, the immunological screening that would begin the following day.

At this stage, Tom Ksiazek had to attend a meeting at USAMRIID. The continuing investigation was now delegated to Pierre Rollin.

3

Pierre Rollin is of medium height and broad muscular build, with dark curly hair. Born in Morocco of French parents, he graduated at the medical school in Nîmes before moving to the Pasteur Institute in Paris, where he trained in virology with the distinguished Pierre Sureau. He met Tom Ksiazek and C. J. Peters, now the director of the Special Pathogens laboratory, during a research fellowship at USAMRIID, when he helped them contain the outbreak of Ebola virus that broke out in a monkey quarantine facility at Reston. Today, as a visiting scientist within the Special Pathogens Branch, he has the title of Chief of Pathogenesis.

When, on Wednesday, June 2, he began the serological testing of those first tissue and blood samples, there was little to suggest that they might be dealing with any of the more familiar BSL-4 viral pathogens. He had a total of eight samples of sera from five patients, including a second, or "convalescent," sample from three suspected cases, including Frank and Dolores. Rollin began testing using the older immunofluorescence antibody test. It proved uniformly negative. On June 3, he moved on to the immunoglobulin capture assay, developed by Ksiazek in those arduous field investigations in Korea and Yugoslavia. Ksiazek had since extended the range to screen for every known hemorrhagic fever virus: Crimean-Congo, Rift Valley fever, the four filoviruses, Marburg, Ebola Zaire, Ebola Sudan, Ebola Reston, and the three arenaviruses, Lassa, Machupo, and Junin. He included the three hantaviruses, Hantaan, Seoul, and Puumala.

The plates were 10 x 15 centimeters of semi-opaque plastic, in columns of twelve wells horizontally by eight vertically. Each vertically running column contained a different serum, with increasing dilution of the serum as you traveled down the wells. To inspect by eye, you hold the plate over a white surface. A positive would show up green.

To his astonishment, when Rollin performed the tests, a number of wells turned green with one or more of the three strains of hantavirus.

If the antibody level is high, if the result is strongly positive, the well turns a deep shade of olive, so dark as to appear black. None of the sera were strongly positive and no positives showed with the tests on any virus other than the hantaviruses. A less experienced investigator might have dismissed the results as too weak and confusing to be significant. But Rollin prepared himself to spend the night in the lab. He restarted the tests, challenging the first batch of results, and extending the lower dilution ranges. The next reading would come through about 3:00 A.M.

Those repeats confirmed the earlier tests: three patients' sera were still showing weakly positive to one or more of the three hantaviruses. They included the sera from Frank and Dolores. All three positives corresponded with the common pattern of "pneumonitis" reported by Sherif Zaki. With the repeat of the test, Rollin was able to confirm that two of the samples showed a rising titre of antibody, perhaps even a conversion from IgM to IgG. Though tired, he felt exhilarated. "I was excited because we had something. I thought it significant because it was the first positive result in all of the tests done at the CDC."

When he read the plates with a densitometer, the values were significant but they were not diagnostic of infection with any of the three viruses. If a patient was infected, say with Puumala, the convalescent sera readings would have gone off the scale.

In the early hours of the morning of Friday, June 4, Rollin drew up a table of the results and sent it by fax to Ksiazek at USAMRIID. The fax would not reach Ksiazek, who was sleeping in the visitors' accommodation, until the following morning. Rollin, meanwhile, took the table to C. J. Peters. Peters was as excited as Rollin, but both men remained cautious. When the results eventually found Ksiazek, he also felt the need for skepticism. "People were dying in the Four Corners region at that very moment. It would have been disastrous to jump into something that might have turned out to be a false diagnosis."

Ksiazek had, of course, seen exactly such a pattern before. These results were a carbon copy of those confusing reactions he had seen in Sarajevo. It seemed likely they were seeing a cross-reaction between antibodies in the victims' blood and all three known hantaviruses. He hardly dared to credit the logical implication: they were dealing with a hantavirus, certainly, but it was not one of the three known to cause human disease. Ksiazek advised Peters to repeat the tests.

Peters took the chart up to the sixth-floor office to discuss it with Dr. James M. Hughes, the director of the National Center for Infectious Diseases. Rollin, meanwhile, repeated the tests. This time he included a fourth hantavirus, the Prospect Hill virus, discovered by the Nobel laureate Carleton Gajdusek. Prospect Hill virus had never been known to cause human illness. But it was an American virus — the only American species of hantavirus. Rollin became even more excited when he obtained a slightly stronger reaction to Prospect Hill than to the other three viruses. But still it was not diagnostic. He did not see the black-green in wells at a high dilution. Another cross-reaction.

Although the serology was now pointing toward a hantavirus, this virus was every bit as mysterious as the plague itself. No hantavirus had ever been associated with epidemic disease in America. This virus followed none of the known patterns, whether on the immunological tests or in its symptoms in people, including the terrifying lethality.

The obvious proof would be to grow the virus in tissue culture or to demonstrate its effects in the inoculated animals. But the scientists were under no illusion that this would be easy. With Hantaan fever, viral isolation had taken thirty years, and although technology had improved greatly in the fourteen years since the first isolation of Hantaan, hantaviruses remained notoriously difficult to grow. A decision had to be made on how to make the best use of the information at hand. The practical implications of a hantavirus diagnosis were spectacular. In every previous outbreak caused by a hantavirus, the virus had infected humans through contact with rodents — a transmission that could be interrupted if the rodent could be identified.

4

On Saturday morning, Jamie Childs, the epidemiology section chief in the Viral Zoonosis Branch, accompanied by the tall, bearded mammalogist John Krebs, stepped off a plane in Albuquerque into a wall of dry heat. The two biologists were accompanied by a tall, dark-haired woman, Dr. Louisa Chapman, who had volunteered to help with the epidemiology in the field. During an extraordinary press briefing in Santa Fe the previous afternoon, and despite the misgivings of many experts, the serological diagnosis of a hantavirus had been announced by the state medical officer, Norton Kalishman, with the support of Mike Burkhardt, Gary Simpson, and C. Mack Sewell. It was into the charged atmosphere following this announcement that the newcomers first arrived, and they were determined to move quickly into their field investigation.

On Sunday they hired a large van to transport the several tons of equipment that had been ferried in from Atlanta by plane. On Monday morning, they picked up their field crew and drove the 135 miles along I-40 to Gallup.

For the two men and the woman sitting across the broad front of the van, the scenery was deeply impressive. Once they had cut through the bowl of mountains that encircles Albuquerque, they emerged onto a desert plain, intersected by magnificent buttes and cliffs. None of them had ever worked in this desert landscape before. John Krebs would remember how the high-sided van was caught in the crosswinds. "Every time we emerged from behind a butte, the wind would toss us around like a ship in a gale." Dark and solitary clouds scampered over a vast dome of sky, trailing ultramarine curtains of shadow. It took them some time to realize that the shadow was in fact rain that evaporated before it hit the ground. Thunderheads would suddenly boil up out of nowhere. Lightning would erupt out of the black underbellies, four or five discharges at once, all from separate foci close to one another, startling zigzags of actinic light that fell onto the highway ahead, evoking lingering afterimages on their startled retinae.

At this time the Navajo were being tormented by the press. The field crew heard reports that people on the reservation were shooting at helicopters. It was not a comfortable place for intruders. By this time, the epidemiologist, Jay Butler, had been heading a small team out in the reservation for more than a week. At the Holiday Inn they met up with the sandy-haired, and mustachioed Butler, already exhausted and sick with what he hoped was no more than food poisoning. Butler had a pained laugh with Childs. "You look so good now. I hope you don't look as bad as I do after a week!" Butler's investigation, made difficult by the paranoia now raging over the reservation, had already suggested that aerosol spread of the mystery virus was very unlikely.

Jay Butler diverted Louisa Chapman into the hurried educational exercise now taking place in Gallup, where large numbers of health visitors employed by the Navajo Nation and Indian Health Service were getting ready to fan out over the entire reservation to begin the epidemiological investigation. A dark-haired, bearded, and blue-eyed Irish American called Pat McConnon, a qualified biologist and able administrator, was establishing the order of a base office and coordination of supplies out of the chaos. The hospital switchboard in Albuquerque and the press information offices in Santa Fe were overwhelmed with panic calls, fielding thousands of inquiries in twenty-four hours. Following two days without sleep, Chapman was diverted to coordinating the trial of the drug ribavirin against the presumed virus. The two biologists, meanwhile, were soon experiencing the intrusive presence of the press for themselves.

News helicopters homed in on their activities, attempting to follow them to places of interest, complicating their attempts to accommodate the culture clash between modern science and the traditional Navajo way of life. Those cultural tensions threatened to undermine the field operations. Before Childs and Krebs could even start their work they had to obtain permission to go into the reservation. Yet they would need to perform intimate inspection of homes and their surroundings for rodents. In the act of doing so, they would need to wear protective clothing that could appear unduly alarming.

With some of the local people already believing they were being punished, in a religious sense, for transgressions, it was obvious that they needed a Navajo representative wherever they traveled. Nowhere was this more important than when they entered those most intimate of arenas, the houses and compounds where people had died. A Navajo, Herman Short, proved invaluable during the coming days and weeks, when they needed increasingly to link up with the Indian Health Service.

Their first destination was the extended family complex at Littlewater, where the outbreak had begun. Even as they drove north on the dusty Route 371, the biologists passed signs pinned to gates or on stakes hammered into the ground that warned away intruders. The arrangements for their inspection were so sensitive they had had to detail in advance exactly how many people would be coming. If they specified four people, then four people, no more and no less, would be allowed in. On the reservation, families were located much as the typical farming communities in the Midwest, in small compound living areas, a half mile to several miles apart. They entered the extended compound of Rosina's family, which comprised three mobile homes, now abandoned, and a small mud structure. There were two or three derelict cars per building that seemed to function as storage sheds, now festooned with full-grown weeds. The cars were sealed, but to the experienced eyes of the biologists, rodents could still get into them through rubber grommet holes for cables and the entry holes for steering levers and foot pedals. There were also outhouses and garbage piles.

It was very early in the morning. The atmosphere was tense. As they arrived, waiting as instructed inside the car, the family representative checked them over, to make sure they had kept to the agreed number of scientists.

One of the homes was a long white trailer. This was Michael's family home, where he had lived with Rosina and their baby son. Now the experienced eyes of the biologists perused this ravaged homestead for the presence of a possible carrier of the virus, a likely familiar and common rodent. For his doctoral thesis, Jamie Childs had tracked rats through the backstreets of Baltimore, astonished by their ability to burrow under the concrete of people's backyards. They would excavate the dirt underneath these huge concrete areas to the point where they would collapse and crack. In those rats, Childs and his collaborator at USAMRIID, an army scientist called James LeDuc, had confirmed the presence of Hantaan and Seoul viruses in American cities.[2] To Childs's practiced vision, there was no way that anyone living here could avoid coming into contact with rodents. Even in Baltimore, well over half the homes were infested with house mice. In this rural setting, the ubiquity of rodents was a fact of life.

The family representative unlocked the doors so they could enter the buildings. The biologists climbed out of the van, donned paper disposable gowns, double rubber gloves, and half-face respirators. They pulled on goggles that fitted separately above the masks. From now on, even the simple act of talking would be difficult. They would have to shout to hear each other above the noise of air being drawn in through their masks' two side filters.

They entered the trailer where Michael and three others of his extended family had first become sick. The family had abandoned the trailer so fast there had been neither time nor the inclination for tidying up. Unwashed dishes were piled in the sink, dirty laundry lay scattered about, family photographs on the walls and sideboards. The anguish of sudden bereavement was poignant. Even for the seasoned biologists, it was distinctly unsettling. There were obvious signs of mouse infestation, and they began to lay their first traps. From the trailer, they moved from building to building, from abandoned car to abandoned car, all the time finding more evidence of rodent infestation, laying more traps. When they had finished with the cars and buildings, they gazed out over the surrounding landscape.

Over the previous weekend the biologists had formulated a standardized protocol with Jay Butler and the epidemiologists that would allow the two groups to collate results. Now they put that protocol into effect around the compound. The index home was the center of their compass. They set more traps at the cardinal points, then they walked along the radii, setting a trap every fifteen to twenty meters for as far as two hundred meters. Pack rats are a common rodent in the New Mexico landscape, with big mounds of twigs marking their burrows. At every third station they put in a large trap to catch pack rats. In all, they set about a hundred traps. They made a special inclusion of wood piles, shade huts, garbage piles. It was an exhausting business. When they returned the next morning, much of their work had been ruined by an unanticipated problem.

There were packs of feral dogs, thirty or forty strong, roaming the area. The dogs had been attracted to the bait laid in the traps and overnight half the traps had been sprung by them. The biologists were forced to abandon the experiment and call in some animal patrol specialists to take some of the dogs away so they could bait the traps all over again.

This time, when they returned in the morning, they found that out of seventy or so smaller traps, forty-five contained rodents. The captured animals were anesthetized, blood samples taken, and they were sacrificed, the internal organs removed at autopsy in the field and the fresh tissues preserved for virus examination. It was clear to the biologists that they were dealing with exceptional rodent densities. Said Childs, "When we had trap successes in the order of fifty to seventy per cent, we knew we were dealing with a major rodent plague."

It was also apparent that they were seeing a great many of one particular rodent, *Peromyscus maniculatus,* commonly known as the deer mouse. This is a wild mouse, whose natural habitat is in fields and landscape, but it will readily enter people's houses. For scientific validity, Childs and Krebs needed to compare the results at the index compound with a "control" home compound. But lack of cooperation from the neighbors forced the biologists to travel farther afield.

Meanwhile, the samples of rodent blood and tissue, frozen in liquid nitrogen, were flown to Atlanta, the first collections arriving on Sunday, June 13, for urgent examination.

The blood was screened using Tom Ksiazek's capture assay for antibodies to hantaviruses. By the early hours of the following day the results were through. From the first twenty-odd blood samples taken from captured deer mice around Michael and Rosina's home compound, eight were positive. The deer mice had antibodies to hantavirus in their blood. The mice appeared to be infected with the same mystery virus.

5

The antibody results Rollin had obtained on those first three patients were very suggestive, but they were not diagnostic. No more were the antibody positives in the rodents. They could not determine which hantavirus was causing the epidemic.

The syndrome of the disease was unlike anything that had ever been seen before. To confirm that people really were infected with a hantavirus, to discover exactly what species of hantavirus was causing the epidemic, and to make absolutely certain that *Peromyscus maniculatus* was indeed the source of the epidemic, the scientists needed to be able to demonstrate the virus in the tissues of infected humans and in deer mice from the locality where people had picked up the infection.

In 1953 the young American James Watson and the Englishman Francis Crick made what is generally believed to be the most important biological breakthrough of the twentieth century when they discovered the stereochemical structure of DNA, the template for heredity and ultimately the master program for life itself.[3] Watson and Crick's elucidation of DNA was the biological equivalent of Einstein's discovery of the interconvertability of mass and energy in physics.

Viruses are living entities, sharing a common template of DNA, or its sister molecule, RNA, with every other form of life on earth. Now it was possible to understand them at their most intricate level, that of their genetic organization. For the proliferating fields of genetics and molecular biology, they afforded a shortcut to the exploration of life.

Many of the most important truths have an elegant simplicity. Darwin's discovery of natural selection is just one example. The essential structure, and therefore the function, of DNA is another that can readily be grasped. To move in these realms is to enter a world akin to magic, that offers a seeming infinity of possibilities.

DNA, or deoxyribonucleic acid, is a coded language. Like all languages, it is made up of words and the words themselves are composed of letters. DNA uses a finite alphabet of just four letters, each of which represents a molecule of nucleic acid. Fantastic as it might seem, all of DNA, the program of heredity within the nucleus of even our human cells, comprises just those four nucleic acids, known by their initials, G for guanine, A for adenine, C for cytosine, and T for thymine (the acronymic GACT). The template of life is literally spelled out using this quaternary code. The individual letters, the words, and the book itself are joined and held together in a double helix of singular beauty by a chemical backbone of sugars and phosphate. Each individual nucleic acid, with its attached sugar molecule and phosphate, is collectively called a nucleotide. The sequences or words spelled by chains of these nucleotides are called genes.

So how could the diversity and complexity of life arise from such a simple coded language? The answer lies in the fact that these genes are thousands or even hundreds of thousands of nucleotides long. Such a gene is the chemical code for a single protein. For every protein in our bodies, as in the bodies of an ant or an oak tree, there is an appropriate gene. When all of these genes — or words — are added together, you have a book, the book of life for any organism on earth. There is a special word for this book of life: it is called the genome.

In the nucleus of the human cell, the genome is so large that it has to be separately bundled up into forty-six separate chapters, which are called chromosomes. These actually comprise twenty-three matching pairs, one from each pair inherited from either the father or the mother. Our human genome, made up of its forty-six chromosomes, contains a hundred thousand genes, themselves coded by billions of nucleotides. This same library of life, written in the language of DNA, is the template for every living creature on earth, from a whale to a kingfisher, from the humble amoeba to Albert Einstein himself. Viruses are no exception, though their genomes are far smaller than that found in the nuclei of human cells.

It was Einstein who declared that the most beautiful and most profound experience anybody can have is the sense of the mysterious. And there are very mysterious properties, remarkable properties, to the behavior of this wonder chemical, DNA. If you put a chain of DNA into a solution — call it chain A — and then add liberal amounts of the four nucleic acids that comprise it, they automatically line up along the chain of DNA. They do not line up each nucleic acid opposite its twin, but adenine will attract only thymine, cytosine only guanine, and vice versa. Now, if an enzyme (a complex protein itself governed by a gene) is added, the nucleotides that are lying adjacent to their mates join together to form a new chain of DNA, say chain B. This chain B is not identical to the parent chain A. It is the chemical equivalent of a mirror image, a perfectly predictable match. Now consider what would happen if we were to repeat the process of creation, using chain B as the parent chain! The result would be a new chain, identical in every way to the original chain A. In the language of genetics, for example when referring to exactly the same process of replication in messenger RNA, chain A is the positive sense and B the negative sense.

Whenever you put that same chain into a solution of the nucleic acids, you get that identical chain forming along its length. In this way the chain of DNA, the word it codes for, has an innate ability to create its word mate. This is how the message of life is passed on, as cells divide to form other cells, and their genomes replicate. In the double strand of DNA that makes up the double helix, one strand follows the pattern A, entwining in reverse parallel with a chain following the pattern B.

Each gene, at the heart of the cell, codes for a single protein, such as the color sense in our vision, the clotting factors in our blood, the mucus in our bronchial tree. If a mistake is made — it might be only a single wrong letter, a nucleotide somewhere along a chain of tens of thousands — this can have catastrophic effects on the function of the protein that is coded by that gene. Such a mistake, which is called a "mutation," is responsible for many hereditary diseases: for example, color blindness, hemophilia, sickle-cell disease, and cystic fibrosis. There are as many possibilities for mutations as there are nucleotides along the string of genes.

In the code of life, DNA has a sister compound called RNA, or ribonucleic acid. RNA is just as easy to understand, comprised again of four nucleic acids, three of which are the same as those spelling out DNA. One of the nucleic acids has been changed. Thymine has been replaced by uracil, denoted by the letter U in the acronym GACU. Despite the single nucleotide change, the precisely coded string of RNA will form opposite its complementary DNA sequence. Human cells do not use RNA as a genomic code but as a messenger molecule, the delegated word that carries the information from the chromosomal DNA out through the nuclear membrane and into the cytoplasm, where granular structures called ribosomes are the factories for the production of proteins. Each gene or word in the nucleus in this way codes for a specific messenger RNA, which in turn codes for a specific protein. Viruses differ from humans — for that matter from every other known form of life on earth — in that while some employ DNA as their code of life, many use RNA instead.

Just as every one of our living cells contains our genome coded in the chromosomes, every virus has a core of DNA or RNA, wrapped in a protective capsule of protein. There may be other layers and other components that play a part in the process of infection, but it is the DNA or RNA that determines the virus, that codes for its species and physical structure, that governs its astonishing life cycle. Where the human genome is made up of a hundred thousand genes or words, a viral genome comprises a very small number, perhaps three, four, eight, or ten. But they are important words, words of immense potency. The language of the genes is devoid of courtesy. It is the jargon of command. When the words are "spoken" — when the DNA sends the messenger RNA to the ribosomes — the proteins that derive from the genes are created. Those proteins fashion the tissues, the enzymes of the blood, and through them every organic component that makes up the living body.

This also means that if you know the words that make up any organism's book of life, if you possess the plan of its DNA or RNA sequences, then you can identify the organism with absolute accuracy.

This knowledge was apparent from the first discovery of the structure of DNA. But there remained one problem in making use of it. The net quantity of DNA present in a single living cell is very small. The genome of a virus is a thousand times smaller. For three decades after Watson and Crick's discovery, the elucidation of the precise sequences of DNA from such minuscule sources was exceedingly difficult.

By coincidence, in 1993, the year of the outbreak of the mystery epidemic in the Four Corners area of the Southwest, the Nobel Prize for chemistry was awarded to an engagingly eccentric scientist called Dr. Kary Mullis for his discovery, in 1983, of a means of chemical manipulation of DNA called the polymerase chain reaction.[4] The polymerase chain reaction made the nearly impossible a practical reality.

Using the polymerase chain reaction it is possible, given a tiny amount of DNA or RNA, to amplify it enormously. The potential of such a discovery was immediately seized upon by molecular biologists throughout the world. Remarkable applications have ranged from the forensic identification of murderers and rapists to tracing the history of man from our hominid ancestors. It was the imaginary basis for the re-creation of dinosaurs from the DNA trapped in mosquitoes in amber in the film *Jurassic Park.* It has a particular attraction within the field of virology, where molecular biologists use it to extract and purify the DNA and RNA of viruses. This research has many applications, including cloning and hybridization, in which viral coding can be inserted, rather like computer programs, into the templates of other life forms, particularly bacteria. This is called genetic engineering. Such genetically engineered bacteria have been used as factories for the production of genes and therapeutic products, such as human insulin.

Another practical application of the polymerase chain reaction (PCR) has been the successive elucidation of the genomic sequences of viruses. Today, every laboratory in the world can obtain access to gene bank databases, where they store the genomic codes for hundreds of different types of viruses. It is the genetic equivalent of the British or American library. Viruses no longer need to be seen or grown in the laboratory to be identified. They can be traced by molecular biologists matching their extracted nucleotide sequences against this standard library of viral genomes.

Stuart Nichol is chief of Molecular Biology in the Special Pathogens Branch at the CDC. When the immunological results tabulated by Pierre Rollin suggested they were dealing with a hantavirus, C. J. Peters asked Nichol to help them confirm and identify the virus.

6

Nichol is young, dark haired, and slightly built. He was born in Newcastle, in the north of England. At school he was interested in biology — not so much in the morphological aspects but in the internal chemical processes that govern the way biological systems work. First he studied biology at Liverpool University, got bored with the gross anatomy of dissections, then he made his own formative realization. "Microbes are very neat models for looking at how biological systems work, how molecules interact. I became very interested in microbes from that aspect."

Viruses based on RNA rather than DNA had begun to fascinate him. Moving on to take a Ph.D. at Cambridge, he found himself working under the directorship of Brian Mahy, an eminent British virologist. Cloning technology was just starting to come on line. The world of molecular biology had just been blown wide open by Mullis's discovery. When Mahy came to Atlanta to head the Virology and Rickettsial

Division at the CDC, Nichol followed him, taking up his present position in Special Pathogens. Nichol's main interest today is the evolution of RNA viruses and the role this has come to play in disease processes. It is very interesting territory. Hantaviruses, for example, are RNA viruses.

The PCR has its limitations. It is far too specific to be used as a broad screening procedure. Without the success of the immunological findings, Nichol would never have had the opportunity of becoming involved. But now, thanks to the work of Ksiazek and Rollin, Nichol and his staff knew exactly what they were hunting for. "As soon as I heard it might be a hantavirus, I went and searched for hantavirus sequences."

The serological results provided Nichol with another important nugget of information. If the epidemic was caused by a hantavirus, it must be antigenically distinct from the other known hantaviruses. The pattern of the illness, the horrific lethality, emphasized the differences all the more. A different strain of hantavirus implied that it would have significant differences in its nucleotide sequences.

The hantaviruses, as noted, are a genus within the family of *Bunyaviridae.* All viruses that belong to the same taxonomic family will be found to have genomic sequences in common, though there will be differences in other parts of the genome between the subdivisions, the genera that make up the family. There would be closer commonalities between the various species of virus that make up the hantavirus genus. Hantaan virus would have these common sequences, as would the Puumala, Seoul, and Primrose Hill viruses. It was those commonalities that now interested Nichol.

He had a small stroke of luck. Several of the hantavirus codes were already loaded into his computer disks because the department had been working with them as part of another project. Now he downloaded these onto his screen to look at them. Using various computer programs, he aligned the genetic codes so he could compare their precise nucleotide sequences. "I asked myself the question: of all the known hantaviruses, which bits are the most highly conserved?"

He found just what he wanted, a common string of nucleic acids just twenty or so units long. Every hantavirus had this string in common. Then he did something that is of almost everyday occurrence in a modern molecular biology laboratory, yet which to me seems truly extraordinary. By typing these nucleotide sequences into a computer linked functionally to a battery of nucleotide wells and empowered with PCR technology, the computer created those nucleotide sequences not on its video screen but in actuality in the solution. This artificially constructed string of nucleic acids would become a DNA probe, termed a "primer," that could be used to hunt for hantavirus genomes. This is the first step in the polymerase chain reaction. Nichol used this to search for matching "target" molecules in the biopsies taken at autopsy, molecules sufficiently different from human material to betray the fact that they were viral in origin.

He took a piece of lung from a victim that had been confirmed to have the mystery hantavirus on serological testing, ground it up, and homogenized it. Using chemical techniques, he collected together all of the RNA that was present in the homogenate. Most of this would of course be human messenger RNA that channeled information from nucleus to ribosomes. If there was no virus in the tissue, all of the extracted RNA would be human. Nichol gambled that there would be some, if very little, hantavirus RNA present.

Next he added his artificially constructed primer to the homogenate. If hantavirus was present, the primer would discover the matching string in the viral genome and line up alongside it. When he then added an RNA polymerase enzyme, this would copy the sequence of the virus beginning at the zone of attachment of his primer and running along the virus chain. It would keep on going, copying as it went, without stopping until it came to the distant end, creating a chain that was very unwieldy. But that could be sorted out later. The next step was to convert this trace amount of extracted viral RNA to its matching DNA. This is made possible using a very unusual enzyme not normally found in human tissues called a reverse transcriptase. The strip of copied viral RNA was now translated into DNA and this became the template for the second step in the cloning process.

He now designed another primer to match another well-conserved area of the viral genome, but this time in its DNA equivalent to start with. The second primer was designed to attach to the extracted DNA strand a distance of say eighty nucleotides from the area of attachment of the first primer. When DNA polymerase was added, it attached to the new primer zone, running in the opposite direction to before and stopping abruptly where the artificial chain began, at the site of attachment of the first primer. In this way, he had not only copied a sequence of viral genome, but he had sharp cutoffs at both ends of his viral extract, coinciding with the zones of attachment of the two primers and giving him a string of DNA that matched the corresponding string of viral RNA and was about eighty nucleotides long.

The DNA polymerase he used for this second step is a very special enzyme, first derived from a bacterium that lives in the hot springs of Yellowstone Park. Its uniqueness lies in the fact it will withstand repeated heating to near boiling temperatures. First, as a polymerase, it makes a copy of the key string of DNA. Now, if you heat up the mixture, the two complementary strands of DNA will automatically separate. The original sequence has now become two. So by allowing the mixture to cool and using the polymerase to reproduce the two strands, and by separating them again by reheating, the two become four. If the mixture is repeatedly heated and cooled, the polymerase enzyme doubles the number of DNA strands with each cycle. Four becomes eight, eight becomes sixteen, and so on. This is the wonder of the polymerase chain reaction: it becomes a factory for the replication of that same strand of DNA. Within just twenty cycles, using this technique, Nichol had accumulated a million copies of the replicating nucleotide sequence. In the next cycle this became two million, then four million, and so on. It is relatively easy to extract these larger quantities and analyze them.

The laboratory has a computer linked to a laser beam that reads off four different dyes that color the four different nucleotide bases that make up DNA. On June 8, eight days from when the samples had first arrived in the Special Pathogens Branch for testing, Nichol was ready to read off his presumed viral sequence. The computer read off the sequences with remarkable rapidity. In Nichol's own words, "Then — wow! We knew we had it. There really was hantavirus RNA back in that original sample of human lung."

Even then the molecular biology had barely begun. He used the sequence he had extracted as a primer to fish out more of the nucleotide sequence of the viral genome. By the next day, he was able to confirm that the sequence, though definitely that of a hantavirus, did not correspond with any of the known hantavirus sequences. As more and more of the genome came to light, Nichol incorporated the information into targeted PCR primers, enabling him to reconstruct increasing proportions of the viral genome. By degrees, he cloned the entire genetic template of the virus. With this sufficiency of information, he could assure his colleagues that they were dealing with a life form that had never been seen before: a new species of hantavirus.

It was a dramatic example of modern science in action, a detective-style solution that would cause headlines in the newspapers and a flurry of more scholarly papers in the formal scientific journals. The distinguished journal *Science* would hail it as remarkable under the heading "Virology Without a Virus."[5] Scientifically, the identity of a virus remains in limbo until it is seen under the electron microscope and grown on tissue culture in the laboratory. In the causation of a new disease, these are the mandatory first steps in a conclusive system of proof first laid down by the great German bacteriologist Robert Koch a century earlier. By June 9, just twenty-six days after Michael had died in the Gallup Medical Center, thanks to the combined efforts of Tom Ksiazek, Pierre Rollin, Sherif Zaki, and now the crowning high-tech glory of the molecular biology by Stuart Nichol, the CDC had established the cause of the epidemic before ever seeing or growing the virus. For Stuart Nichol "that was the most exciting moment I have had in my scientific life."

But even with the further discovery on June 14 that the rodent specimens from New Mexico had antibodies to hantavirus in their blood, important questions needed to be answered. They, as scientists, needed to know a great deal more about the deer mouse. They needed to know where the mouse and the virus came from in their evolution. They needed answers to why the virus was so commonly found in mice that were not in themselves sick. What was the nature of this curious relationship the virus had with the mouse? As with many of the lessons of science, the initial discovery of the hantavirus had been no more than the first step in a fascinating odyssey into new realms of the unknown.

SIX

The Struggle for Life

The most wonderful mystery of life
may well be the means
by which it created so much diversity
from so little physical matter. . . .
Yet life has divided into millions of species,
the fundamental units, each playing
a unique role in relation to the whole.

EDWARD O. WILSON[1]

1

Three quarters of New Mexico enjoys an upper Sonoran habitat, including most of the plains, foothills, and valleys above 4,500 feet. These higher altitudes are cooler and benefit from a little more rainfall. Prairie grasses thrive there, with low-growing piñon pines and juniper bushes. Much of this land is wilderness, little changed by the advent of man. The commonest mammals are small rodents. Deer mice flourish, along with harvest, house, and pocket mice. There are several varieties of rats, including browns, pack rats, and kangaroos. Deer mice are easy to distinguish: they are larger than the other mice, with outstanding ears and brown fur, white on the underbody. It is the color of their fur that gives the species its name: they have the same biological camouflage as deer.

You could not ask for a worse carrier of a deadly disease in America than the deer mouse. With various subspecies and cousins, such as the white-footed mouse, it thrives in an incredible variety of climates, from the cold thin air of Canada's arctic mountains to the humid heat of the southern swamplands. Deer mice will gather whatever food they can, nuts, grass seeds, flowerheads, and insects, hoarding their larders in hollow logs. Their very abundance is a factor in the local ecology. They are the dominant rodent species in America, inhabiting a vast territory, from Hudson Bay to California, widely dispersed throughout most states, with the exception of the Southeast, and extending in a long thick tongue into the heart of Mexico.

It was the local wildlife that attracted Robert Parmenter to New Mexico. Bob Parmenter grew up in rural Virginia before it was devoured by the expanding beltway of Washington. "Then," he reminisced, "it was much like England, with deciduous forests. As a child I was always down there poking around in the forests, the swamps and creeks." From those boyish adventures he came to love nature. It made him want to become a biologist.

He took his B.A. in Colorado Springs, where the Great Plains meet the Rocky Mountains. During his master's studies at the University of Georgia, he studied freshwater turtles, adding ecology during his Ph.D. at Utah. Ecology has been his major interest since. In 1988 Parmenter started work on the long-term ecological research at the Sevilletta in southern New Mexico. The Sevilletta is Professor Terry Yates's brainchild, part of a network of eighteen sites through America, two extending to Antarctica, where the flora and fauna are studied long term in relation to environmental and natural forces. This is the American branch of Biodiversity 2000, the most ambitious ecological project ever mounted. Animals are harmlessly trapped to monitor their numbers, birds counted in the air or monitored in their nests, plants on the ground, even the very insects that crawl, fly, and burrow into the soil: all life is observed, the interaction of one species with another in their biological niches, their numbers and distribution plotted on computers, their survival measured against predation, food sources, illness, and climate.

In mid-June 1993 Parmenter was approached by a television producer who had heard that he was an expert on deer mice. This was true: he had been fascinated with the small rodent for many years and it was one of the most important animals under study in his research at the Sevilletta. But when the producer raised the subject of the hantavirus epidemic, Parmenter was surprised. He knew nothing about the epidemic. Here was a different kind of scientist, one who was not normally troubled by human diseases. He found himself being interviewed live on the television news that same evening.

"It's hard to believe," he told the viewers, "that such an adorable little animal could cause so much trouble." With tawny fur, big protruding ears, lustrous black eyes, and an inquisitive rounded snout with prominent black whiskers, the mouse looked more like a friendly little personality from the tales of Beatrix Potter than a threat to people. He had been bare-handling deer mice for forty years, capturing and coming into close contact with thousands every year. His wife, Cheryl, herself a virologist, helped him conduct behavioral research on them. They kept cages of mice on their back porch at home, and when they took off for a week or so of camping, some trapped mice kept them company in their sleeping tent.

But Parmenter also knew that the deer mouse is exceptionally tough and enduring. He had seen a startling example of this when studying the return of life into the devastated area around the Mount Saint Helens volcano, when the first animal to recolonize the landscape after the catastrophic eruption had been this mouse. Overcoming any privation and prepared to eat whatever is going, it never hibernates, produces as many as five litters a year, and can conceive while nursing the previous litter. He was impressed by the endurance of such a tiny animal in the tooth and claw of evolutionary survival.

Parmenter's ecological studies were suddenly very relevant. This last year he had noticed massive increases in mouse populations, as much as thirty-fold in some areas, before the start of the human epidemic: it seemed more than fortuitous that the same geographical area was the epicenter of both the explosion in the mouse population and the outbreak of the lethal virus. The two had to be causally related. Why then had the mouse population seen such a massive increase?

From the start of the epidemic, the Navajo medicine men and women had pointed to changes in the weather as the key factor in its origins, but nobody took much notice. Now the scientists were also wondering if the weather had played some part in triggering the epidemic.

After seven years of drought, New Mexico had enjoyed two mild winters with more rain and snow than usual. Now Parmenter's data showed that this mild weather had nurtured a heavy crop of piñon nuts and insects such as grasshoppers, the favorite food of the mice. Perhaps it was this abundance of food that had given rise to the massive increase in the deer mouse population during the spring and early summer? This appeared to fit very closely with the traditional wariness of the Navajo toward mild weather. One of Bruce Tempest's colleagues is married to a Navajo woman who tells a very remarkable story. In 1919 her grandmother had survived a fever epidemic that had wiped out the rest of her family. They had all died following the spring-cleaning of their hogan. Since then, whenever they decided to spring-clean the hogan, her grandmother would insist that they cover their noses and mouths with a cloth and that they burn their clothes when they had finished. Like 1993, 1919 was a year when the traditional elders recalled an epidemic of human deaths following mild weather and an increase in mice.

If mild weather was the crucial link, why then had there been such marked changes in the weather?

A possible explanation lay with the El Niño effect. For decades scientists have been investigating a curious natural phenomenon. Every three to seven years a huge current of warm surface water moves over the Pacific to the west coast of South America, dramatically altering rainfall and wind force and direction throughout the Pacific Basin. This warming of the coast is called El Niño, the Christ child, because it arrives at Christmas. El Niño can spawn droughts, floods, and storms. In the southwestern states of America it heralds wet winters and mild springs.

Although some of the intricate mechanisms of real cause and effect remain obscure, it seems likely that ecological factors played a key role in triggering the epidemic. And here was a clue that might help us understand viral epidemics in other parts of the world. Ecological factors, such as environmental pollution and despoliation, is an increasing worry to scientists. In the words of Dr. Abraham Verghese, "As humankind continues to disrupt its delicately balanced and amazingly complex ecosystem, the shock waves from the tears in nature's fabric reverberate at increasingly microscopic levels. The recent outbreaks of rare pathogens such as hantavirus may be an example of such a reverberation."[2]

This is an important observation. But the Navajo were not recent invaders of the mouse's ecosphere — not in the timespan of human history — though, perhaps relevantly, the presence of humans in the New Mexico desert is very recent from an evolutionary perspective.

By summer Parmenter went on to record a sharp decline in the number of deer mice, from twenty per hectare in May to four per hectare in August.[3] This fall in numbers was as mysterious as the original dramatic increase. Tellingly, as mouse numbers fell so also did the incidence of human infection. A number of possibilities could explain the subsequent crash in the mouse population, from exhaustion of food to the attrition of predators, perhaps even the intervention of disease. But though Parmenter could not yet provide an answer for the decline in numbers of the mice, he was sure they were not dying from infection with the hantavirus. It was one of several growing mysteries.

For example, given the causal relationship of El Niño to mild winters in the Southwest, and from there to the explosion in the population of deer mice, we still cannot explain why this resulted in a virus soon to be discovered to have long infected deer mice now crossing species to humans. And there was another, equally fundamental mystery that demanded an answer. This was a question that had been posed by the Navajo traditional healers when the virus was discovered.

Why were the deer mice not themselves sick from a virus that was lethal to humans?

Infected mice scurried about their rugged lives in perfect health, all the while shedding copious quantities of virus in their urine, saliva, and feces. How very strange that the virus seemed not to harm the mice at all! Why would a viral illness contracted from a common rodent behave in this manner? Some astonishing findings were now emerging from the continuing field inquiries.

2

In the parched heat of high summer, the biologists were busy from daybreak to sunset, even with the addition of locally trained volunteers from the Indian Health Service. Jamie Childs and John Krebs were forced to work into the night so they could cover two or more sites in one day. Necessity had compelled them to streamline their operating patterns.

During those first three frantic weeks, the biologists spent nineteen days in the field. The workload seemed to mushroom at an exponential rate. They were helped by colleagues from Arizona. Soon they were training anybody they could get hold of, not only personnel from the Indian Health Service but from all three of the surrounding states. Over the two months of June and July, the biologists performed an astonishing 1,696 autopsies in the field, from small mammals of thirty-one different species. Almost half the total were deer mice, the rest made up of piñon mice, brush mice, rock mice, canyon mice, several species of chipmunks and squirrels, and whitetail prairie dogs, together with many species of vole and rat. In all they would collect ten thousand samples of blood and tissue. Back at the CDC, Tom Ksiazek, Pierre Rollin, and technicians such as Mary Lane Martin took on the gargantuan workload of screening them all for antibodies to the virus. The results were invariably supportive. In every batch of samples, antibodies would be found in the rodent sera. Although other species would show some positives, by far the commonest association was the deer mouse, 30 percent of which were infected with the virus.[4]

The field inquiry was linked closely with that of the epidemiologists, led by Jay Butler, so they could pool results in a meaningful way, site by site, area by area.

People and rodents from the same areas would be screened using identical tests, serum first, because it was easier. The positives could then be subjected to fine analysis, using the polymerase chain reaction to compare viral genome in the tissues of people who had died with that extracted from the organs of sacrificed rodents from the same homes, the same yards, the same landscape area.

As the huge volume of serum samples from people and rodents deluged back to Atlanta for lengthy investigation, Jay Butler and his colleagues were plotting out the profile of the epidemic. The case control study was designed to identify risk factors, but it did not plot the true attack rate. They needed a cohort of people who had had exposure to infected people.

Experience in other settings had shown that laboratory personnel exposed for instance to Hantaan virus without taking protective precautions contracted the illness within minutes of bringing the rodent carriers into the lab. Suddenly a dreadful thought occurred to the epidemiologists: what if the disease might be transmitted to staff from patients — so-called nosocomial spread. This did not need to imply aerosol contagion. Specimens were being handled at a stage before infection was even suspected, saliva and body secretions might result in contagion.

Chuck Vitek, one of Butler's team, began a retrospective study of health care and laboratory worker exposure. The findings proved very interesting. Almost three hundred people had experienced some exposure to infected cases. This included doctors, nurses, clinic staff, the people who collected blood, urine, and fecal samples, the technicians and other people who worked in the mortuaries, the fire service and other emergency staff involved in the emergency transport of patients. Although they found no evidence that any member of staff had contracted the virus from patients, the implication for epidemics generally — notably one that might spread by aerosol contagion — was stark.

Two of Fred Koster's research colleagues at the UNM hospital, Steve Jenison and Brian Hjelle, perfected a quick diagnostic test for hantavirus infection, based on the Western blot technique, which proved invaluable in the rapid diagnosis of patients entering the intensive care unit.[5] Jim Cheek, the IHS epidemiologist, took blood samples from the people flooding into the hospitals and clinics and tested them for antibodies. Out of the four hundred people screened in this way, he found antibodies to the virus in four. All four had high levels of IgG only. The normal antibody pattern to a newly acquired infection is to respond with IgM-class antibodies within days and to follow on with IgG weeks later. Raised levels of IgG could signify recent infection, but it could also signify infection many years in the past. If the antibody studies could be believed, there were scattered survivors that the doctors knew nothing about. Among those presenting with clinical symptoms, a small number also had a mild pattern of illness.

This was a new insight: the disease did not always prove serious. Yet for most victims it remained highly lethal. In those early days the mortality from the virus reached an all-time high of 77 percent, then fell to about 70 percent. It remained a very disquieting pattern.

In Mescalero, which is in south central New Mexico, there is an IHS hospital run for the local Indians, who are related to the Apache. A physician called Charles Wilson, while examining a fifty-year-old cowboy with a painful knee, realized that the man had been admitted to the hospital in the past with a condition very suggestive of hantavirus infection. Wilson sent a fresh specimen of serum to Steve Jenison, who found antibody activity both to nuclear capsid and to the surface glycoprotein of the new virus. This was hard evidence for infection with the hantavirus long before the present epidemic.

At the UNM, Brian Hjelle also began to screen people and rodents using PCR. When a new patient was confirmed to have the virus, Hjelle at the UNM, or Stuart Nichol at the CDC, would use a razor or scalpel to cut out the diagnostic band of PCR-amplified DNA on the agarose gel. The agarose was melted away, giving them purified snippets of DNA that matched the nucleotide sequences of the virus RNA. Using recombination, these would then be cloned in bacterial cultures, when the bacteria would in effect become factories for the viral genome. If things went well, the time from receiving a blood sample to reading the genetic sequences was five days. They would print out the virus sequence from the infected patient, then compare this to the sequences from locally trapped mice until they discovered the exact group of deer mice that had led to the infection.

By June 14, Nichol and his staff had extracted the first PCR band from rodents. When on June 16 the seventh human case was confirmed by serology, they could show that the sequence from the patient exactly matched that extracted from the local deer mice. In ways that bacteriologist Robert Koch would never have dreamed, they could now prove beyond a shadow of a doubt the complex cycle of virus infection in mice and its transmission to human victims.

Not only could they determine what species of mice had given infection to what person, they could tell the exact location where a person had picked up their infection. Even the imperturbable Jamie Childs was amazed. Viruses from towns or settlements no more than twenty or thirty miles apart were sufficiently different — the scientists call it genetic drift — to distinguish nucleotide changes. That not only made it possible for Jay Butler and his epidemiological team to track down exactly where infections came from, it gave them an important clue to the evolution of the virus itself.

When examining the tissues of a man who died in Snowflake, Arizona, Nichol was surprised to find that the genetic sequence of the virus that killed him was different from the Arizona strain. The man had only just moved to Arizona for work. His home was in Hesperus, Colorado. "When we tried to match his virus sequences from the rodents from the house he had been staying in Arizona, they didn't match. But when we got the rodents from his house in Hesperus — bingo! We got an exact match." The virus that was killing people in

Arizona was a slightly different strain from that killing people in Colorado, and this was different again from that in New Mexico, or Utah.

Another case, investigated by Brian Hjelle, had legal significance. A woman had died from hantavirus infection and her family believed she must have picked it up from her place of work in Gallup. If so, they intended to sue her employers. Hjelle tested the genetic coding of her virus and compared it to the sequences from mice found at her workplace. They did not match. Next he compared her virus to that found in the mice around her home, thirty miles to the north of Gallup. But these also failed to match. In bafflement, they turned to her car. This had stood unused for several months and was then put back into use, ferrying her to and from work. In the car ventilation system they found evidence of mouse droppings. When Hjelle tested the viral sequences found in the mouse droppings he obtained a perfect match to the virus that had killed her.

In the words of Stuart Nichol: "It was like fingerprinting in a crime whodunnit. The scientists could be presented with the body of a victim and all the attendant evidence. Who then was the culprit? The PCR sequence provided the fingerprint identification of the culprit."[6]

But there was more to be learned from this than just an astonishingly precise application of high-tech science. Stimulated to major efforts by an epidemic in its own heartlands, scientific America began to probe the wider ramifications. Why should the virus change so much when you moved a mere twenty or thirty miles? We knew that viruses, and particularly RNA viruses, had a phenomenal mutation capacity. But this biological phenomenon appeared to go beyond mere mutational capacity.

The genome of the new hantavirus, emerging with surprising variations from the ongoing PCR research, was much closer in its sequences to Prospect Hill virus than Puumala, and closer to Puumala than Hantaan. This hinted at a slow evolution over aeons, as rodents colonized whole continents. Viewed from the narrow perspective of human infection, science would never understand the evolution of the hantavirus. The clue to the mystery lay in a wider biological perspective, in its relationship with its rodent host. There was accumulating evidence that the mice and virus had lived together over great spans of time. The virus had infected the local rodents for tens of thousands if not millions of years. This suggested that the field studies, showing striking evolution of the viral genome with the geography of the rodent were the key to an extraordinary virus-host relationship — information that in turn might provide a basis for understanding viruses and their relationship with their natural hosts throughout all of nature.

3

At the UNM hospital the clinicians were also beginning to make progress in their investigations of why people were dying from the infection.

The most poignant clue lay in the fact victims died so rapidly, often within hours of the onset of breathlessness. Steve Simpson, a specialist working with Howard Levy in the intensive care unit, could only compare this to the pattern he had previously seen with fulminating meningococcal septicemia. This is a condition where the virulent meningitis germ causes massive shock as a result of blood-borne infection, overwhelming not only the immunological army sent to combat it but even the adrenal stress glands. Death results within hours, the adrenal glands exhausted, their tissues shrunken to husks. It was obvious that people were not getting intensive medical attention quickly enough. More lives could be saved if that message could be conveyed to doctors manning the peripheral clinics and hospitals.

But what about the manner of death itself? The victims were often fit young people. The pathological mechanisms must surely be devastating to kill them so quickly.

The primary cause of death was long established: a deluge of fluid from the blood into the lungs of the victims. Sherif Zaki had found virus in the capillary walls of the lungs. This seemed to fit a pattern. Hantaviruses are hemorrhagic fever viruses. These cause bleeding when an immune reaction triggered by the virus damages the walls of blood vessels. It showed up as bruises in the skin or bleeding from the bowel, kidneys, nose, or eyes. When they sucked fluid from the lungs of these patients, they knew that it was identical to serum. Now they figured out the answer. It was caused by a phenomenon called "capillary leak."

The immune damage in the capillaries was so fine, the holes created in the blood vessels so small, it allowed fluid but not cells to leak out. They were dealing with an internal hemorrhage, but the leaking blood was being sieved through the holes, as if through the finest of nets, to swamp the alveoli with its serum component. This was what had drowned Michael and Rosina.

This made sense to the intensive care specialists, whose first step in treatment was high-quality supportive therapy. In early June they had rescued Dolores in this way. Howard Levy was going around telling people that the only thing he knew about this disease was what he had learned from Dolores. Levy felt encouraged because she had been very sick and yet, with good support therapy, she had survived. This meant they were on the right track. "What," he asked himself, "did I do right with her? What was it that saved her?"

Despite her very severe lung disease, he had been able to oxygenate her lungs without insurmountable difficulty. Simpson helped Levy treat another victim with aggressive resuscitation measures, and once again the patient survived. It confirmed exactly the line of their thinking: if they could just get patients into the hospital quickly enough and treat the lung flooding very aggressively, they could rescue most of the victims. Then, in Simpson's words, "Next thing that I remember is feeling very humble."

On June 20, a thirty-year-old Navajo woman arrived in the ICU with the disease. She could not have come to him earlier in the course of her disease. She had actually gone into work that morning, but her colleagues had taken one look at her and driven her straight to the hospital. Simpson saw her within minutes of the gurney coming through the admission doors. He stayed with her every moment of her treatment and resuscitation. Tragically, in spite of every intensive support measure, she deteriorated and died just two hours after admission.

While early admission and treatment remained vital, her death was irrefutable evidence that there must be another pathology operating in some patients.

He had ventilated her, oxygenated her, kept perfectly good arterial oxygenation. In most aspects her case resembled the successfully resuscitated Dolores. She had, however, differed from Dolores in one noticeable respect. There had been marked irregularities of her heartbeat. The electrocardiographic traces on the monitor and the readouts had been grossly abnormal. Before her death she had suffered a series of cardiac arrests. It looked as if her disease had focused not so much on her lungs but her heart. Howard Levy was reminded of an earlier case, the fifty-eight-year-old Icelandic woman who had come into the intensive care unit on May 18, before anybody was aware of the epidemic. She had died so rapidly it was only in retrospect that her illness had even been attributed to the hantavirus. Yet he recalled similarly devastating cardiac complications.

That woman had also been ventilated. They had managed to restore perfectly normal oxygen levels in her arteries. They had inserted a very fine measuring tube, a Swan-Ganz catheter, through a vein and guided it through the heart, and wedged the tip in her pulmonary vascular bed. This is a means of finding out more about the working pressures in the left side of the heart, where the powerful left ventricle, stronger and thicker-walled than the right ventricle, has to pump blood under high pressure to the entire body. They were able to calculate the work rate of her heart, a mathematical derivative called the "cardiac index" and this had registered almost no left ventricular function.

It was obvious that they needed to analyze the clinical patterns of the growing list of hantavirus patients passing through their hands. A bright young doctor, Gustav Hallen, was delegated the responsibility of compiling the information on the hantavirus cases. At this point some ten or eleven confirmed victims had been treated in the department. His instructions were to chart everything, the patterns of presentations, the clinical findings, the pulmonary and cardiac findings, the responses and failures of responses to various treatments.

Young Hallen collected together all of the case notes, X rays, investigation findings. He plotted details on large charts. Very soon the intensivists could discern a pattern of how these patients presented. There were some surprises. The first surprise was that so many patients presented not with a cough or breathlessness but with abdominal pain, nausea, and diarrhea. Three people had been discharged home with symptoms of "gastroenteritis" only to be rushed back to the discharging hospitals rapidly worsening, with the telltale shortness of breath. One had been admitted to a surgical ward and had been lying there for three days, during which time the surgeons debated removing her appendix. The pattern was so striking it made Levy and Simpson wonder if the route of contagion was not through inhalation at all but ingestion.

Were people contracting the virus from swallowing dust and not inhaling it? Pathologists looked for virus in the stomach and intestinal walls at the autopsies, but they did not find it.

Hallen's charts carried an important practical lesson. Doctors with less experience of the presentation would assume a dramatic pulmonary presentation. They needed to be warned about the prominent vomiting and diarrhea, otherwise they would turn away people thinking they had gastroenteritis. The detailed case histories and charts also drew attention to a puzzling yet apparently vital factor. At the end of June a cardiologist had drawn people's attention to the results of a noninvasive investigation called echocardiography in one of the fatal cases. The echocardiogram involves passing sound waves through the chest in order to show the movement of the walls of the heart: the alternate contraction and relaxation of the two ventricles and the opening and closing of the four valves. "Somebody should look at the cardiac findings on this patient," he cautioned. "This echo looks like a horrible cardiomyopathy."

Cardiomyopathy is a disease of the heart muscle itself, which seriously impairs its ability to function. In two of the cases on Hallen's table there was clear evidence of cardiac involvement. Levy already knew that when the blood lactate levels were high people tended to die. Now Hallen's charts correlated high lactates with heart problems. The pattern was consistent with a cardiomyopathy.

Until this moment every death had been attributed to the lung pathology: the victims were drowning. Now Levy, Simpson, and Koster knew this could not be the whole story. Some people died in spite of good arterial oxygenation.

The medical name given to the new disease was the hantavirus pulmonary syndrome, or HPS. Now it appeared that the cause of death, at least in patients whose course was modified by intensive therapy, was often not pulmonary at all but cardiac. With growing confidence, the intensive care specialists monitored all future patients for these cardiac complications. They found people who would go into cardiac failure, they would develop disturbances of the electrical control mechanisms of the heart, called arrhythmias. In the terminal stages of some people's illnesses, they would see a catastrophic manifestation, called electromechanical dissociation, where the heart registered electrical activity while there was no pulse at all. They had echocardiograms on these patients that, in Levy's words, showed "the picture of a dying heart."

Continuing ultrastructural examinations performed by the CDC chief of pathology, Sherif Zaki, were showing virus in the capillaries of the heart muscle itself. It varied from focal immunostaining in some cases to diffuse and extensive staining in others.[7] This was direct confirmation of virus invading the heart, though it did not prove that the cardiac deaths were due to this alone. Preliminary research would suggest an additional mechanism. There was some toxic factor in the blood of hantavirus patients that caused a profound depression of the heart muscle. But search as they would, Levy, Simpson, and Koster could not pinpoint it.[8]

The intensivists knew they could effectively treat the lung whiteouts, but they needed to improve on their treatment of the failing hearts. They decided they would attempt to rescue these patients with a dramatic interventional therapy called "extracorporeal membrane oxygenation," or ECMO.

This involves placing a tube in a major vein, such as the jugular, taking the venous blood out of the body, oxygenating it using mechanical equipment, and then returning it through another tube in a major artery, such as the femoral. In effect, you bypass the lungs. ECMO was initially designed for patients with overwhelming respiratory failure, but Howard Levy knew that it also supported the heart. Easing the burden of work on the failing ventricles, it could give these patients an artificial cardiac output of five liters per minute.

The first patient they tried to help with this dramatic measure was a man in his thirties who had already had a cardiac arrest. His position appeared hopeless. He was being kept alive using cardiopulmonary resuscitation at the bedside, with artificial ventilation and manual chest compression. He was put on the ECMO machine, stabilized and survived another eight hours. Though it ended in failure, to Levy it seemed to offer hope of assistance in catastrophic circumstances.

Soon afterward, they were faced with another desperately sick man who had progressed to intractable heart failure. Not quite so desperate as the first, he was kept alive on the ECMO machine for forty-seven hours. At this stage, his heart seemed to rally and he made a good recovery. No other patient who had developed a similar degree of heart failure from the hantavirus survived.

By July 1994, twenty out of the national total of eighty hantavirus victims had been treated in the intensive care unit at the UNM hospital. Nationwide, with patients presenting to doctors lacking this depth of experience, the mortality was still 60 percent or more. Here in New Mexico, the local experience and evolution of treatment had reduced the mortality of their patients from its initial 70 percent to about 35 percent. It was an heroic achievement with a struggle for survival in every new case that would draw upon every morsel of courage on the part of patients and their families and that would continue to test to the very limit the experience, determination, and skills of this dedicated team of doctors and staff.

4

Yet the hantavirus pulmonary syndrome did not go away. Even as the outbreak was being investigated, people were still contracting the disease, and tragically they were still dying from it. By October 1993, although half the reported cases had been registered in the Four Corners states, other cases had been recorded in California, Louisiana, Texas, Idaho, Nevada, the Dakotas, and Montana.

On January 14, 1994, a twenty-two-year-old New York man developed the hantavirus. A student at the Rhode Island School of Design, he went to his doctor with fever and chills, put down to a flulike illness. Three days later he could not breathe. His doctor called an ambulance, which ferried him to the emergency room at Miriam Hospital, in Providence, Rhode Island. He died there four hours later. There were no deer mice in Rhode Island. The student had cleaned out an old warehouse in Queens, New York City, which was infested with mice. CDC biologists soon arrived in the warehouse and the surrounding area, where they found a new rodent carrier, the white-footed mouse, long known to be a risk for the bacterially mediated Lyme disease.

Cases now being diagnosed in Texas and Louisiana were equally puzzling, since these states were thought to be outside the range of the deer mouse. Even more intriguing, the viral sequences extracted and amplified from the lungs of a fifty-eight-year-old bridge inspector in Louisiana showed a previously unrecognized hantavirus, "closely related to but distinct from both the Prospect Hill virus and the virus circulating in the four-corners area."[9] In Florida, where there are few deer mice, the cotton rat was found to be the carrier. Even the relatively few parts of America that had no deer mice did not escape the plague: other strains of the virus, carried by alternative host rodents, were equally capable of killing people.

In January 1994, C. J. Peters took a call from colleagues working in São Paulo, Brazil. They had been dealing with the cases of three young men, all brothers living in the same household. All three had contracted the same febrile illness.

The first to fall sick suffered a mild pattern of illness, then recovered. The second brother died with classical symptoms of hantavirus pulmonary syndrome. By the time the third brother arrived in the hospital of the small town of Juquitiba, near São Paulo, the local doctors hurried him into the big city. Luiza de Souza from the public offices of health and the virologist Lygia Iversson recognized the pattern from the CDC weekly reports of morbidity and mortality.[10] They sent sera to Peters, who confirmed the nature of the infectious agent. Using PCR, Stuart Nichol demonstrated a new strain of hantavirus in the dead brother's tissues.

Soon evidence was coming through for hantavirus pulmonary syndrome infections in Argentina. The first cases were also being recognized in Canada. It seemed likely that in time the extent of the virus geography would encompass the whole of the American continent.

In the third week of November, some six months after the epidemic had first been recognized, the virus that caused it was successfully grown in tissue cultures. It had not been an easy isolation. And the natural competition between CDC and USAMRIID had resulted in not one but two separate and independent isolations of related strains of the virus.[11]

After all this labor, how curious it was at last to conjure up its image in the green eye of the electron microscope. The culprit for so much terror and lethality appeared as a tiny sphere, no more than 107 thousandths of a millionth of a meter in diameter. It looked as innocent as a tiny ball of cotton wool. In taxonomic terms, it was a typical hantavirus, a new member of the family of *Bunyaviridae.* Named after a prototype first isolated at Bumyamwere in Uganda, the Bunyaviruses form the largest of all the viral families, including more than two hundred separate species. Most are transmitted to animals by arthropods, the majority by mosquito. Their books of life are all coded by a genome based on a single "negative" strand of RNA.

The isolations made possible the development of more refined diagnostic tests. Drugs such as ribavirin could now be tested against the virus in tissue culture.[12] It also made a vaccine possible, if unlikely given the cost implications and the relatively small numbers of dead.

At USAMRIID, Alan Schmaljohn and his colleagues in the Virology Division decided to name their isolate the Mammoth's Lakes, California, virus, intending this to be a substrain of Four Corners virus. At the CDC they decided to call their isolate Muerto Canyon virus, because a canyon of that name lay within the territory where the virus had first shown itself to be epidemic. It was an unfortunate choice of name. The CDC did not realize that there was another Muerto Canyon within the Navajo reservation, of tragic historical significance. This was the alternative name for the Canyon de Chelly, where in 1805 a hundred Navajo, including women and children, had been massacred by Mexican soldiers.

The controversy set up a furious correspondence with the Navajo representatives. In New Mexico, the alternative title of Four Corners virus was the sensible choice of the local experts. But they had not considered the sensibilities of local civic authorities: the thrill of your own named plague virus held little appeal for the tourist industry. To C. J. Peters there was only one prudent option: the virus was given a name so artfully contrived it could offend nobody. The hantavirus that made its explosive first appearance in the Four Corners states in 1993 is now known as *Sin nombre:* the virus without a name.

5

With the growing understanding of the American hantavirus and its relationship with the deer mouse, scientists now made the disturbing realization that it would never go away. The virus would remain a hazard for people as long as there were deer mice in the landscape. And eradication of all the rodents in America is impractical. "To try to eliminate mice from thousands of square miles," Bob Parmenter explains, "would be an impossible task. There are just too many of them.

Short of a nuclear strike that would take everything else out with it, we simply do not have the funds to do it within the entire federal budget. They would bounce back faster than you could eliminate them." The best that could be hoped for was that the hantavirus would continue its present occasional rate of infection.

By August 1995, such occasional infections ran to a total of 115 confirmed cases, collected from twenty-three states. The heaviest rates of infection were still in the Southwest, with forty-six from this total in New Mexico and Arizona. There was an additional ironic twist. Two thirds of the cases were now white, with just 30 percent Native American and 12 percent Hispanic. The mean case fatality rate was as alarming as ever at 51.3 percent, despite the lowering effect of the UNM hospital contribution.

It came as a shock to American doctors to acknowledge the existence of a new endemic disease in their clinical practice. In Peters's words, "It will be a new chapter in the medicine textbooks. People are going to diagnose it regularly. It will be on their lists of differential diagnoses — and not just in this country."

The ribavirin study conducted by Louisa Chapman was hampered by its open nature — the drug was released on compassionate grounds rather than as a controlled scientific trial — and by the very rapidly progressive nature of the infection. Few patients had the benefit of receiving the drug early. In consequence, the results proved difficult to interpret. Chapman would conclude that while there might well be benefit from the drug, it proved too marginal during the epidemic experience to show significance on statistical analysis.[13] Longer-term controlled studies are pending at NIH.

Although the total number of human infections and deaths caused by the *Sin nombre* hantavirus remains modest when compared with the great plagues of history, in terms of mortality rate for those infected, the virus is one of the most deadly microbes ever encountered. Even after the shock wave of its first emergence, it left fear in the air, coupled with a heated intellectual curiosity that comes from the challenge of important mysteries. There were large chasms in our understanding of what had really happened in New Mexico on that fateful May 4, 1993, when the first person fell ill. And understanding, one sensed, was important.

Nobody had yet answered the question why a virus that caused no disease in the mice should cause such lethality in humans.

Was it really true, as most people still assumed, that humans had become infected by accident through coming into contact with a parasitic cycle primarily directed at rodents? Was the horrific lethality no more than the biological equivalent of bad luck? Could it be that within this admittedly difficult question there might be a commonality — a mystery that once explained might apply to a great many other emerging plague viruses?

The hantavirus epidemic was more than an example for America. In a curiously timely warning, exactly a year before the hantavirus outbreak began, a distinguished panel of American microbiologists had gathered under the aegis of the U.S. Institute of Medicine, to be jointly chaired by Robert E. Shope, Professor of Epidemiology and Director of the Yale Arbovirus Research Unit, and Joshua Lederberg, the Nobel laureate and President of the Rockefeller University in New York. The purpose of the meeting seemed at the time iconoclastic: to assess the threat of emerging infections to the United States.

Their conclusions were published in a booklet that sent a wave of shock through the prevailing establishment complacency. The very opening lines of that small booklet belied the purely American focus of their title. "As the human immunodeficiency virus (HIV) disease pandemic surely should have taught us, in the context of infectious diseases, there is nowhere in the world from which we are remote and no one from whom we are disconnected."[14]

Emerging infections, and emerging viruses in particular, were everybody's concern.

SEVEN

Deadly Returns

Plague: from the Latin, plaga:
a blow, stroke, wound;
an affliction, calamity or evil,
especially through divine visitation;
a general term for any malignant disease or
pestilence with which men or beasts
are stricken, especially an epidemic attended
with great mortality.

Oxford English Dictionary[1]

1

In 1961, six years before the Surgeon General's speech celebrating the conquest of infectious diseases, the seventh world pandemic of cholera began in the Celebes, in Indonesia, from where, as in the six great pandemics of the nineteenth and early twentieth centuries, it slowly began to spread.[2] It moved through Asia and into southern Russia, reaching western Africa by 1970.

Cholera is a waterborne disease, as was first discovered by two British physicians, Snow and Budd, in the mid-nineteenth century. The comma-shaped germ that causes it, highly mobile through a single polar flagellum, is called the *Vibrio cholerae.* Taken into the body through contaminated water, or food such as shellfish, the germ attaches to the lining of the human bowel, producing a poison, called an enterotoxin, which causes the gut to leak fluid. The leak is massive, in excess of a liter per hour. The profuse diarrhea resembles rice water, teeming with germs. This contaminates soil and local water supplies, continuing the germ's cycle of infection to others.

From its first appearance on the west coast of Africa, the epidemic meandered along coastlines and waterways so that, a year later, it had reached the southeastern Mediterranean and, through it, eastern Europe.[3] The world reacted with something less than alarm.

Until the nineteenth century, cholera had been endemic to Asia, confined largely to the great deltas of the Ganges and Brahmaputra rivers in India. The new pandemic was caused by a vibrio called the "01-El Tor" strain. El Tor replaced classical cholera as the dominant biotype of the germ in the Ganges River delta. There was a lesson here that went unnoticed outside the circles of experts. The cholera vibrio was adapting to circumstances: it was demonstrating a chilling capacity to evolve. El Tor cholera is less virulent than the classic germ of ages past but is more likely to lead to long-term excretion by human carriers. In this new epidemic health authorities were finding between thirty and one hundred asymptomatic infections for every severe case. This makes it much more difficult to eradicate since carriers have more time to contaminate sewage and water.

The World Health Organization and the United Nations fought hard to control the new epidemic. Their efforts, in conjunction with those of the public health authorities in afflicted countries, met with hard-won success. During the 1980s the numbers of cases worldwide declined; meanwhile, the actual numbers of involved countries increased threefold. Then, without warning, in January 1991, an outbreak of cholera was reported in the city of Chancay, 60 kilometers north of Lima, Peru.[4] Simultaneously cholera was reported in Chimbote, a seaport 400 kilometers north of Chancay. Over the first two weeks in February 12,000 cases were confirmed in the two cities. Suddenly the plague showed its teeth. Spreading, in the words of the formal WHO report, with "unexpected speed and intensity," the epidemic traveled 2,000 kilometers along the Peruvian coast, involving every coastal health department, and quickly extending to Ecuador. By March and April,

Colombia and Chile were struggling with cholera in their coastal cities.

Why had this ancient terror appeared off the South American coast? What was its link to the Asian epicenter?

For thirty years, Rita Colwell, a Maryland-based scientist, had been studying the microbial populations in Chesapeake Bay. Colwell had gathered some compelling evidence for the harmful effects of marine pollution. A century or more of despoilment with human waste, pesticides, fertilizers, and unwanted chemicals was having unpredictable and potentially disastrous effects on bacterial and virus populations around the shorelines. The evolution of new strains of cholera was a typical example. Colwell showed that the vibrio could metamorphose into a minute encysted form, resistant to injury. In this form the germ could survive for months or even years inside the living cells of algae. It should have been a timely warning, but nobody listened to Colwell. No more did they listen when she warned that thanks to the mixing with viruses and other bacteria that teemed around untreated sewage, germs such as the cholera vibrio were picking up genes that made it resistant to most antibiotics and even chlorine.[5] In Thailand, the El Tor strain was found to be resistant to eight different antibiotics. In Lima, the vibrio was also resistant to chlorination of water.

The cholera invading the coast of South America had arrived in the ballast tanks of a ship from China that discharged its pestilential cargo into Peruvian coastal waters. When the outbreak was investigated, it confirmed Colwell's timely warnings. The vibrio was now infecting the local algae that were food for shellfish, crabs, lobsters, and fish, which in their turn were eaten by people. Once it infected people, a huge epidemic began to spread throughout Latin America. Rapid urbanization, foreign debts, and political unrest strained the economies of the afflicted countries, making the expensive necessities of public hygiene and sanitation low priorities. Three years later, Dr. Robert V. Tauxe, of the CDC, thought a million reported cases a gross underestimate, with 10,000 reported deaths from nineteen countries throughout Latin America.[6] There were additional lessons to be derived from the South

American cholera outbreak. One was the potential of modern global travel and transportation in the propagation of plagues. Another was the possibility that a plague might evolve into something else: an emerging infection need not imply a new germ or virus; it could equally imply a change in pattern of one that was all too familiar. Such microbes needed only minimal genomic change, for they had already discovered a path through human defenses to pandemic vulnerability.

In a series of articles, Dr. Paul R. Epstein drew attention to the level of the impending catastrophe. It is easy to forget the vital role played by simple hygiene in the war against epidemics. Yet good clean water was the way cholera had originally been controlled. In spite of vigorous efforts on the part of the United Nations, which had declared the 1980s the banner decade for clean water, in the 1990s, 2.7 billion people, or half the world's population, still lacked safe water and sanitary toilet facilities. Epstein quoted the director of the Pan American Health Organization, Caryle Guerra de Macedo, who estimated that 6 million people in the region could become infected in the next three years.[7]

In July 1991, L. K. Altman of the *New York Times* added his voice to the growing chorus in an article entitled "'Catastrophic' Cholera Is Sweeping Africa."[8] Cholera, like so very many plagues, has a behavioral pattern that is intrinsically interwoven with that of human society.

Before its ethnic conflict, Rwanda was one of the least known countries in Africa, small in African terms, impoverished, yet fertile enough to feed its burgeoning population of 7.5 million. In 1994 it erupted onto the world's news and television screens when three months of war between the majority Hutu government and the mainly Tutsi rebels of the Rwandan Patriotic Front left at least half a million dead, slaughtered in an appalling genocide of civilians. The Hutu perpetrators lost the war, more than two million of them fleeing the country. At least a million fled northwest, across the Zairian border and into the countryside around the town of Goma. Goma had been a quiet town of 80,000 people, nestled by the beautiful Lake Kivu in the lee of a volcano. Suddenly the town was invaded by a desperate torrent of refugees, carrying everything from blankets to their meager rations of yams and beans. Two hundred thousand arrived in a single day, confused, thirsty, hungry, and homeless. They camped on doorsteps, in schoolyards and cemeteries, in fields so crowded that people slept standing up. There was an urgent need for shelter, food, and water.

Nancy Gibbs, writing in *Time* magazine, estimated that it would take an extra million gallons of purified water a day just to counter the refugees' thirst; the rescue services, meanwhile, were managing no more than 50,000. People foraged for fresh water, scrabbling hopelessly in a hard volcanic soil that needed heavy mechanical equipment to dig a well or a latrine. For many the only water available was "a thick, slimy brew already fouled by human waste."

In these relief camps filled with refugees the circumstances were perfect for cholera to show its timeworn face. "On Wednesday the first confirmed case of cholera appeared; within 24 hours 800 people were dead. Then it became too hard to keep count."[9]

Cholera kills quickly, from massive dehydration as a result of the fulminant diarrhea. An adult can lose thirty liters of fluid and electrolytes in a single day. Within the space of hours, the victims go into a lethargic shock and die from heart failure. Without treatment most cholera victims die, yet given good quality medical treatment, including very rapid replacement of fluid and electrolytes, less than one percent of its victims will die from it.

The cholera in the camps was now confirmed as the 01-El Tor pandemic strain, resistant to many of the standard antibiotics including tetracycline and co-trimoxazole. This presented immense problems for the medical staff from local health ministries and the World Health Organization who were attempting to fight it. The tragedy invoked one of the largest relief efforts in history, involving the Zairian armed forces, every major global relief agency, and French and American army units. But the spread of the plague was too rapid for the combined forces to take effect. Within just days of the arrival of cholera, the camps were so littered with the dead and the dying it became hard to tell the sleeping from the dead. Three weeks after the plague had first erupted, it was believed to have infected everybody in the camp. An estimated one million people had contracted it from the waters of Lake Kivu itself, which had become contaminated with infected sewage. The disease would kill 50,000 people before wells could be sunk to provide clean water and before medical aid, including massive transfusions of intravenous fluids and the second-line antibiotic furazolidone could be efficiently dispensed among the teeming victims.[10]

Meanwhile the pandemic continued its meandering global assault. On October 1, 1994, the pandemic was reported from many different republics from the former Soviet Union. It had begun in Dagestan, an isolated and backward rural area in Russia's part of the Caucasus Mountains, where more than a thousand people had contracted it and thirty-five had died. An outbreak in the Ukraine caused ten deaths and threatened a population of more than 50 million people.[11] Daytime temperatures as high as 99°F combined with drought and poor hygiene were contributory factors. Ten Ukranian cities were reported to be affected and half those sick with the disease were reported to be homeless vagrants. Six different strains of cholera germ had been found in the Salfir River, which flows through Simferopol, the Crimean capital, where the health authorities were openly admitting that they lacked the essential skills to contain the epidemic.

"The spread of cholera and other infectious diseases is the calling card of an economy in trouble," declared Alexander Moroz, head of the Ukrainian parliament.

Even as health authorities around the world struggled to contain the El Tor pandemic, a new strain of the classic cholera germ, dubbed *Vibrio cholerae* non-01 CT^+ was emerging in Bangladesh and India. In the words of Paul Epstein, this new germ "has the potential of becoming the agent of an eighth pandemic of human cholera."[12] A year later, a second, and even more infamous "man of death" was making its fanfare return.

2

On September 20, 1994, bubonic plague broke out in the city of Surat, in the Indian state of Gujarat. Reporters arriving at the civil hospital found moaning victims on torn mattresses without sheets. Doctors were said to be as rare as vital drugs. "Nobody comes to clean the ward at all," cried an eighteen-year-old patient, one of 219 on the isolation wards, where the most critically ill were separated from the others by screens of cotton sheets. Food refuse, discarded paper, and pillows were strewn about the floor.[13]

The remainder of the hospital had been evacuated in a climate of fear, triggered by rumors of more than a hundred deaths, though official reports put the total at a more modest thirty-four. A hospital doctor admitted to reporters that every hour, eight to ten patients were being admitted. In the previous twenty-four hours at least thirteen had died. Panic spread through the streets as rumors shook the city's population of 2 million people. Local supplies of the antibiotic tetracycline disappeared from the market, reputedly being hoarded by the rich. Ambulances were looted for drugs and doctors were attacked by the relatives of dying victims, outraged that treatment was no longer available for the poor. Media photographers captured the scenes as masked paramilitary units in camouflage gear with riot sticks guarded the corridors in the Surat hospital to prevent victims' leaving. Hundreds of frightened people, wearing handkerchiefs over their faces, packed the local railway stations as a panic-stricken exodus of 300,000 people erupted outward from the city. A shudder embraced the world.

What is the nature of a disease that could cause such a terrified reaction?

Bubonic plague, from which the generic term plague (the Latin *plaga*) is derived, has devastated humanity for at least one and a half millennia. The first recognizable manifestation was in the reign of the Roman emperor Justinian, between the years A.D. 541 and 544. The plague recurred many times during the succeeding two centuries, causing epidemics that are believed to have killed 40 million people. In 1346, after a gap of some six hundred years, the second great pandemic of bubonic plague arrived in Europe, to become the most notorious human catastrophe in history.

Although there is controversy about its origins, it is thought to have arrived in early October 1347 when twelve Genoese galleys, fleeing from a siege at Kaffa on the Black Sea, arrived at the Sicilian port of Messina. Some ten years later a Franciscan friar, Michael of Piazza, described the frightful arrival of the moribund sailors as they hove into the port, with "sickness clinging to their very bones."

In fact, during the preceding year, there were rumors of a plague of unprecedented ferocity ravaging the East. The pestilence had taken birth in the Tatar lands of Asia Minor, from where it had scourged the cities and plains of India, Syria, Armenia, and Mesopotamia, leaving the streets littered with unburied corpses. In China, then called Cathay, the dead were said to number millions. The Tatars, blaming the Christians in their midst, had besieged the Genoese traders in the fortified port of Kaffa, now called Feodosiya, on the Crimean coast. But while the town withstood the human assault, plague broke out among the attacking army. Before withdrawing in a collapsing rabble of fever and death, the Tatars had flung corpses over the walls using their giant catapults. Whether as a result of this or through a rather more likely influx of rodents through their porous walls, plague erupted in Kaffa and from there, through the fleeing galleys, to Europe.

It seems likely that the pandemic also spread along other fronts, along the historic land routes, the caravan trails, from village to village, and farm to farm. Whatever its portals of entry, from the day of its arrival a pestilence smothered Sicily, taking three months to reach the Italian mainland, and from there, over the succeeding years, it invaded Europe from Mongolia to Ireland and from the Mediterranean to Iceland. During its destructive passage it was called "the great dying." Only two centuries later would chroniclers, for reasons that remain obscure, give it the title it has since adopted, "the Black Death."

The Black Death killed between a third and a half of the populations of the known world of the Middle Ages. The epidemic in Justinian's reign is believed to have contributed to the fall of the Roman Empire. In the Middle Ages it brought down the Byzantine Empire, terminating in its passage the reign of the Mongol emperors of China, which had begun with the bloody campaigns of Genghis Khan.[14]

Today we know that bubonic plague is caused by a bacterium called *Yersinia pestis.* The bacillus has a natural reservoir in wild rodents, such as rats, squirrels, gerbils — in New Mexico it can be prairie dogs — some species of which it has infected for thousands of years. In its rodent host the infection is interestingly "mild," spreading from animal to animal by the bite of an infected flea. Major outbreaks in human history have resulted from exposure to infected rats, spreading to humans through the bite of the rat flea, *Xeopsylla cheopis.* When the rat flea becomes infected with the plague bacilli, they multiply in its stomach until they erupt outward through a gullet obstructed by a mass of bacterial bodies. The flea bites in a frenzy, attempting to relieve the sense of obstruction, transmitting its deadly cargo through the skin at the site of the bite. Often this is on the leg, or less likely the arm, where the bacilli secrete poisons that enable the rapidly multiplying germs to ascend the draining lymphatics. Within two to six days the glands under the arm or in the groin swell enormously — the "pocketful of posies" of the popular nursery rhyme.

These engorged and painful lymph glands, or buboes (from the Greek *boubōn,* for groin), give the plague its name. Most people at this stage of the disease will respond to antibiotics, and transmission can be interrupted by rodent control and by isolation of the infected. In the majority, however, the germs break through into the bloodstream, where they cause a virulent blood poisoning, or septicemia, resulting in widespread bleeding and clotting disorders, meningitis, and involvement of the lungs, giving rise to pneumonia. Without drugs, virtually all such cases will die.

It is this involvement of the lungs that bestows on plague a much more sinister form of contagion. The pneumonia causes the victim to cough and sneeze repeatedly — the "atishoo, atishoo" of that same rhyme — each cough ejecting millions of germs into the air. Inhalation of this deadly aerosol infects others, giving rise to the most fearful epidemic potential. In this form, the rodent carrier and its infected flea are no longer necessary, the plague spreading directly from person to person. Transmission is rapid, a single infected sufferer transmitting the germ to dozens of contacts. Pneumonic plague can therefore threaten anywhere in the world, even in the most hygienic and previously healthy populations.

Pandemics in history have often accompanied changes in human population or behavior, such as the decline of the Roman Empire and, in the case of the Black Death, the ravages of the Mongols. Significantly, in 1994 the city of Surat had seen a fourfold population expansion in the past decade, triggered by an expansion of the local textile and diamond polishing industries. Over a million migrant workers were crowded into squalid makeshift accommodations heavily infested with *Rattus rattus,* the black rat, an infamous plague carrier.

On September 27, barely a week after the onset, and while the Indian health authorities were claiming that the plague had been brought under control, scores of victims were still surfacing hundreds of miles from Surat. The Gujarat state government refused to seal off the city, from where the exodus was estimated at 600,000 people fleeing to surrounding cities and states hundreds of miles away. Indian health authorities' confirmation that some of the cases were acquiring the disease in a pneumonic fashion was potentially a very dangerous development. It was this realization that triggered a wave of alarm among public health authorities around the world.

By September 28 immigration controls were stepped up in every country to prevent the plague being imported beyond India. Passengers attempting to travel out of India were subjected to medical inspection. Meanwhile the Indian medical authorities were reassuring foreign journalists that nobody would be allowed to board any outgoing flight with even the slightest evidence of illness. Even as those same authorities issued bulletins claiming that the plague was contained, television news cameras filmed the shantytowns and slums of Bombay, where thousands still slept in the streets, with sewer rats darting between the sleeping bodies. Those same cameras recorded desperately ill victims still being helped into the local hospitals and teams of health officials wearing masks and spraying dense clouds of insecticide through the rat-infested streets and public places.

Qatar, Bahrain, and Kuwait, all Arab states employing large Indian labor forces, had now banned flights to and from India until further notice. The United Arab Emirates were spraying pesticides in ships and aircraft from India. In Germany, doctors boarded flights from India, checking for signs of illness among passengers. Singapore, Malaysia, Thailand, and the Philippines set up screening of travelers. Pakistan suspended flights between Karachi and Bombay and refused visas to people from the affected states. In Britain, where the plague was receiving daily coverage in the national television news, Dr. Kenneth Calman, the government's chief medical officer, advised all port medical officers on diagnostic and containment measures.

By September 29, plague had spread to at least eight states, with 330 cases now confirmed in Maharashtra, immediately south of Gujarat. Two hundred were believed to be bubonic, with 130 thought to have arisen through pneumonic spread. Other afflicted states now included Delhi, where the capital was located, together with Rajasthan and Uttar Pradesh in the north, and Bengal and Orissa in the east. In a little over a week, the plague had traveled over a thousand miles, covering the landmass from west to east. Health officials in Delhi, who had confidently claimed to have controlled the plague, were now admitting to profound disquiet about the half million migrant workers who had fled Surat over the previous week. Two people had now tested positive for pneumonic plague in the capital and ten more in Bengal. People in New Delhi were "busily killing rats," even though health officials warned that dead rodents might be more dangerous than living ones, since it forced the fleas to change hosts. The chief epidemiologist from the National Institute for Communicable Diseases was himself isolated in the hospital, suspected of acquiring the plague.

On September 30, in Britain's *Daily Telegraph,* the science editor, Roger Highfield, quoted Professor Alexander Tomasz of Rockefeller University, who said the era of antibiotics was ending as age-old infectious diseases staged a comeback. While discussing the recent worry about antibiotic resistance in a number of common germs, the article raised the specter of a bubonic plague bacillus that could also develop wide-ranging antibiotic resistance. Such a prospect would indeed be appalling. The expert went on to reassure people, "There has been no report of significant resistance in the plague organism to date. But understanding of the organism is extremely limited in the West, where it has only been of interest to germ warfare laboratories."[15]

Health experts from the European Union met hastily to coordinate measures, worried that with their newly opened internal borders, if one European country became infected, plague would quickly spread throughout the Community. France became the first European country to recommend that its citizens avoid India altogether. Similar urgent meetings took place at the CDC in America. The WHO, meanwhile, was appealing for calm.

By October 2, the toll of infected in the Indian plague had reached 2,500, with the authorities admitting to 58 deaths. In Delhi, where the air was acrid with the smell of burning rubbish and where thousands of the citizens were walking the streets wearing face masks, the authorities felt the disease was now contained.

As with the El Tor cholera pandemic, there were several profound lessons to be learned from the Indian plague outbreak. Surat and India in general were fortunate that the strain of germ was of relatively low virulence and sensitive to tetracycline. Otherwise the loss of life might have been much greater. Biochemical studies of the germ, at Fort Collins in Arizona and at the Pasteur Institute in Paris, confirmed that the germ was different from any strain seen previously: like the cholera vibrio, the plague bacillus was evolving, metamorphosing. It was a warning that new variants might be seen in the future, including some that might not prove as benign.[16]

3

If the conquest of a single great disease epitomized the hubris of the world, it was the defeat of tuberculosis. This single plague, which spreads in humans through inhalation, can affect any organ but mainly destroys the lungs. It has caused the death of approximately a billion people during the nineteenth and twentieth centuries. Even with the discovery of the antibiotic cure, the battle to control it had been nothing less than a global war, involving all of the manifest arms of medicine.[17]

In 1960, war was declared against tuberculosis in every major developed country and, on behalf of poorer nations, at the WHO in Geneva. An effective triple therapy had been discovered that both cured patients and allowed a much safer and more effective surgery where it was needed. These advances created the potential to eradicate tuberculosis globally. Tragically, even as the battle was won in developed countries, it was lost in the Third World. And in time that failure would return to haunt the planet.

In the mid-1980s, what had been an unrelenting decline in the incidence of tuberculosis ground to a halt. The setback was global, alarming the few remaining experts. The causes, as we now know, were complex. The arrival of AIDS was a vital factor, its pervasive damage to human immunity mimicking the triggering effects of war or famine in other arenas. There were other, more predictable causes: for example, immigration into industrialized countries from the developing world, where the disease was literally epidemic, was a factor in Europe. In America, poverty and lack of adequate medical supervision, which increased under the Reagan administration, was very significant in inner-city areas. In the United Kingdom, the numbers of new cases of tuberculosis continued to fall until 1988, after which they remained virtually stationary for two years before rising by 5 percent in 1991. Today, one in fifty of London's homeless population has tuberculosis. In 1992 and 1993 combined, there were an estimated 8,000 more cases than had been anticipated. From then on they rose year by year until a belated response from the authorities, from 1992 onward, saw its first benefits in 1994–95.

In a press release entitled "TUBERCULOSIS IS RISING IN INDUSTRIALIZED COUNTRIES," dated June 17, 1992, the World Health Organization listed no less than ten countries in which tuberculosis was dramatically increasing. It surprised experts to see such countries as Switzerland, Denmark, the Netherlands, Sweden, Norway, and Austria included, in addition to Ireland, Finland, Italy, the United Kingdom, and the United States.

This strange and frightening recurrence of tuberculosis in the United States and Europe was only a faint reflection of an alarming pandemic. In 1991, Professor John Murray, of San Francisco General Hospital, drew attention to the fact that during the previous five years, there had been an ominous worldwide resurgence of tuberculosis, with a veritable explosion in sub-Saharan Africa. Eight million or more people worldwide still contracted open tuberculosis every year. It is estimated that tuberculosis caused 2.9 million deaths in 1990, making this disease the largest cause of death from a single pathogen in the world. While the largest numbers of deaths occurred in the developing world, a surprising 400,000 people were still contracting it in industrialized countries, where 40,000 people were still dying from it.

For the tuberculosis germ, too, was metamorphosing. Throughout America and many of the hardest hit areas in developing countries, there was a strong association between AIDS and multi-drug-resistant tuberculosis. In the past, when the mycobacterium that causes tuberculosis developed multiple resistance, its virulence — its capacity to infect others — waned. The MDR germs associated with AIDS had developed new levels of virulence, with a high degree of infectivity, best seen in the proliferation of nosocomial outbreaks. There were 600 such outbreaks in the United States alone in recent years, introducing a potential hazard for staff not only in the wards dealing with patients but for microbiologists and pathologists and their staff when dealing with specimens and postmortems.

In sub-Saharan Africa, where 171 million people are infected with latent tuberculosis, the catastrophe has already happened. In 1992, an estimated 6.5 million Africans are infected with the HIV virus, of which 3.12 million are additionally infected with life-threatening tuberculosis.

This resurrection of tuberculosis has given the developed world a considerable fright. It has caused a massive turnaround in our thinking. Millions of dollars, francs, marks, pounds sterling, and yen are now being poured into a belated effort against a disease long considered dead and buried.

4

The return of such plagues soon rekindled interest in the threat of infections, both among the researchers and the media, and those who investigated the trends were shocked by what they discovered. In an article in *Time International* with the provocative title "The Killers All Around," Michael D. Lemonick highlighted the growing fears concerning infectious microbes. "They can strike anywhere, anytime. On a cruise ship, in the corner restaurant, in the grass just outside the back door. And anybody can be a carrier: the stranger coughing in the next seat on the bus, the classmate from a far-off place, even the sweetheart who seems perfect in every way."

John Burke Davies is a journalist with a natty gray beard and thinning hair that hangs back over his collar. He wears thick black sunglasses and writes for the *Daily Sport*, the British equivalent of the U.S. *National Enquirer.* "Normally," he confessed, "we would have a sex story on the front page." On May 11, 1994, while skimming through press agency reports from various regions in Britain, a brief from the West country caught his attention. A mysterious infection had appeared out of the blue in the rural county of Gloucestershire. The opening paragraph described two people who had died from this killer bug and three others who were fighting for their lives. Then he read the second paragraph, which described how this bug crept along the flesh, eating it at a foot an hour. This seized his imagination.

The story had begun some three months earlier when, in the village of Chalford near the town of Stroud, Les, a retired engineer, had approached his doctor with two hernias.[18] On Friday, February 4, he went into the Stroud hospital for elective surgery. The operation went well and he appeared to be recovering normally. At 7:30 the following evening he was found collapsed on the floor next to his bed. He was vomiting, had a high temperature, and the tissues at the top of his leg were swollen and discolored. Rushed to the Gloucester Royal Hospital, whose intensive care facilities are superior to those at Stroud, he was diagnosed with necrotizing fasciitis. This is a type of localized gangrene. Suddenly his condition warranted an emergency operation, which involved removal of all the infected skin. By Sunday morning, he was lying on the intensive care unit following major surgery.

Helen, a middle-aged woman, also lived in Chalford. The day after Les was admitted to intensive care, she arrived in the Stroud hospital for an operation to strip her varicose veins. It was a routine operation and she expected to return home the following day. But Helen also developed necrotizing fasciitis. Her next memory was waking up in the Gloucester Royal Hospital, nine days later. Two people from the same village, who had gone into Stroud hospital for minor surgery, were now lying in beds seriously ill in the intensive care unit. Dr. Keith Cartwright, the hospital bacteriologist, was called in to see Helen that evening.

Setting up urgent platelet and blood cultures, he also studied pus swabs from the gangrenous wound. Under the microscope, he saw chains of bacteria, the typical appearance of a germ called the streptococcus. The cultures confirmed the presence of a bacterial infection caused by an unusually virulent form of streptococcus, known as the Lancefield's Group A beta-hemolytic streptococcus. This puzzled Dr. Cartwright. Most cases of necrotizing fasciitis were caused by a combination of germs rather than just this one. He could remember no previous reports of the streptococcus affecting more than one patient after recent surgery. Over the next few days, he studied the medical journals, but found nothing.

Dr. Cartwright closed the Stroud hospital operating room for all surgery. The hospital set up an internal committee to investigate the strange infections, taking swabs from hospital and OR personnel and undertaking a major sterilization of floors, walls, and equipment. Two bacteriologists, Dr. Cliodna McNulty and Margaret Logan, from the Gloucestershire Health Authority, arrived to help with the investigation. Local physicians were put on alert.

The emergency room was reopened on Tuesday, February 15. Three days later a third case of necrotizing fasciitis turned up, this time the victim a retired general practitioner who lived just one mile from Les and Helen. He had no connection with the operating room prior to his infection, though he now had to have his leg amputated. The Gloucestershire Health Authority injected a new urgency into their investigation, worried about the threat of a community-based epidemic. The communicable diseases center at Colindale, North London, was notified. This is the central register for serious infections in Britain. They issue weekly CDR bulletins, rather as the CDC issues MMWRs in the United States.

No bacteriologist takes the beta-hemolytic streptococcus anything other than very seriously. There is a laboratory at Colindale, the Public Health Laboratory Service (PHLS) Reference Laboratory, which constantly monitors infections in Britain. Necrotizing fasciitis was so rare, and a cluster of three cases so unique, that Dr. Cartwright called his colleague at Colindale, Dr. Norman Begg, at his home on Friday evening. Cartwright's anxiety was soon justified. On March 3, a woman coming home to Gloucestershire from a trip to Scotland died on board the train. In April the deadly infection killed two more people in Gloucester Royal Hospital. Six people, all of whom lived in or near Stroud, had now contracted necrotizing fasciitis. A forty-six-year-old woman, who worked as a travel agent, was the seventh victim. She recovered on massive doses of antibiotics in the Gloucester Royal Hospital.

The South-West News Service circulated the story to see if any nationals were interested. It was at this stage that it caught the attention of John Burke Davies. Davies felt simultaneously revolted and fascinated by what he was reading. He decided he would re-christen the germ. He called it the "flesh-eating bug."

His dramatic story took the front page of the *Daily Sport* opposite a half-page color photograph of two buxomy young women wearing nothing more than the red and blue football shirts of the national cup final teams.

A week or so later, the picturesque market town of Stroud hit the front page of every national newspaper. Journalists were excited by the public abhorrence of a bug that devoured living human flesh. On May 24, the *Daily Telegraph* carried the headline "KILLER BUG CLAIMS ITS FOURTH VICTIM." Eschewing the more lurid name for the germ, the article described the spread of the disease to a fatal case involving a man in London. A day later, the same newspaper threw aside its previous caution when, with a banner headline across the entire front page, it proclaimed, "FLESH-EATING BUG KILLS YOUNG MOTHER AFTER 'COMPLETE CURE.'" The victim was in her mid-twenties and had contracted the infection three weeks before, after having a baby.

So began a different kind of epidemic: the battle between newspapers for increasingly lurid headlines. On Thursday, May 26, the same paper carried a new headline. "MOTHER SAW HER CHILD DESTROYED WITHIN 24 HOURS." The tabloids were having a field day in carmine-dominated Technicolor. Adjacent to a full color photograph of a mutilation, the *Daily Star* carried the headline "KILLER BUG ATE MY FACE." The *Sun* challenged with "FLESH BUG ATE MY BROTHER IN 18 HOURS." The *Daily Mirror* topped that with "FLESH-EATING BUG KILLED MY MOTHER IN 20 MINUTES." In the *Sunday Times*, Sean Ryan, Lois Rogers, and Margaret Driscoll debunked the story under the heading "THE BUG THAT ATE INTO OUR IMAGINATION AND SENT US ALL MAD."[19]

The effect on general practitioners' emergency calls was all too predictable. Sky News parked in the grounds of the Gloucester Royal

Hospital, putting out hourly updates on a story the hospital had been handling for months previously. One such request was from the *Daily Sport* inquiring if there was some connection between the outbreak and the victims of the worst serial killer in British history, Fred West, whose bodies had recently been disinterred in the same town.

A comprehensive ongoing surveillance and investigation of the disease was now implemented, bringing together the Gloucestershire Public Health Laboratory, the Gloucestershire Department of Public Health Medicine, the PHLS Communicable Diseases Surveillance Centre (CDSC), and the PHLS Streptococcus Reference Laboratory. An "Action Group" for necrotizing fasciitis was set up. Drs. Marian McEvoy and Sarah Harrison at the CDSC issued a case definition, together with a telephone contact number for information on any possible cases. While reassuring colleagues that there was no reported increase in the national reporting of serious Group A streptococcal infections, they gave instructions for the urgent treatment with huge doses of penicillin, augmented perhaps with clindamycin, for all clinically suspected cases.

Medical reassurances seemed merely to fan the flames. The press were now accusing the British Health Secretary, Virginia Bottomley, of a cover-up, while the World Health Organization, siding with Bottomley, was accusing the press of making a mountain out of a molehill. In the words of Sean Ryan and his fellow journalists, "The killer bacteria could have been created by Stephen Spielberg . . . poised to eat into the hearts and minds of the public as surely — and as fast — as it had eaten its way through its victims."[20]

That year the flesh-eating bug was among Britain's more successful exports to the United States, where the media and public were every bit as entranced. In the *Weekly World News* the successful import was duly acknowledged with an entire front page taken up with the banner headline "FLESH-EATING VIRUS INVADES THE U.S." The streptococcus had now evolved into a killer sci-fi virus. Newspaper artistry had improved upon nature with viruses upstanding on insect-like legs, baring teeth reminiscent of the killer shark of the film *Jaws.* Under flashbacks of other headlines, such as "EATEN ALIVE" and "I WATCHED KILLER BUG EAT MY BODY," the *Weekly World News* warned that 50 million could die from the mystery bacteria that kills in minutes. No American was safe — according to a statement attributed to, and even more vigorously denied by, the Centers for Disease Control. While the official line in America as in Britain was to suggest there was no problem, in Canada, where authorities had recorded a fivefold increase in cases in one year, Lucien Bouchard, the fifty-five-year-old leader of the Quebec separatist movement, contracted the disease and had to have his leg amputated.

Back in Britain, the hysteria had evolved to an age-old solution: looking for somebody to blame. Hearing a hint that two nurses at the Stroud hospital had been found to carry the streptococcus, journalists descended on the hospital in a determined flurry of investigative reporting, only to discover that when the local bacteriologist tested fifty of the journalists themselves, one in ten were carrying the streptococcus. The "epidemic" settled spontaneously to a trickle of cases.

In retrospect, if part of the reaction was voyeurism, part was also a genuine and deep-seated horror of an invisible entity that killed people in such a horrible fashion. Why was the germ behaving in this bizarre fashion? Less severe infections with the same streptococcus are so common they are not reported. With the easy availability of antibiotics, the germ is virtually always susceptible to penicillin and its derivatives and treatment usually goes no further than the family doctor's office. In the first half of this century, it was a very different story.

In those days the beta-hemolytic streptococcus was the most notorious of all the common acute germs. Spreading as epidemics of sore throat, it diffused poisons into the bloodstream that caused fearful epidemics of scarlet fever, killing as many as 35,000 people each year in Victorian Britain. Even after recovery it could leave its victim with damaged heart valves through rheumatic fever or the kidney disease nephritis. It ravaged obstetric wards as the notorious puerperal fever, feeding upon the richly engorged lining of the recently delivered womb. Once the germ gained entry even to a minor cut or wound, its progress would soon be highlighted by the livid red tramlines that ascended the skin, or by the scarlet reefs of spreading cellulitis, brushing aside the body's defenses, until it killed its victim through a fulminating blood poisoning. It was so feared that a surgeon contacting an infection in his finger would have the finger amputated.

While public health authorities were dismissing the scare as nonsense, some experts welcomed it if it would "wake people up to the continuing danger of infections." The experts had recent cause for worry. In Britain, salmonella food poisoning was still afflicting 30,000 each year. Meningococcal meningitis caused 1,500 cases and 50 deaths. Most worrying of all to bacteriologists around the world, there was growing concern that we were losing the life-saving efficacy of those "miracle drugs," the antibiotics.

5

Microbes were the earliest forms of life on earth to evolve under the pressures of Darwinian evolution. Their existence over aeons has been dictated by natural selection, and consequently they owe their remarkable endurance to their ability to adapt to the most injurious of prevailing circumstances. Perhaps it should not surprise us that today many of the familiar, if notorious, germs of history are becoming resistant to the drugs commonly used to treat them. These include the second of the blitzkrieg twins of acute infections, the *Staphylococcus aureus.* The "staph" is the cause of boils, disastrous infection in bone, and even blood-borne spread. It was the first germ to show resistance to penicillin. Scientists got around this problem by discovering a derivative of penicillin, called methicillin, that kills the resistant strains. But for more than a decade, doctors have seen a proliferation of staph, resistant even to methicillin. By 1995, this "superbug" was sweeping through hospitals throughout the affluent world. Known by its acronym, MRSA (methicillin-resistant *Staphylococcus aureus*), the germ is carried on the bodies of one in three of the healthy population. Since the mid-1980s, it has become a common cause of serious hospital infection. The elderly and people recovering from surgery are most at risk. In 11 hospitals in the West Midlands region of England, a recent study found 346 cases of infection in just three months, and at least 60 people had died.[21]

Today vancomycin is the only antibiotic to which all staphylococci are sensitive. Languishing in a deep freeze at St. Thomas' Hospital in London are strains of staph that are resistant even to this final course of treatment.[22] This germ was discovered by experiment on human skin. But if such a "superbug" were to emerge as a spreading epidemic in the community, there would be no effective antibiotic treatment. Professor Tomasz, a world expert at the Rockefeller University, believes that this is only a matter of time.

Another germ, called the *Enterococcus,* is already resistant to vancomycin in 20 percent of cases found in American hospitals. Rarely infecting healthy people in the past, it is now an increasing cause of blood poisoning in immunologically compromised patients and those with artificial heart valves and surgical prostheses. In Britain, a report from the PHLS has recently reported strains resistant to every antibiotic in forty-four hospitals, giving rise to serious problems in intensive care units. Professor David Speller, head of the antibiotic reference unit at the PHLS, has reported a worrying increase in infection, with patients dying from infection of artificial heart valves.[23] Yet another germ, the pneumococcus, appears to have first acquired resistance to penicillin in Spain in the mid-1980s. Ten years later, it has added resistance to the very latest cephalosporin antibiotics in America. In consequence, this germ, which is one of the causes of bacterial meningitis in children, can only be treated with vancomycin given by unpleasant injections directly into the spinal canal.

Globally, malaria is second only to tuberculosis among the worst of the killer infections. More than a hundred million people worldwide contract malaria in 102 countries each year, and approximately one and a half million people die from it. Caused by a protozoan called a plasmodium, malaria too is becoming increasingly resistant to antimalarial drugs, such as chloroquine and Fansidar. The mosquitoes that carry it are also becoming resistant to the commonly used pesticides that formerly suppressed them. The result is a dangerous resurrection of this ancient peril in East Africa, South America, and Asia, wiping out much of the wonderful gains made in the eradication programs of the 1960s. Tens of thousands, for example, are dying in Madagascar, where malaria was previously well controlled. Sadly, in many parts of the world, the situation is actually worse than it was at the very beginning of these early pioneering attempts.

And what we see in the poorer world has a tendency to arrive unexpectedly into our own backyards. In America, between 1960 and 1991, there was a tenfold increase in the number of people with malaria traveling from abroad. Recently it was reported in migrant communities in Southern California, and in 1994, an isolated outbreak was reported in New Jersey, apparently transmitted by local mosquitoes. With global warming, this problem could rapidly worsen, with movement of the vector mosquito northward into many Western countries.

In 1992, 4,000 cases of diphtheria, a disease entirely preventable by vaccination, were reported from the former Soviet Union, where vaccination programs had declined as a result of political and social instability. The following year it had grown to epidemic proportions, with 15,211 cases in Russia alone and almost 20,000 cases in Eastern Europe as a whole. That same year, the CDSC reported two cases of diphtheria in British children with no history of foreign travel. Both had been fully immunized as part of the customary triple vaccine. In 1994, the British Department of Health announced that an extra booster dose of diphtheria vaccine for children aged 15 to 19 would now be included in the regular immunization program.[24]

If it appears surprising that at the same time in Britain there was a scarlet fever alert, how incredible that in this same year a single page of the *British Medical Journal* contained reports of new outbreaks of bubonic plague and cholera at the same time that microbiologists worried about the rising incidence of whooping cough, dysentery, and salmonella.[25]

These observations are far from the media hype of the flesh-eating bug. A series of bulletins from highly professional sources warned colleagues about the reality of the danger.

In fact there is nothing really new in the appearance of bacterial resistance. When isoniazid, one of the wonder discoveries against tuberculosis, was first being assessed for its efficacy against the *Mycobacterium* in 1952, before the results of the first clinical testing of patients were published, no less than six different papers appeared, all reporting astonishment at the effortless ease with which the tuberculosis germ developed resistance to this wonder drug.[26] Such drug resistance is facilitated by the flagrant abuse of every antibiotic ever discovered.

In developing countries today, children hawk antibiotics at truck stops and charlatans sell antibiotics with little idea of diagnosis or dosage as a panacea for all ills from piles to sexual impotence. And the abuse is not restricted to developing countries. In the industrialized West we have grown equally blasé. Parents pressure doctors into giving antibiotics to children with mild sore throats, often caused by viruses that are totally resistant, with the result that common germs, which are often found in small numbers in the child's gut, or in the nose or throat, and which are resistant to the antibiotic, now proliferate. Perhaps most reprehensible of all, farmers administer antibiotics to meat animals because it has been discovered by trial and error that the animals develop more flesh.

Problems arising from such behavior are now commonplace in the emergency medicine and intensive therapy arenas of big city hospitals, where very sick patients, often with compromised immune defenses, face infection with unusual microorganisms. In the complex interplay of resistant germs and very sick patients, doctors are tempted to prescribe uncommon and relatively toxic drugs — what Dr. J. Fisher, author of the book *The Plague Makers,* terms "drugs of fear." Even worse still, he warns, some germs, capable of epidemic spread, are now resistant to all of the available antibiotics.[27]

Among such disquieting trends is the rise in sexually transmitted diseases. Common strains of the *Gonococcus* that causes gonorrhea have developed resistance to penicillins and several other antibiotics. The *Haemophilus influenzae,* one of the causes of meningitis, ear infections, and pneumonia, is often resistant to five different categories of antibiotics.

Tragically, in the world at large, 4.3 million people every year die from these ordinary infections in the form of acute chest infections, often brought on by flu. A surprising 3.2 million, many of them children, actually die from diarrheal illness. In the salutory words of Hughes and Berkelman, "Our once seemingly invincible array of antimicrobial drugs is declining in effectiveness for many hospital and community-acquired infections."[28]

The antibiotics are so vital in everyday medical practice that it would be disastrous if we were to lose even some of their benefit. It is very important indeed to try to understand how germs develop antibiotic resistance, and a vast amount of research is looking into this worrying trend in laboratories throughout the world. That research is turning up surprises.

Bacteria reproduce very quickly, commonly with a new generation every twenty minutes or so. Mutations arise as a natural phenomenon of their genomic chemistry. The presence of an antibiotic in their environment will exact an evolutionary pressure through mutation to evolve new strains, resistant to the antibiotic. Inside every person infected with tuberculosis, for example, there are clones of mycobacteria resistant to every known antibiotic, perhaps every antibiotic we will ever devise. These have already arisen by mutation. The presence of an antibiotic kills all of the others but permits the strains that have mutated resistance to that antibiotic to proliferate and take over the infection.

Mutation does not, however, explain everything. All complex forms of life have sexual reproduction, when huge amounts of genetic material is interchanged and married to form the new offspring. In the early decades of this century, it was the common perception that bacteria could not exchange genetic material: bacteria, reproducing by simple binary fission, did not enjoy the advantages of such sex lives. Now we know that this is not the case. Bacteria, it seems, are quite promiscuous.

In 1946, Lederberg and Tatum radically altered the old perceptions with the discovery that bacteria could readily pass genetic information between them, not only between members of the same species but even bacteria of very different species.[29] Lederberg called these mobile packages of genetic information "plasmids." It was a vision of bacterial behavior that was both iconoclastic and threatening: bacteria sometimes linked to each other, reaching out penile bridges, through which they exchanged genetic messages. Those messages could contain the genetic code for antibiotic resistance.

In fact, bacteria move their genes around in a surprising variety of ways. Another way in which they do so is to join up physically (as in sexual mating) to swap genes. This is the mechanism of evolution of the staphylococcal superbug. The drug of last appeal was vancomycin, which destroys the bacteria's cell wall, causing it to blow up like an overstretched balloon. Nothing other than a re-creation of the entire cell wall of the germ would stop this action. Yet this was exactly what happened at St. Thomas' Hospital when Professor Noble and his team in the Institute of Dermatology deliberately mixed two different species of germ on living human skin. To their astonishment a harmless germ, normally resident in the bowel, passed on a package of genes giving the genetic coding of an entirely new bacterial cell wall to the deadly staphylococcus.

Viruses, called bacteriophages, or "phages," infect a huge variety of different bacteria, importing vital genomic information on antibiotic resistance information. Germs can even cannibalize their own dead to the same effect, as Oswald Avery showed in his wonderful experiment that demonstrated for the first time that DNA was the chemical of heredity.[30] Experts believe that the group A streptococcus, popularly known as the flesh-eating bug, is infected with exactly such a virulence-enhancing virus.

6

It is an altogether normal reaction when the ordinary man or woman retracts in horror at the thought of microbes invading and destroying their body. Such microbes threaten us. Every parent knows the terror when a child develops a high temperature and the fear of loss strikes deeply into the heart. The answer to such fears does not lie with empty reassurances but with an honest assessment and the understanding that comes from patient scientific appraisal.

For the moment, mercifully, most germs are still sensitive to antibiotics, and that means most bacterial infections remain eminently curable. But this cannot be taken for granted. The abuses of antibiotics lie within our human control: they must be tackled urgently before the situation becomes uncontrollable.

A new epidemic, caused by an emerging virus, is potentially more threatening because it does not lie within human control. Although there are some effective antiviral drugs, most lack the curative power of antibiotics in bacterial treatment, and viral resistance to the few drugs we already possess is fast appearing. Vaccines, the mainstay of prevention against viral diseases, are not always effective and take too long to develop in the case of a new or emerging virus. As we have seen with the hantavirus epidemic, in the case of a new plague caused by an emerging virus there is usually no vaccine and no treatment.

There has been a significant increase in the emergence of such new plague viruses over the past forty years. That increase makes it mandatory that we look very hard at what is happening, that we study every aspect of it so we can come to an understanding of why such plagues emerge.

In his excellent review article, which took its theme from the underlying sociological factors of the great plagues of history, the Harvard-based Paul Epstein underlines the observation that pandemics tend to happen during times of great social transition.[31] At no time in our history has human society been going through a greater period of change, or with such unpredictable acceleration.

The First World remains utterly beyond the means of the Third. The Second World, formerly the communist bloc of the East, is embroiled in a chaos of transformation, with economic catastrophe, food lines, and a wave of organized crime amid falling standards of personal and public health. In our world at large, it is estimated that AIDS, almost uniformly fatal, may infect one in every hundred of the human race by the end of the millennium.

Why, we ask ourselves, should our world be unduly prone to plagues at this late stage of the twentieth century? Where do such plagues come from? Nowhere is it more urgent to address this question than to the origins of plague viruses. If we can understand the source of such viruses, then we have made an almighty step toward reducing their threat.

To understand the riddles of origin of any form of life, one has to look deeper, to gaze into the profounder mysteries of evolution, to explain the paradoxes of life itself. And those explanations lie not in the hospitals or research laboratories but in the fields, in the oceans, and perhaps most pertinently in the rain forests — territories that are more the province of biologists and of ecologists than of doctors.

EIGHT

The Coming of Ebola

Ex Africa semper aliquid novi.
There is always something new out of Africa.

PLINY[1]

1

At its southeastern edges, the African rain forest does not end abruptly but tapers into a hinterland of lush savannahs and riverine valleys, tongue-like extrusions of the forest that encourage the more inquisitive of its wildlife to come out and explore. Nzara is a small town in the far south of the Sudan, close to the Zairian border. It occupies land in the Yambio district of the Western Equatorial province that has only recently been carved out of this junctional hinterland. Colobus and Cercopithecus monkeys live in the trees, and colonies of baboons hold fierce territorial squabbles in the high grass.

Nzara has a population of 20,000, mainly of the Azande tribe, with its powerful family taboos, polygamy, and ancestor worship. There are a few brick buildings, roofed in corrugated iron, but most people live in dense communities of mud-walled and thatched "tukels." In 1976,

Yusia lived about ten miles outside Nzara in a homestead of three circular tukels, one for himself and one each for his two wives. He enjoyed the rural setting, hacked out of dense forest on the old Yambio road. It pleased him that he could eat fresh eggs laid by his own chickens or grow maize in the bloodred soil, storing the crop in a wooden box covered with its own thatched roof — like a capacious covered coffin — which he had lifted up on stout poles with collars on the poles to keep out the rats. His wives would sing to themselves as they ground the maize on long hollowed quern stones, like the *metate* of the Maya.

The main employer in the town was the cotton factory, which employed 455 people. Cropped in the local fields, the cotton was carted to the factory, where it was put through every stage of production to cloth. Yusia worked in the cotton factory, cycling the ten miles into work, where he had his own small brown table in an office next to the storeroom. On the left of the table was a battered and worn box of files in which he would tally the bales of cloth that were stacked and sorted in the storeroom before being loaded through its wide corrugated iron doors onto lorries for export. Unassuming and hardworking, Yusia kept to himself and his two families, avoiding contact with his working colleagues outside factory hours. He knew about the good times at John's Jazz Bar in the center of the town, but he never joined in the drinking and dancing or in the wild parties that went on there from time to time.

On June 27 Yusia fell sick. A pain began over his forehead and spread until it involved the whole of his head. It was nauseatingly severe. Next he developed an extremely sore throat, which he described to his brother Yasona as like a ball of fire. Slim and fit, Yusia had never been seriously ill in his life. But now his tongue was as dry as rope, crops of tiny painful ulcers tormented his cheeks, and it was agony even to swallow saliva. Next came severe muscle pains, in his chest, his neck, and the small of his back, from where the pain lancinated down both his legs. His face became sunken and drained of expression. He just lay groaning on his bed. Yasona stayed with him in his mud and thatch home, nursing him through an illness that was progressing with a mind-numbing intensity and rapidity.

No measure seemed to alleviate Yusia's suffering. By June 30, his brother was so alarmed he arranged for him to be taken to the local hospital.

The hospital in Nzara was a shack with a few iron-frame beds. There was a dispensary run by a nurse and a doctor who spent much of his time hunting monkeys. By now Yusia was suffering cramping abdominal pain, diarrhea, vomiting, and prostration. Two days after his admission, he started to bleed profusely from his nose and mouth and the diarrhea became heavily bloodstained. The flesh shrank about his bones until his face resembled a skull-like mask, with eyes that were sunken and staring. Death came as a merciful release on July 6.

On July 13, Yasona also fell sick. He suffered an identical cycle of harrowing symptoms, but his illness lacked the lethal severity of his brother's and he survived after two weeks of high temperature, headaches, chest pains, diarrhea, and vomiting. There it might have rested, the death of one quiet man, poignant but not so very remarkable.

Mortality from a febrile illness is commonplace in the tropics. Malaria, typhoid, tuberculosis, and sleeping sickness have a firm clutch on the people, and any one of these might cause a similar presentation and fatal outcome. But soon other people also began to fall sick. On July 12, another storekeeper from the cotton factory was admitted to Nzara hospital with identical symptoms. Bullen had worked at a table adjacent to Yusia's in the records office, though there was no other family or social connection between them. Whereas Yusia had lived in the country, Bullen lived three kilometers east of the factory, in the organic warren of urban streets. Within just two days Bullen was dead. Then one of his two wives was also sick, with identical symptoms. She also died, five days later, on July 19. In their terminal decline, Bullen and his wife had shown signs of bleeding, from the nose, the mouth, in their vomit, and in their diarrheal stools.

Suddenly a fourth man reported sick. He also worked at the factory, in the cloth room beside the office where Yusia and Bullen had kept their tally books and records. His name was Paul, and his symptoms began about July 18. His illness would have a much wider impact than that of either of the quiet brothers or the second storekeeper.

Paul was a bachelor who lived in the densely populated heart of the town, where he was active in a number of business enterprises. He was well known to everybody in Nzara, a young man who enjoyed the good life. His home was next to that of a merchant, Mohammed, and he helped out in Mohammed's shop. Paul was useful to Mohammed, acting as interpreter for visiting tradesmen, including people traveling across the border from neighboring Zaire. He was friendly with Mohammed's two assistants, Samir and Sallah, who lived with the merchant's family. The brothers played in a local jazz band. All three would entertain the customers in the cool of evening at a local hotspot. This comprised a few shady lean-tos roofed in thatch inside a protective palisade of dry sticks to keep the leopards out. There was a handwritten sign over the entrance that read "John's Jazz and Dancing Bar." Inside the palisade was a bare dirt yard for dancing. Then there was the bar itself, open-fronted and thatched, with its few frames of plank shelving, stocked with Makasi beer, brought in from Uganda by riverboat, and rice wine from Zaire, smuggled in along with the routine imports through the factory. Over the bottles was a photograph of a smiling woman and an amateur oil painting of a yawning hippopotamus in a river landscape. A handwritten sign, in crimson letters, read, "Do drink but love."

It would appear that the conviviality also ran to an easy intimacy with the women. The extroverted and fun-loving Paul was adept at organizing women for such parties, even from as far afield as Zaire. The women genuinely liked him and he had many girlfriends. During Paul's illness, he was comforted by many such girlfriends, who nursed him in his home before he became so ill he had to go into the Nzara hospital. Paul died in the hospital, shocked, exhausted, bleeding, his face expressionless, his eyes sunken, their whites injected with fever.

It is a local tradition that the female members of the deceased's family clean and bathe the corpse after death. This ritual takes place in the home or compound where the dead person lived. The bathing ceremony is accompanied by an open display of grief, weeping and crying, with frequent fondling of the body and kissing of the face of the dead. The corpse is then buried in a grave next to the abandoned home. Many of the victims of the mystery fever were washed and mourned in this way, their bodies buried in traditional graves, with mounds of rocks and earth raised over them. The mound would be capped with a touching little thatched roof, as if the spirit of the dead lived on, in a pastiche of its former home. Following Paul's death, and the cleansing and bathing of his body, several of his female friends, including his fiancée, Hawa, contracted the illness. From this point onward there was an explosion of the fever in the town.

Samir, one of the two jazz-playing brothers, became ill on July 26, not long after Paul's funeral. His illness progressed slowly, but by August 6, his brother, Sallah, removed him to Maridi, a town with a large and well-equipped hospital, eighty miles east of Nzara, where on August 7 he was admitted to the "Class 1" pavilion. This was a hut with a corrugated iron roof that housed a total of four private patients, including Samir. Here the tragic young man suffered a protracted decline, dying ten days after his admission. Sallah helped to nurse his brother, hitherto so fun-loving, throughout the ten days of his final torment. Returning to Nzara on August 18, Sallah took sick himself and was visited by a male nurse from the Nzara hospital. The frightened nurse gave him injections of chloroquine and antibiotics, but it made no difference and Sallah subsequently died in the merchant's home, where his death was soon followed by that of one of the merchant's sons.

Subsequent investigation would suggest that there had been sexual spread between Paul and several of the women who helped to nurse him. Certainly Hawa herself soon became ill, being nursed in turn by her three sisters and her stepmother, all of whom eventually became sick and died. In this extending vortex of sickness and death, the friendly merchant, Mohammed, was himself taken ill on August 21. Terrified by what he was witnessing locally, he traveled to the provincial capital of Juba, was admitted briefly to the Juba hospital, discharged himself, and then traveled on to Khartoum, where, after four days of progressive decline, he died in neighboring Omdurman on August 30. Shortly after he left Nzara, several others of his family and employees fell sick and died. The nurse who had treated Sallah at his home was eventually taken to Maridi hospital, where he too died on September 3.

Epidemiologists would be able to relate no less than forty-eight cases of illness and twenty-seven deaths in Nzara to contact with Paul, or with people who had contracted the infection from him. Meanwhile, other cases would continue to arise in cotton factory employees, for which no direct contact with previously sick people could be established. These individuals would in turn infect members of their own families.

In Maridi, Samir's illness began to spread among the patients in the Class 1 pavilion. Within ten days of his arrival, all three of those who shared the private facility with him were either sick or dead. Then the nurses who had taken care of Samir themselves contracted the fever and were admitted to other wards in the hospital, where they in their turn were nursed by their colleagues. Sallah would never realize the time bomb he had left behind him, as, after the death of his brother, he had made his grieving way back to Nzara, himself to die there.

2

Eight hundred and twenty-five kilometers southwest of Maridi, across the Zairian border in a small mission town called Yambuku, a forty-four-year-old man called Mabolo woke up early on the morning of Thursday, August 26, with a fever and a headache.

The mission was based in the Yandongi district, a collectivity within the Bumba zone, which straddles the equator. In 1976 the zone enclosed a population of 275,000 people. The area is part of the Zaire river basin, in the heart of the tropical rain forest. The main local tribes are the Budsa and Bagensa, who speak a trading language called Lingala. Yandongi district had a population of 35,000 people, living by subsistence farming in a scattering of hundreds of villages linked by laterite tracks hacked out of the forest during the Belgian colonial period.

Mabolo was a teacher. But today he felt too sick to go into the school, and instead he joined the queue outside the dispensary, which was in the temporary charge of the Belgian nun Sister Beata. The dispensary was a single large room, two-tone walls of upper light and lower dark blue, with a large table laden with medicines at the center, where a pharmacy assistant prepared the medicines. There were two enamel trays. One held some glass syringes and needles, a metal tongue depressor, and some vials of medicines. The other tray contained some stainless steel kidney bowls, with disinfectant, cotton wool, and clean water.

Sister Beata, who ran the busy obstetric unit, was sitting at a desk with a register book, which listed the bare particulars of each patient in a single line entry. She must have glanced up at Mabolo in surprise as he took his turn to come in from the queue on the verandah.

Mabolo was an important contributor to the mission. In addition to teaching, he served as president of the Parish Committee. Beata, at forty-two years old, was very close to Mabolo's own age. She was smallish, with a rounded face and a gap between her two front teeth that gave her a girlish innocence when she smiled.[2] Now she listened as he described his headache, fever, and listlessness, and she performed a cursory examination. Between them, they agreed it was a relapse of malaria. She drew some Nivaquine into a syringe and injected it into his arm. Following the injection, Mabolo appeared to improve, but on September 1 he returned with a worsening of his symptoms. This time he was given an injection of quinine, in case he had malignant tertian malaria, which could be resistant to Nivaquine alone. But this time the injection did not help him. Over the next three days, his fever escalated, he became progressively more nauseated, and he passed clotted blood in his stool. He was carried into the hospital, where he was admitted to a twin-bedded side room, off one of the medical wards.

That night he deteriorated further, the abdominal pains came in agonizing cramps that doubled him up and made him scream with pain, and he vomited altered blood. His wife, Mbunzu, nursed him, changing the soiled bedclothes with her unprotected hands. Mabolo died in the early hours of September 9, in the arms of his wife and his mother.

The funeral customs in Yambuku are very similar to those in Nzara, with the same intimacy in the funeral preparations. Mbunzu washed the body, kissing, fondling, and hugging it, before dressing it in clean clothes. Others who had been close to him showed a similar affection for his memory — and Mabolo had a lot of friends.

Soon, as in Nzara, the deadly fever began to spread. By the day following Mabolo's death, there were six new victims on the medical wards and reports were coming in of identical cases out in the villages.

Just prior to his illness, Mabolo had toured the Mobaye-Bongo zone in the northern part of the Equateur region. He had started out on August 10 and returned home some twelve days later. Six other mission employees had accompanied him, all traveling together in a mission vehicle. Along the way they had stopped off at a number of towns, aiming to reach the village of Bandolité, close to the border with the Central African Empire. They had almost reached Bandolité, failing just a few miles away because a bridge had been washed away by floods. On his return home, Mabolo had stopped at a roadside market fifty kilometers north of Yambuku, where he bought meat for the family, some freshly killed antelope. One of his companions had bought some freshly killed monkey. When he arrived home, Mbunzu had dried and stewed the antelope meat and it was eaten by his entire family. Mbunzu was insistent that they had eaten none of the monkey meat.

A baffled Sister Marcella looked elsewhere, gathering the hospital registers going back more than a year, looking for similar cases. All she could find, prior to Mabolo, were a few fatal postpartum hemorrhages. As day by day more and more cases were pouring into the clinic or being carried into the hospital wards with what was now being described as "the fever," Marcella became increasingly alarmed. Mabolo's widow was already sick with the same symptoms her late husband had.

Within days, the frightened nuns and their Zairian staff at the Yambuku Mission Hospital were struggling to cope with a growing avalanche of desperately sick patients, with more and more of the nursing staff collapsing with the same deadly pattern of symptoms. The three nursing sisters were exhausted with sleepless nights and the unremitting demands of the sick. On Sunday, September 12, the exhausted Sister Beata developed a headache and a fever. She was so frightened she dared not measure her temperature. By Monday, she was retching with nausea, her headache was unbearable, and heavy pains scourged every muscle and bone of her body. When she did now measure her temperature, it was 103°F. Sister Myriam came to see her in her room. Myriam brought her a little food and drink and arranged to take her place, covering the dispensary. Beata's illness frightened the other nuns more than any previous development.

The nature of the fever itself was utterly baffling. There had been talk of yellow fever, but few if any of the patients were jaundiced. While the disease had many features that resembled any severe fever, there were strange manifestations, signs that were different in scale or bizarre altogether. The patients were often terrified. Some became confused and agitated, as if their brains were involved. As the sickness deepened, their faces became strangely expressionless, their eyes sunken and glazed, like a mask of impending death. Nothing the sisters could do seemed to give relief. And the reports filtering through from the surrounding villages were now alarming.

Several times each day the nuns would gather in the tiny convent chapel to pray, a single room barely eight by six meters and as personal as a soul. Each sister would kneel down on her individual prie-dieu of dark-stained wood, with its individual hymnals and missals. A window with olive-green curtains fanned the strong light past the altar, draped in an African wax decorated with the monogram "P" and "X" combined. Around the walls, the little stations of the cross were distributed at intervals, various stages of Christ's journey on the road to

Calvary, intaglio figures of suffering in a black hammered copper in little wooden frames.[3] In the growing anguish of their prayers, they implored God to spare Beata.

On September 15, Doctor Ngoï-Mushola, a young Zairian and the official director of the hospitals scattered throughout the Bumba zone, arrived at Yambuku, where he examined Sister Beata, who was progressively deteriorating. The doctor thought the most likely diagnosis, given the high fever and the prominent diarrhea with blood, was malignant typhoid fever. He prescribed intravenous fluids to combat the dehydration and injections of an antibiotic called chloramphenicol. Dr. Ngoï took blood from Sister Beata so it could be tested for antibodies to the typhoid bacillus.

Beata could no longer swallow food or even liquids. Her tongue and palate were matted with festering ulcers. There was a measles-like rash over her neck and upper body. She was exhausted from bloody diarrhea. The whites of her eyes had turned red with conjunctivitis and blood was oozing from her eyes and nose. Doubts were now entering Dr. Ngoï's mind about the diagnosis of typhoid. When Beata died, a wave of fear swept through every ward in the hospital. Within hours there was a panic-stricken mass exodus. Surgical patients too feeble to walk were carried into the bush on the backs of their relatives.

During the day, Dr. Ngoï examined the nine patients who still remained in the hospital. Though he was uncertain what was really causing this lethal fever, he drew up the first official report of the events in Yambuku. Prefaced with the heading "An Inquiry into the Alarming Cases in the Collectivity of Yandongi, in the Zone of Bumba, 15–17 September, 1976," the young African doctor described with perfect lucidity and accuracy all of the clinical features of the epidemic. He identified Mabolo as the index case, described his four-day automobile journey prior to the illness, then went on, case by case, identifying every victim by name, age, gender, village of origin, dates of hospitalization and outcome to date. He went on to list the negatives on simple examination of stool, urine, and blood. He concluded, correctly, that Yambuku was the epicenter of the epidemic. The paucity of investigations available at the hospital were not sufficient to arrive at a diagnosis. Finally he drew attention to the need for more radical measures to protect the population, appealing to the health authorities in the region for help. He signed off: *"Fait à Yambuku, le 19 septembre 1976. Signé; Le Médecin Directeur des Hôpitaux Dr Ngoï-Mushola."*[4]

The following day, another Bumba-based doctor, Rurangawa, wrote to Dr. J. Focquet, the Lever's Plantation Medical Director in Kinshasa, passing on what he had heard from Dr. Ngoï concerning Yambuku. The letter was ferried to the capital by plane and arrived the same evening. Dr. Focquet called Dr. Ngwété, the Minister of Health, at his home and read the contents of the letter to him over the telephone. This was the first intimation anybody in the capital had of the disaster.

3

Early in the morning of October 1, Dr. Ruppol, Chief of the Belgian Medical Cooperation, an organization run by the Belgian government to assist Zaire with medical problems, received a call in his office in Kinshasa. The caller was a missionary sister from Marcella's order.

"Doctor, you have to do something. We have heard news over the radio system that our sisters are dying in Yambuku."[5] At first Ruppol could not believe it was anything so very serious. Minor epidemics were common in Africa, though it was less common for missionaries to die. Ruppol left his office that morning and went to see the Christian brother, Le Frère René, who was in charge of the mission link for the radio network in Kinshasa. Demanding to be put through to the missionaries in the Bumba zone, he learned for the first time that hundreds of people were dying of the mystery epidemic. He heard that Sister Beata was already dead and that another of the Belgian nuns, Sister Myriam, had been brought back to Kinshasa from Yambuku by two doctors sent out by the Minister of Health and was being supported by her colleague, Sister Edmonda, at the Ngaliema hospital.

Myriam, it seemed, had already spelled out the course of her illness to Dr. Courteille, the doctor ministering to her, advising the staff to adopt barrier nursing precautions and predicting her own death. Ruppol was both astonished and furious at being kept in the dark. Now, listening to the main symptoms of headache, vomiting, high fever, and gastrointestinal bleeding, he suspected a hemorrhagic fever.

Jean François Ruppol is of medium height, fair-haired, passionately interested in medicine. He was born in Zaire of Belgian parentage, where his father, a distinguished missionary doctor, was of the first generation born in Africa. In essence, he was a general physician, with a special interest in public health. He had never been involved in a viral epidemic before. Returning to his office, he called Brussels, speaking first to Dr. Kivits, the director of FOMECO (Fondation Médical Coopératif), and then to the Ministry of Health in the Belgian government. He warned them that this was a serious epidemic, and, since he now intended to travel to Yambuku, that he would need protective clothing, masks, and diagnostic support.

Next he spoke to the Zairian Minister of Health, Dr. Ngwété. The minister granted him permission to go to Yambuku. It was agreed that he would be accompanied by Dr. Raffier of the French Medical Mission and a Zairian doctor called Krubwa from the University of Kinshasa. Ruppol also called the Catholic Communications Center and asked Frère René to pass a message to the sisters at the Yambuku mission. They were to stay at the mission until he arrived to talk to them. He hoped to arrive at Yambuku the following day.

After thirty-six hours spent arranging protective clothing and making other preparations, Ruppol, Raffier, and Krubwa flew out of Kinshasa in a C-140 army plane, arranged for them by General Bumba, Commander in Chief of the Zairian army. The expedition arrived in Bumba in the late afternoon, where they were met by Father Carlos of the Bumba mission and Dr. Ngoï. Carlos and Ngoï estimated that there were about two hundred people still sick with the fever in the surrounding villages. An unknown number had already died. The group walked the dirt street to the office of the commissioner, where they instructed him to place all of Bumba zone into strict quarantine. The army would set up roadblocks. Nobody would be allowed in or out.

Ruppol explained to Dr. Ngoï that he suspected a hemorrhagic fever virus. The most likely possibility was Lassa fever, caused by a potentially lethal virus discovered just four years earlier in Nigeria. Lassa had killed a lot of people and frightened a good many of the doctors who had had to deal with it. If this present epidemic was caused by such a virus, they needed to identify it, discover its origins and the means by which it was transmitted from person to person. If it proved to be a different virus, possibly even a new, or "emerging," virus, then there were many dangerous unknowns that must be evaluated with extreme urgency.

Dr. Ngoï told Ruppol about a man and his wife who had fled Yambuku after the death of their eighteen-month-old son, arriving in Bumba that same day with a surviving two-year-old. The man was also named Mabolo, the head of the mission school in Yambuku. He and his wife, Molombe, had been admitted to a pavilion in the hospital grounds, while the child, who appeared to be well, had been taken into care by relatives. Ruppol was compelled to reinforce the brutal need for the complete isolation of this pavilion. It was now enforced by a military cordon sanitaire.

The tragic couple occupied the end of a large empty ward. Their food was left for them at the entrance. Nobody was allowed to nurse or minister to them. Early the morning following their arrival, Ruppol and Raffier visited the pavilion. Wearing protective clothing, they stood outside the doors and shouted to the people inside. But there was no reply. As soon as they entered, they could see that the couple were already dead. In a family photograph, the school director Mabolo, who was twenty-six years old, is dressed neatly in a lightweight suit and Molombe, a pretty woman, is wearing a green top and a pagne with matching scarf.[6] This morning, as Drs. Ruppol and Raffier entered the pavilion, everywhere in the small cubicle was the pervasive evidence of the suffering this couple had gone through. The room was disorganized and clearly had not been cleaned since their arrival. There was a nauseating stench of death. Molombe was lying on a bed wearing the soiled clothes in which she had fled Yambuku. There was no linen on the bed. Bloody tracks ran from her mouth, nose, eyes, and even her ears. Her husband was half reclining in a chair, his head thrown back, and one of his arms, resting on the arm of the chair, was raised. Rigor mortis had set it in this bizarre position. Once again there was evidence of bleeding from his mouth, nose, and eyes, and soiling of the chair from his excretions.

Forcing themselves to proceed against their own mounting panic, the two doctors attempted to obtain blood from the man's arms, but the blood had clotted in the veins. In desperation, they attempted to extract urine by direct needle puncture of the bladder, above the pubic bone. They extracted very little, no more than a few drops.

Saturated in sweat, they emerged from the pavilion and called for volunteers to bury the bodies. Nobody would volunteer. Even the soldiers refused, murmuring, "We are married men. We have children at home." Prisoners from the town jail were brought to the hospital, where they dug a hole in the hospital grounds a mere five meters from the entrance to the pavilion. They were promised their freedom if, wearing protective clothing, they would bring the bodies out, but they refused. Ruppol and Raffier carried them to the makeshift grave, wearing full protection.

The woman's body was put in first and then the man's, hampered by the rigor mortis, which kept his arm elevated. Before filling the grave, they poured petrol over the corpses and set them afire. As they filled in the hole with earth, the hand of the schoolmaster, engulfed in flames, protruded ominously through the disturbed earth.[7]

Yambuku is about ninety kilometers from Bumba. The road was considered dangerous in the rainy season, so the medical team needed the army support helicopter to take them. But first they had to persuade the reluctant pilots, who had been drinking in a brothel the night before, to take them into the plague zone. The protective clothing, ferried in by C-140 transport, was now transferred from the plane to a helicopter, as were surgical gowns of green paper, paper bootees, goggles, masks, and gloves, together with the necessary equipment to take blood and tissue samples.

Dr. Massamba, a regional Zairian doctor, volunteered to go with them. In Ruppol's estimation, he proved very helpful and courageous. The pilots did not know the way to Yambuku, so Massamba directed them. They stopped at villages along the way to ask for directions, when the wind from the rotors blew the thatch off the huts. Some time in the early afternoon of October 6, the large gray Puma helicopter arrived at Yambuku, landing on the mission football pitch. The sun was shining directly overhead in a cloudless sky and it seemed that everybody still living disgorged from the mission and the nearby huts and villages to see what was happening. Dr. Ruppol climbed out, still dressed in his casual clothes. He called out through a megaphone to the gathering crowd to keep their distance. "You must come no closer than that tree," indicating a tree about thirty meters from where he was standing. The three surviving nuns, Marcella, Genoveva, and Mariette, stood at the front of the crowd, struggling to hear him, bewildered and uncomprehending as this fellow Belgian asked them to tell him what was now happening.

For Jean François Ruppol, this was the single most distressing moment, standing beneath the helicopter, seeing the tears in the eyes of the white-clad Belgian sisters, who thought he had come to take them away from there. "I had to tell them that they must stay." He would remember staring up with his own hazel eyes into that cloudless tropical sky: "La soleil et la silence!"

Ruppol did his best to explain what a hemorrhagic fever was. He told them he must collect samples from people who were still sick or who had survived the illness. Still speaking through the megaphone, he also advised the nuns on the measures they must take from now on to prevent its spread. The nuns told him that they had no masks or goggles and had virtually run out of surgical gloves. He promised he would come back the next day and he would bring these protective items for them. In a gesture of sympathy, he had brought their letters from home, which, careful still to keep his distance, he placed at the bole of a nearby tree. He had to disappoint their wishes to take homebound mail, however, because they had licked the stamps.

Now began the task of entering the thatched huts of sick villagers to obtain samples. Everywhere the team went, they were followed at a distance by the crowd of nuns and Africans. In front of the entire gathering, they stripped naked, donned surgical gowns, masks, goggles, and bootees. Inside the huts, even the cinderblock houses, they could hardly see the sick patients in the gloom. They simply took blood samples, then departed. They would come out of the houses, eyes blinking in the dazzle of the sun, remove the protective clothing until naked, pile the discarded clothing into a bundle and burn it, don new clothes, start again. The samples were transferred for preserving into a flask of dry ice. They had no plastic tubes, so they were forced to use glass, though they realized this was dangerous. Glass has a tendency to shatter, and those sharp fragments would effortlessly penetrate surgical gloves, injecting their contagion into anybody handling them.

In all it took them about three hours to collect samples from seven people. It was extremely hot and the doctors soon felt thirsty. People offered them water, but they had to refuse. Repeating their promise to return, the doctors returned to the helicopter and returned via Bumba to Kinshasa, where the Minister of Health, Dr. Ngwété, was waiting for them when they alighted at a small airport on the outskirts of the capital. Ruppol took the thermos containing the samples to the FOMETRO (Fondation Médical Tropicale) offices, where, in the little kitchen, he again stripped bare, put on protective clothing, gloves, mask, and goggles. A curious group of onlookers watched his patient preparations through the glazed door. When he opened the flask on the kitchen table, he found that some of the glass had shattered.

There were splinters of glass mixed with the frozen blood. There had been five milliliters of blood in each tube, blood he did not feel he could afford to discard. "I did not wait for it to thaw. I picked up the solid pieces of frozen blood between two needles and transferred it to another container. I was wearing gloves, but the worry was that a piece of broken glass might penetrate the gloves."

The samples were divided into three lots, one for Dr. Raffier, to be sent, with an accompanying letter, to the Pasteur Institute; a second for the World Health Organization in Geneva; and one for Professor Pattyn's laboratory at the Institute of Tropical Medicine in Antwerp.

4

On October 3, as Ruppol arrived in Bumba, a British doctor, Cenydd Jones, who was Chief of the Epidemiological Surveillance Center in Nairobi, traveled to the Sudan at the instigation of the World Health Organization, flying by civilian air charter directly into the southern capital of Juba, from where, a day later, he made his way by army helicopter into the plague-hit Maridi.[8] In the hospital in Maridi the fever had exploded in all three ward pavilions, where the sick and the dying were still accommodated in open-plan "Florence Nightingale" wards. At the peak of the epidemic, there had been no attempt at barrier nursing or isolation. Touring the wards, Jones saw chamber pots containing unprocessed urine and feces. The investigating doctors from Khartoum, though they put on gowns, masks, and gloves to examine patients, forgot to remove them afterward, when they returned to smoke and drink coffee in the doctor's office. Even the family relatives walked in and out of the contagious areas without gown, gloves, or masks. At the same time, the massive mortality among the staff had led to a profound demoralization, and the maternity ward was now a mortuary, piled with bodies.

Two days later, Jones was joined by another British doctor, David Smith, a tropical diseases expert working in the Ministry of Health in Nairobi.[9] Experienced in working with dangerous viruses, Smith brought some key diagnostic materials with him, including plastic sample tubes, flasks, and several boxes full of dry ice for preserving viruses. On October 7, Jones, Smith, and a local WHO representative, Dr. P. L. Giocometti, traveled the eighty miles of arduous road to Nzara, where they compiled the first detailed epidemiological studies of the fever, tracing it back to Yusia and the cotton factory. They also plotted the epi-line of deaths, which graphically outlined the two separate explosions, peaking in the middle of the second week of August in Nzara and, massively, in the final days of September in Maridi.

Smith took liver biopsy samples from some of the bodies in the makeshift mortuary, using a percutaneous instrument called a viscerotome. He also took blood samples from a dozen patients in both Nzara and Maridi, which were packed, with the liver samples, in double-insulated flasks and preserved in dry ice. On October 9, after their return to Nairobi, Dr. Wilfred Koinange Keruga, Director of Communicable Diseases Control in the Ministry of Health, allowed them to go into voluntary quarantine in Jones's home, while also enabling Smith to dispatch his frozen samples of liver and blood on a British Airways flight to London that same night. Those precious diagnostic samples were addressed to Porton Down in England, the leading viral diagnostic laboratory in Europe.

On October 10, the samples taken by Jones and Smith arrived on the desk of Ernie Bowen, the director of the maximum security virus laboratory at Porton, some six days after the earliest samples from Zaire arrived at the Institute of Tropical Medicine in Belgium. In 1993, some of the American media would criticize the CDC for taking just nineteen days to find the cause of the Four Corners region hantavirus epidemic. Yet only now, more than three months after the death of Yusia, on July 6, had these appropriately equipped laboratories even begun the investigation of the African epidemics.

NINE

The Hairs Stood Up on My Neck

An African epidemic
might be compared to a giant tree
falling in the forest.
Nobody notices it has fallen.
When the first white person dies,
the epidemic begins.

DR. JOEL BREMAN
personal communication

1

In the late summer of 1976, western Europe basked in a heat wave. Millions of people living in the Southwest went without tap water and the crowds cheered when rain stopped the cricket at Lords. The situation in Ulster was worsening, and on August 31, the district of Notting Hill, London, was the epicenter of race riots, echoing a much more vicious situation two months earlier, when South African police opened fire on rioters in Soweto, killing a hundred and wounding a thousand. Americans, while scrutinizing the credentials of the Democratic presidential nominee, Jimmy Carter, were preoccupied with celebrating the bicentennial with sky-bound lasers declaring "Happy Birthday, USA!" In Japan, people were keeping a ready eye on China, where, following the death of Mao Tse-tung on September 9, a nation still reeling from the turmoil of Chiang Chin and her fellow radicals was planning to explode the most powerful nuclear bomb to date. The developed world was preoccupied with its own diversions and problems. Nobody was overly concerned about an emerging plague in Africa.

On Monday, October 4, Dr. Luc Eykmans, the newly appointed Director of the Institute of Tropical Medicine in Antwerp, arrived at work, ascending the magnificent marble staircase, impressed by the somewhat idealized murals: Africans tilling soil, aiming bows over lakes and hills, or gazing wistfully into tropical distances. He had no sooner entered his oak-lined office than his secretary rushed him across to the laboratory of Professor Stefan Pattyn. Here Dr. Eykmans, himself a microbiologist, watched as the small man, wearing gold-rimmed bifocals perched upon a patrician nose, examined the contents of a package. The package, it seemed, had been forwarded by a Belgian doctor, Dr. Courteille, who worked in Kinshasa, and it contained a letter wrapped around a jam jar containing several glass tubes. The jar and its specimens had been packed with dry ice and the glass had shattered in the mail. Pattyn opened the shattered package wearing surgical gloves.

Eykmans watched his colleague hold aloft a letter that was dripping with blood. Pattyn read the words of Dr. Courteille aloud:

"I have enclosed liver biopsies and blood samples from the autopsy of Sister Myriam, née Louise Ecran, aged forty-two years, who just died here in the Ngaliema Hospital. She comes from the Mission Hospital of Yambuku. We have received reports of hundreds of cases in the Yambuku area and in the Bumba zone generally. Apparently there are no survivors."

Pattyn dropped the bloody letter into a bag for disposal, unpeeled his gloves and dropped them into the same bag. Then, his irascible blue eyes sparkling, he turned to Eykmans. "It seems we have a serious problem!"[1]

After a lifetime of experience, most of it as Professor at the Institute of Tropical Medicine in Antwerp, Stefan Pattyn was one of the few eminent virologists in Europe. From the information in Courteille's letter, he assumed he was dealing with a viral illness. His first thought was yellow fever. But without exception everyone who left Belgium for central Africa was vaccinated against yellow fever. Sister Myriam could not, therefore, have died from yellow fever.

In 1976, viruses could be difficult to diagnose. There were three ways in which you could proceed. You could look for specific antibodies to the virus in the blood of infected patients. The tests were fairly rudimentary, but they were accurate enough if you had a good virus antigen to test against. The problem with a newly emerging virus is you had no antigen and therefore the standard serological tests would prove negative. There were two ways of hunting for a more elusive virus. You could inoculate animals or you could attempt to grow the virus in cell cultures grown in flasks. When he began to investigate the blood and liver samples from Sister Myriam, Stefan Pattyn had a first-line hypothesis. He thought they might be dealing with Lassa fever. So he instituted the appropriate investigation, which was to inoculate some of the serum into adult mice.

There was another rather more worrying possibility: the agent might be a hemorrhagic fever virus but not Lassa. If so, Pattyn was convinced it would prove to be an arbovirus. Arboviruses, or viruses transmitted in the main by biting insects, interested Pattyn greatly. For years, he had been experimenting with ways of isolating them. Now he inoculated some of the serum into suckling mice, while homogenizing some of the liver tissue and adding it to cultures of Vero cells.

Next he fixed a sample of the liver in formalin and sent it to Dr. Gigase, the institute pathologist, for routine histological examination. Within twenty-fours hours, the pathologist called him to say the liver showed a pattern consistent with hepatitis. Pattyn's heart jumped when the pathologist told him that he could see "inclusion bodies" in the liver cells: inclusion bodies confirmed the presence of a virus. But many viruses that attacked the liver would cause similar inclusions within the cytoplasm of infected cells. It was a pointer, but hardly diagnostic.

On October 5, he called up Dr. Paul Brès, Chief of Virus Diseases at the World Health Organization in Geneva. Pattyn was surprised to find that Brès already knew about the mystery epidemic in Zaire — in fact Brès made it clear to Pattyn that he was extremely worried. Despite his experience and enthusiasm, Pattyn was no better equipped to deal with a dangerous virus than an ordinary hospital diagnostic laboratory. The flask had been opened on a bench with no more protection than a lab coat and a pair of gloves. Brès thought this was much too dangerous a situation and he instructed the disappointed Pattyn to send on all of his specimens to Porton Down in England, where they had facilities to deal with lethal viruses. Over the next few days, Pattyn reluctantly bundled up most of the serum, the liver biopsies, some brains he had removed from the infected mice, and his cultures. He sent most of these on to Porton, but he did not send them all. He retained some of the inoculated tissue cultures and some of the suckling mice.

Myriam's serum and her liver biopsy samples arrived at Porton from Antwerp on October 8. This would be followed three days later by the specimens from Ruppol in Yambuku and from Smith in the Sudan. The late Ernie Bowen was in charge at Porton — in the words of David Simpson, "a stocky Welsh pit prop with a broken nose." Bowen had worked his way up the hard way, from a job as laboratory technician to take charge of the high-grade pathogens laboratory, with its glove-port cabinets and secure animal room. It was the only such facility in Europe equipped to work with BSL-4 agents. Bowen also enjoyed a close working relationship with David Simpson, one of the most experienced virologists in Britain, who was currently based at the London School of Hygiene and Tropical Medicine.

Bowen organized a wide range of serological screens, cell cultures, and animal inoculations on the specimens from both African epidemics, hunting for a variety of hemorrhagic fever viruses, including yellow fever, Crimean-Congo fever, Rift Valley fever, and Lassa fever. Bowen also picked up the phone and called Simpson to talk about the possibility of a virus called Marburg. They had worked together in the first diagnosis of the Marburg virus some nine years earlier when the most lethal hemorrhagic fever to date had exploded into medical and public consciousness.

2

Marburg is a small and prosperous city at the heart of Germany, an hour by car from Frankfurt. A Gothic castle looms over the timbered and ornately carved stone streets of the medieval center, with a Knights' Hall and historic associations with Martin Luther and the turbulent birth of Protestantism. In 1967, this picturesque holiday city became the reluctant name for a mysterious and deadly illness that seemed to arrive out of the blue into its summer streets.

In August several employees of the Behring monkey factory fell sick, suffering from fever, prostration, vomiting, and diarrhea. They were admitted to the University Department of Medicine, under the care of Professor G. A. Martini. The pattern of their illnesses suggested a virulent form of dysentery. Dysentery has two very different forms, one caused by a germ called shigella and the other a tropical variety, caused by an amoeba, but neither germs nor amoebae were found in the patients' stools. Instead, and despite every attempt at support therapy, the patients suffered a fulminant decline and died. By August 22, the public officer of health in Marburg was instituting urgent measures to contain a lethal outbreak in the city.

Identical cases were soon turning up in Frankfurt, and a few weeks later in Belgrade, Yugoslavia. The numbers of affected people had already risen to thirty-one, twenty-three of these in Marburg alone. As the illness progressed, and particularly in the most severely ill, there was a vivid rash that tended to coalesce into a livid reddening of the skin over the face, trunk, and extremities. After the sixteenth day, the patients' skin peeled off, their hair fell out, and they lost their nails. Seven of the infected people suffered from severe hemorrhages from the nose, from the gums, and from sites where blood had been drawn from veins or where intravenous drips had been set up. These patients were also bleeding from the stomach and bowel, with vomiting of blood and passing of blood *per rectum.* These proved very difficult to manage. In the most severe cases, particularly in those who were to die from it, confusion and coma, with restless or disturbed behavior, signaled a dangerous involvement of the brain.

There was a common factor in every one of what now amounted to a total of thirty-one cases: they had all had contact with African green monkeys. Twenty worked in the Behring monkey factory, while two others, a doctor and a nurse, had come into contact with patients in the hospital itself. The green monkeys involved, *Cercopithecus aethiops,* had been imported from Uganda, after a period of quarantine at a London airport. Uganda borders both Sudan and Zaire, and the area where the monkeys were trapped was five hundred miles from Nzara, Sudan.

Cells grown from the kidneys of green monkeys would grow very well in monolayers in flasks, making them a suitable culture medium for the growth and testing of disease-causing viruses. The Behring factory was harvesting monkey kidney cells for the production of polio virus vaccine. Ten of the infected people had assisted in the killing or postmortem examination of monkeys, three had performed operations that involved opening up the monkeys' skulls, one woman had dissected monkey kidneys, and another had handled cell cultures derived from the kidneys. Five others had cleaned glassware contaminated with monkey blood. Handling rules had been broken. Some had worked with monkeys before they had been kept in quarantine, ignoring the international rules of safety. Others had handled monkey tissues and blood without the protection of surgical gloves. Two women had contracted the infection from their husbands, one after she had cut his skin while shaving him and another through sexual intercourse after he had seemingly recovered from the disease.[2]

Failing to isolate any causative bacterium or protozoan, the medical authorities in Germany were convinced they were dealing with a virus. But attempts to isolate the virus were hampered by the extremely infectious nature of the mystery agent. It was felt that the disease was so contagious and lethal that all future work needed to be carried out in laboratories with high biosafety-level precautions. The hunt for the cause of the green monkey disease moved to Porton Down in England.

* * *

Late one Saturday afternoon, toward the end of August 1967, ten or so thin-walled glass tubes, most containing frozen serum, one or two frank blood, appeared on the desk of David Simpson, then working as senior assistant to the eminent virologist C. E. Gordon Smith. They had been forwarded by Professor Dick, a British microbiologist then assisting his colleagues in Germany, who had packed them in a large thermos, preserved in dry ice. When, following full BSL-4 safety precautions, Simpson opened the flask in the safety cabinet, he found that most of the tubes had cracked.[3]

Simpson resolved to place each of the shattered tubes into a sterile thick-walled universal container, where the specimens would be allowed to thaw out at room temperature inside the cabinet. He inoculated some tissue cultures of monkey kidney cells and others grown from mice and chick embryos. Then he put the tissue culture flasks in the incubator to see what developed.

Over subsequent weeks of protracted experiment, all of the screens for the known hemorrhagic fever viruses proved negative. Simpson was now hunting for unusual viruses. He began by inoculating serum from the Marburg patients into guinea pigs, and both intracerebrally and intraperitoneally into mice. Simultaneously he instituted many different lines of tissue cultures. The tissue cultures again drew a blank, but strangely there was a change in the color of the medium. Normally it was pink. Now, it was seen to turn yellow, a change toward acidity in pH. Suddenly, and very dramatically, the guinea pigs were becoming very hot. A week or so after being injected with the infectious serum, their temperature peaked at about 106°F.

Simpson tried passing blood from some of these hot guinea pigs into other guinea pigs. They too spiked enormously high temperatures. And the incubation period had now shortened to three days. This second generation of test animals became very sick and died. The team performed postmortems on these dead animals, when they discovered a horrifying devastation of the internal organs. "The livers, in particular, seemed to be shot to pieces. That was our first real breakthrough."

In the guinea pig tissue, Smith, Simpson, and their colleagues began to see some large and strange-looking inclusion bodies inside the cells. By late September they had reached the stage of putting the tissues and serum from the guinea pigs under the electron microscope when they saw enormously long and filamentous particles. The appearances were so bizarre that Simpson was frankly startled. In some views the strange new entity looked like horribly writhing snakes or worms. In other views they formed rings, like doughnuts, or alphabetical shapes, like question marks or commas. He turned to the technician working the electron microscope. "What in the hell is this?"

Today, David Simpson is the Professor of Microbiology at the Royal Victoria Hospital in Belfast. He is white-haired, a quiet-spoken man with cornflower-blue eyes. He was born in Northern Ireland, graduating in medicine at Queens University in Belfast, from where, in 1959, a growing interest in microbiology had taken him to work with Smith at the London School of Hygiene and Tropical Medicine. He had worked for many years at the East African Virus Research Institute in Entebbe, Uganda, where he gained extensive experience investigating hemorrhagic fevers, including yellow fever, Crimean-Congo fever, and Rift Valley fever. Yet in 1967, as he gazed into the viewer of the electron microscope, no infectious agent he had ever encountered had remotely resembled this.

Simpson had discovered a totally new family of viruses. Today they are called the *Filoviridae,* from the Latin *filum,* which means a thread. They are among the most dangerous pathogens on earth.

Perplexed by the bizarre appearances, Simpson called up June Almeida, the leading electron microscopist in London, then a senior lecturer at the Hammersmith Hospital in London. "June," he cried, "I've seen these damned things!" Almeida prepared her own electron microscopic grids and took more pictures. Her reaction was one of equal shock: she too had never seen anything like it. The bizarre viruses were about 80 nanometers thick but as much as 14,000 nanometers in length in their serpentine coils. Serological testing now confirmed that these strange entities, now named the Marburg virus, were the cause of the fatal epidemic. They also confirmed that the source of the strange viruses was the green monkeys. The Behrings factory was ordered to kill every monkey.

It is difficult for us today to realize what a shock the appearance of the Marburg virus created a generation ago. In 1967, the world of microbiology took comfort in the conceit that most, if not all, new forms of infection were known. Now, with startling implications, a totally new life form had emerged from a rain forest primate and hopped species to mount an overwhelmingly lethal attack on humanity. In the words of Professor S. Halter, a senior Belgian public health physician writing a decade later: "When Marburg disease appeared . . . the scientific world was struck by the suddenness and the brutality of the viral assault. The fear was great that such a new type of infectious disease could disseminate and quickly become a new and frightful world health problem."[4]

Nine years later, with the arrival of each batch of specimens from Zaire and the Sudan, Ernie Bowen found himself repeating all of the series of tests they had employed during the Marburg epidemic. In particular he inoculated guinea pigs with Sister Myriam's serum. Soon, to his horror, in an identical manner as with Marburg nine years earlier, the guinea pigs were taken violently sick, collapsing with hugely elevated temperatures.

3

At the Centers for Disease Control, a team of scientists, then working in Hot Lab No. 8, a tin structure stuck onto the end of the virology building, became keenly interested in what was happening in Africa. They included the tall, dark-haired Fred Murphy, forty-two years old, who was Chief of the Branch of Viral Pathology. Karl Johnson, an M.D., was running the Special Pathogens Branch, while his wife, Patricia Webb, worked with Murphy as his assistant.

Johnson was already a legendary figure in virology. All of his working life had been spent working with dangerous viruses. He had only recently returned from Central America to take up the newly created post as Chief of the Special Pathogens Branch. The first he knew of the two African epidemics was when he received a call from a doctor at the Institute of Tropical Medicine in Hamburg at the beginning of the second week in October. The German doctor told him about the outbreak in the Sudan. He painted the picture of devastation, with medical staff either dead or having fled the local hospital. Surely the CDC should investigate the outbreak.

In fact, there was little Johnson could do. The CDC was powerless to intervene unless invited in by the two African governments. Johnson tried calling Dr. Bryan and Dr. Brachman, who were directors in the CDC epidemiology services, to see if anybody knew anything about the two epidemics. Through various rumors he heard that Porton Down was already involved. But when Johnson made contact, Bowen was reluctant to send on any of the samples he was currently testing. In fact he was under an embargo from the WHO. Bowen did, however, tell Johnson about the guinea pigs' pyrexial reaction. He suspected Marburg but could not as yet definitively prove it. Johnson knew that the standard test for virus identification involved complement fixation, which was clumsy and slow, taking two weeks. Immunofluorescence, using reagents he had specially designed for use with the fluorescence microscope, had the advantage of a quick and accurate test. The time saving might be important in these desperate circumstances.

Johnson offered to send on freeze-dried slides and immunofluorescent antisera, so Bowen could do the tests himself. Bowen agreed to try them. They arrived in England on October 9. A day later, Bowen stuck his neck out and sent Johnson serum derived from Myriam, together with some material from her liver biopsies, which arrived in Atlanta on October 10.

The samples of serum and liver were examined by Patricia Webb, who inoculated representative samples into a series of tissue culture flasks, essentially a monolayer of monkey kidney cells covering one internal face of a flask. On the same day the specimens arrived from

Bowen, Johnson received a telephone call from Pierre Sureau at the Pasteur Institute in Paris, telling him that some more specimens, sent to him by Raffier from the expedition to Yambuku, were also on their way to the CDC. These specimens arrived on October 13. Johnson unwrapped them personally, wearing full protection with gloves, goggles, and face mask under the protection of a hood. He was particularly interested in the convalescent blood sample, taken from Mbunzu, Mabolo's wife, who had recovered from the illness in spite of the death of her husband. This convalescent sample was invaluable, since it provided not only high levels of antibodies to the mystery virus but would also show a conversion from IgM to IgG class immunoglobulin. Such a conversion, if clearly demonstrated, would be absolutely diagnostic.

Johnson pried the tubes apart with great care. With a sigh, he registered the fact that two of the glass tubes had ruptured in transit. The experienced Sureau had wrapped each tube individually in cotton wool anticipating such an eventuality, when the contained blood would soak into the cotton wool surrounding it. Mbunzu's serum was in one of the ruptured tubes. Patricia Webb squeezed a few drops from the soaked cotton into a fresh test tube. This would have to suffice for preparation of the fluorescent antibody.

Within two or three days, the cells in the cultures she had incubated with Myriam's serum were showing cytopathic change: curling of cells, opacification, shrinking away from each other and from the glass wall of the tiny flasks that contained them. It pointed toward the presence of a pathogenic virus. She passed on a drop of the supernatant from the culture flasks to Fred Murphy, so he could examine it under the electron microscope.

The electron microscope was housed in its own room within the virus building. Murphy needed to prepare a grid, which is like a small flat disc, over which any suspected microbe might be deposited so it could be viewed for identification through the immense magnification of the electron microscope. Knowing a lot of people had died from contact with this agent, he wore surgical gown, mask, and gloves. This was before the days of the space suit, so he stood in front of the large glass plate of the hood with his hands inserted into big black rubber gloves that extended into the working area. With these routine precautions, Fred Murphy took a drop of the supernatant and watched it ooze over the tiny copper grid. Carefully, he placed the prepared grid into a small box and carried this over to the room housing the electron microscope. "I stuck it into the microscope. Virtually immediately I saw long curled filaments, absolutely unique among all the viruses. It looked exactly like Marburg. Nothing else looks like that. The hairs stood up on my neck."[5]

In Belgium, Saturday, October 16, was election day. Stefan Pattyn went to vote and came back home to receive a call from a journalist who had a late-morning slot on Flemish national radio. By this time the epidemic in Zaire was hot news in all of the leading Belgian media. Pattyn had nothing new to tell him.

The Monday previously, the baby mice he had inoculated four days earlier were dead. There had been an accident when the bottle of water linked to their cage had leaked. He did not know if they had drowned or died from the mystery disease. Nevertheless he harvested the brains, and then, on that same Monday morning, he also looked at the tissue cultures. The cells looked sick. Pattyn needed to see what was happening in the mice and he called the electron microscopist at the University of Antwerp. The microscopist agreed to examine material from the cell cultures. Pattyn passed on the samples and at five in the afternoon he received a call from the microscopist. He simply said, "You should come."

Pattyn and his colleague were now gazing at exactly the same appearances as Fred Murphy had seen in Atlanta. "We saw these long threads of virus. There was only one virus that looked like this at that time. This was Marburg." Pattyn's reaction was identical to Murphy's. "I was extremely scared. I realized then that all those things the journalists had been writing about, of people dying like flies in the Congo, all of that became real to me at that moment. I remembered the terrible lethality of Marburg. Now, suddenly, it was like the big bang."

Pattyn returned to the Institute to call Paul Brès in Geneva. Brès was stunned at Pattyn's finding and furious that Pattyn was still manipulating the virus in his low-security laboratory. He said, "You are to stop working with this material immediately and transfer all remaining specimens to Porton Down or the CDC."

For Fred Murphy, as for Stefan Pattyn, there was no doubt at all in his mind that he was looking at Marburg virus. "What I did was just to shut down the scope, quietly walk out of the room, lock it carefully, then go back to where I had done the preparatory work. I had left the specimen in the scope and told my technician not to go in. Now I decontaminated the hell out of that area with Clorox without ever telling anybody. I swabbed out the working surface, cleaned out the hood. The regular procedure involved removing my gown and gloves, and putting them in a discard pan. I carried that pan right to the autoclave and sterilized it there and then. I went back to the lab, called Patricia and Karl. Then I took some pictures."

One of those pictures taken by Fred Murphy on that first inspection down the electron microscope would become famous in the world of virology. On that picture the virus looks like a bizarrely distorted shepherd's crook — a biblical metaphor of doom that might have decorated the seventh seal in the Book of Revelations.

The microscope produces a negative on a small glass plate. Murphy went on to print some of the negatives and from these he used the enlarger to develop some wet photographic prints. Meanwhile Patricia Webb was back in her lab, doing immunofluorescence on the cells, using Marburg reagents. Fred left the hot lab to walk through the winding corridors up to the administration building, then up the six floors to the office of David Senser, the director. When Fred Murphy came looking for him, Senser was in the conference room just across the corridor from his office. Fred moved across the hallway, with his two or three films still dripping fixative on the floor. The conference room was walnut paneled, long and narrow. Fred Murphy sat down at one end of the long conference room table, which would have seated forty people.

There were three scientists at the end of the table by this time, including Murphy and Johnson, talking animatedly about what a Marburg epidemic in Africa might imply. Several more colleagues drifted in, alerted by the spreading excitement within the building. The conversation grew a little louder. At the other end of the table, David Senser was in conference with another man, clearly discussing something important, so they couldn't be interrupted. By degrees, more and more people were drifting in and the conversation became even louder and more impassioned. After a while Senser and the other fellow walked down the table to find out what was creating all the excitement. It turned out that the man with Senser was Sam Nunn, then the U.S. senator from Georgia. He too joined in.

While Murphy was preparing to examine his grids under the electron microscope, Patricia Webb conducted a parallel experiment, adding sera that might contain antibodies to certain viruses to the cell layer in her culture flasks. The sera were labeled with a fluorescent dye. When antibody recognizes a virus, it attaches and will not wash off. The virus within the cells lights up like a yellow-green starry sky.

She was very glad now that Pierre Sureau had had the vision to anticipate the breakage of the glass, for those drops of Mbunzu's convalescent serum, containing high levels of antibody to the mystery virus, were precious. She used this serum to check for virus in the cells. Simultaneously she added serum known to contain antibodies to the Marburg virus to other flasks containing infected cells. She read the results by examining the plates through a fluorescence microscope under a security hood.

When she added Mbunzu's serum to those same cells infected with Myriam's virus, the cells lit up dramatically: it was clear evidence that the antibody was reacting with a virus growing in the cell cultures. Marburg antibody also lit up when she tested it against standard preparations of cells containing this virus. So there was a virus in the cells and her test system was working. But when she added Marburg antibody to cells infected with Myriam's virus, they did not light up at all. The conclusion was devastating.

Patricia Webb hurried to join her colleagues, gathered in the charged atmosphere of the conference room. Fred Murphy watched her face as she walked into the room. It puzzled him that she didn't just blurt out, "It's Marburg!" Instead she went up to her husband, Karl Johnson, and whispered urgently into his ear. Then she turned to the group and announced, to everybody's astonishment, "It's not Marburg."

In Murphy's words, "We had a virus that looked like Marburg but it wasn't." They had discovered a new plague virus that is now classed as a separate member of the *Filoviridae.* That virus is now called Ebola virus, after an obscure river that runs through Zaire many miles north of Yambuku.

4

In a draft plan of action, proposed at a hurried meeting at WHO headquarters in Geneva on October 18, it was agreed that the exceptional severity and spread of the two African epidemics necessitated urgent measures.[6] Paul Brès, chairing the meeting, decided that the WHO should send two separate teams into the plague zones. David Simpson would head the Sudan investigation, assisted by David Smith from the Medical Research Council unit in Nairobi. They would be joined by a young epidemiologist from New York who had just arrived in Geneva with Karl Johnson's party. A British entomologist from the MRC unit in Kenya, Barney Highton, would look for the viral animal reservoir. Karl Johnson would head the team in Zaire, with Joel Breman as epidemiologist. Pierre Sureau from the Pasteur Institute and his Belgian colleague Professor Pattyn, together with his junior colleague, Peter Piot, would complete the Zairian team.

Background laboratory support for Sudan would come from Porton Down, and for Zaire from the CDC. At this stage the authorities in Khartoum had given their agreement for the arrival of a team, under the auspices of the WHO, to go to the stricken area. To the frustration of Johnson and Breman, the Zairian government had neither asked for nor given permission for a WHO investigation on their soil. There were no maps of the countryside, with its roads and terrain, that would help them to plan ahead. Johnson's team would be forced to operate under the awkward umbrella of an "International Commission." That evening there was a startling complication. The young epidemiologist from New York confessed to Johnson that he could not bring himself to go to the Sudan. He was absolutely terrified. His place would be taken by a thirty-four-year-old medical graduate called Don Francis.

Pierre Sureau was the first of the international commission to arrive, landing in Kinshasa on Wednesday, October 13. He knew nothing of the virus isolation. That same day, he went to the Ngaliema hospital, where Sister Edmonda, another Belgian nun originally from Yambuku, was now sick after having nursed Myriam. She was lying in a small private room in Pavilion 5. The pavilion, which had been a private ward before Zairian independence, was now fully given over to isolation. It took the form of a U-shaped single-story facility, with private rooms running around a courtyard of dirt, grass, and scrabbly bushes. A solitary palmyra, six feet high, provided an ornamental reminder of more comfortable days.

When the two nuns had first arrived from Yambuku, the cause of the epidemic was assumed to be typhoid. Barrier nursing precautions were now in operation: cotton gowns, masks, and plastic overshoes. But shortage of supplies meant they were not destroyed after each attendance; instead, the gowns and overshoes were left to hang outside the door for reuse. Edmonda had refused to wear them, comforting her friend Myriam in her ordinary habit up to the moment of her death on September 30.[7] Edmonda had herself contracted the fever five days before Sureau's arrival and was now suffering the inexorable progression of dreadful symptoms. Dr. Courteille and Dr. Ruppol were treating her with aspirin and antibiotics.

Arriving at the hospital about five in the afternoon, Sureau stripped off his jeans, short-sleeved shirt, and sandals and donned surgical gown, cap, boots, gloves, goggles, and mask. Accompanied by the elderly matron, Sister Donatienne, who donned the same protection, though she had nursed Edmonda previously without it, Sureau entered the sickroom. Edmonda, whose photographs would show her round-faced and smiling, was now bedridden, emaciated and unable to swallow. Her temperature was 103°F. Proliferating over her chest, arms, and groin was an angry rash of raised red papules. What stools she was passing were essentially altered blood. The gentle Sureau comforted her with, in his words, "des bonnes paroles." But Edmonda, like the brave Myriam and Beata before her, waved away such niceties. She told him that she was under no illusion as to what lay in store for her. Her calm repose in the face of death impressed him deeply. Sureau took a throat swab and blood sample for viral cultures.

The hospital was now refusing to disinfect any material from Pavilion 5, so Sureau burned his discarded gown and disposables on the lawned courtyard. When he returned to the hospital the following morning, sister Donatienne came out to tell him that Edmonda had died during the night. "Merciful God," he remarked, "how this virus kills!" Nobody was willing to perform an autopsy. Returning to the little laboratory at FOMETRO, where there was neither a laminar flow hood nor an isolator, he separated the serum from the blood in a simple centrifuge before making arrangements for its shipment to Karl Johnson, who was still in Atlanta, at the CDC.

The following day, the Ngaliema hospital was enveloped in crisis. On October 15, a pretty Zairian nurse called Mayinga, who had nursed Myriam and Edmonda without barrier precautions, was admitted to the hospital with the infection. Her case had alarming implications.

In the week before her illness, Mayinga had been preparing to travel abroad for overseas study. While incubating the virus, she had traveled all over the city, making numerous contacts. On October 12 she had even visited the Ministry for Foreign Affairs, where she had spent several hours waiting in the company of numerous strangers. The following day her illness broke with severe headaches. On October 14 she traveled by taxi to the Mama Yemo Hospital, the giant municipal hospital named after Zairian president Mobutu's mother. Here, after waiting her turn in the overcrowded emergency waiting area, she was examined and blood was taken for malaria testing. She was given antimalarials and referred to the isolation wing, but not admitted. Mayinga decided she would be better treated at the university hospital and took herself across the city. Here, after a long wait in similarly crowded areas, she was again examined, diagnosed as a presumptive malaria, and sent home. On October 15 she reported to the Ngaliema to be immediately admitted to Pavilion 5.

That same day, Sureau was alarmed to receive a telegram from Pattyn, warning the local people about the virus he had seen under the electron microscope, which had "la morphologie du virus de Marburg."[8] Sureau was worried that an event much feared among virologists worldwide had come to pass: the first great African epidemic of Marburg. But Pattyn's message now gave him an idea. He knew that a small outbreak of Marburg had been contained some years before at the Johannesburg General Hospital. Now he kept his fingers crossed that they had saved some convalescent serum that could be used to treat Mayinga.

Even getting hold of information presented a major logistical problem. South Africa was currently ostracized because of apartheid. The presidential office could not be seen to deal with South Africa. On October 14, Dr. Ruppol called the director of the presidential office and told him they desperately needed Marburg serum if they were to save Mayinga. The director worked clandestinely to contact Dr. Kozewski, the director of the Institute of Medical Research in Johannesburg. Excitement rose as Dr. Kozewski told them that convalescent serum had indeed been saved. But Kozewski did not have it. It was in the possession of Dr. Isaacson.

One of the colorful figures of international virology, Margaretha

Isaacson was forty-seven years old, five feet five, blonde-haired and blue-eyed — in her own words, "a tiny person with small bones." Those slender bones housed a formidable determination.

In 1975 it was she who had treated a young Australian couple who had contracted Marburg disease while backpacking through Rhodesia. The man had died in the Johannesburg hospital, but during his hospitalization a nursing sister had contracted the virus from him and she had eventually recovered. Over a period of eighteen months or so, Margaretha had collected a total six units of convalescent serum from her nursing colleague. On October 15, a day after she received the request, the South African doctor strode out of an Alitalia plane in the sunshine of Kinshasa, carrying four of those six units of Marburg serum in her hands. She was astonished to find herself promptly whisked off to the VIP lounge. "Of course I didn't have a visa — the Zairians didn't want it known that I was there. It was one of those cloak-and-dagger things. I never got to see passport control or customs. My passport was taken away from me. They brought me tea and biscuits and the protocol officer from the president's office was there to keep me amused."

As soon as she could extricate herself, she went to the Ngliema hospital, bringing with her a large quantity of protective clothing and gear, donated by the South African government. She found Pavilion 5 in a state of panic. They had collected everybody with even the remotest contact with the nuns or Mayinga together in a second pavilion and surrounded the place with armed troops.

In Pavilion 5 she discovered Mayinga in the fourth day of her illness. The first three days had been marked by nothing more specific than fever and tiredness. Today had seen a marked deterioration. Her throat had become severely inflamed and the profound lack of appetite had begun. She obtained relief from sucking ice cubes and was receiving intravenous fluids. Dr. Isaacson added a unit of precious Marburg serum to the intravenous infusion.

Margaretha Isaacson took over the management of Mayinga from Sureau and the Belgian doctors. She also assumed responsibility for the containment within the Ngaliema hospital. Her first priority was to sort out the physical protection of staff and the quarantine arrangements for the contacts. The staff were still following the simple regime of surgical gown, gloves, and mask, reusing the gowns and overshoes. She put a stop to this, supplying high-quality disposables. From now on they wore balaclava-style helmets and their own full-face respirators, which provided an airtight seal over the face, allowing the nurse or doctor to breathe through a filter canister. Nondisposables were left inside the patient's room and disposables were burned in a makeshift incinerator fashioned from an oil drum. They set up an iron bath over a fire in the courtyard that was used to boil up bed linen. Dettol was replaced as an antiseptic by sodium hypochlorite, proven to kill viruses.

In a situation of petrified horror, the rooms where Myriam and Edmonda had died had been left untouched. Both nuns, and especially Myriam, had suffered bleeding, from their mouths, from injection points, and through vomiting and diarrhea. Bloody excretions stained bedclothes, the bedside furniture, and even the floors and walls. Scared out of their wits, the staff were refusing to go near them, yet Margaretha Isaacson had need of those rooms.

She talked it over with Sister Donatienne. This redoubtable woman was famous among the Belgian expatriate community in Kinshasa. She had a pet gray parrot called Coco, who was said to make his own rounds of the wards. But Donatienne was herself physically disabled with a clubfoot. She was also seventy years old and in no physical state to get down and scrub the floors. "Sister," Margaretha now urged her, "you don't have to clean the floors. You can just pretend you are cleaning them." The two women picked up buckets and brushes and, wearing full protective gear of masks and goggles, began to swab and scrub out the residua of death. In a few minutes the other nurses, inspired by their courage and leadership, arrived and took over the duty.[9]

There remained the problem of the safe disposal of bodies. Cremation would have been ideal, but African custom, and the prevailing rules of the Catholic Church, prohibited it. Meanwhile, as day followed day in relentless progression, Mayinga was deteriorating. She was given a second precious unit of the Marburg serum. They tried heparin to prevent the widespread internal blood clotting, believed to underlie the bleeding manifestations. Still this unfortunate young woman, who only a week earlier had been preparing to fly abroad to a new life filled with promise, was now reduced to a pitiless death sentence in an isolated room, surrounded by armed guards and comforted by colleagues wearing the frightening array of face and full-body protection.

5.

On Friday, October 15, when first setting out for Geneva, in anticipation of taking charge of the expedition to Zaire, Karl Johnson, with dark brown curly hair and a much lighter beard, met Joel Breman for the first time at the Pan Am terminal at JFK Airport in New York. Breman was six feet two, black-haired and -bearded, a CDC epidemiologist who had been diverted to Africa from an investigation of the influenza epidemic then sweeping through America.

In a televised interview many years later, Karl Johnson would explain his thoughts at that uncertain moment. Not only was he one of the most experienced virologists in the world, but he also had the ethical rigor, coupled with the necessary vision and courage, to articulate what others were too cautious or conservative to perceive, perhaps even to imagine.

In the near decade since the discovery of Marburg the virus had been subjected to a good deal of experiment in high-security laboratories around the world. Nobody had found a single antiviral drug that was of any value in treating it. Nor had anyone managed to produce a vaccine. From the scraps of information he had been able to obtain about the situation in Sudan and Zaire, they were dealing with a plague virus that was "tremendously lethal and that could be spreading from person to person in a manner not previously seen with such a lethal virus." The cardinal issue was now lucid, if starkly frightening: "Is this virus going to turn out to be transmitted by sneezing and coughing in a manner similar to influenza. If you have an infection that is virtually fatal, as this turned out to be, and transmitted by that mechanism, now, suddenly, you have a threat to the entire human species."[10]

TEN

The World Was Lucky

It shows you how small the world is.
It shows how Ebola could get out — how HIV
did manage to get out.
I mean, it's nothing at all — no problem!

DON FRANCIS
conversation with the author

1

Don Francis was the first to arrive into the Sudan, stepping out into a wall of heat at the airport in Khartoum. The WHO representative took him to the ramshackle Saraha Hotel, devoid of air conditioning and baking in the heat. Here he met Paul Brès for the first time. The WHO virologist proved to be a precise, mild-mannered Frenchman, bespectacled, gray-haired, and balding. Francis knew the city well from a previous posting here during the smallpox eradication ("It was like coming home"). That same night, David Simpson flew in from London. When he arrived at the hotel, Simpson was in time for an early working breakfast with Brès, planning their operation over a table in the dining room with a creaky "punkah louver" fan in the ceiling disturbing the flies. "It doesn't matter what you tell this waiter." Brès laughed curtly. "You can wave your hand all you like. He will decide when he wants to serve you and what you are going to eat."

It was a propitious introduction to Khartoum. Francis introduced his colleagues to the dusty streets, lined for the most part with two- and three-story buildings of unrelieved concrete, blazing white in the sun, and through which trundled huge Bedfords trucks with hyperinflated desert tires. Moored by the yacht club near the confluence of the Blue and White Nile was the historic gunboat that had come to relieve General Gordon at the siege of Mafeking.

On their arrival, the three doctors had assumed they would travel next day to the plague zone. Now they discovered that there was no possibility of a commercial flight. The country was in a state of panic. Khartoum had bolted the doors on the poorer south and a cordon sanitaire was in force around Maridi and Nzara. All travel to Juba, the regional capital of the south, was forbidden. David Simpson had brought six boxes of equipment with him, including a centrifuge, animal traps, and protective clothing, from Porton Down. These lay gathering dust at the airport. When they approached the Ministry of Health, begging a military plane, they were promised a helicopter, but the promised transport never arrived. October 20 began five days of fruitless rounds in search of help, with repeated visits to the Ministry of Health and the WHO offices.

The WHO representative in Khartoum was a Pakistani called Raffiq Khan, in Francis's words, "built like Superman and an incredible racquetball player." In effect he functioned as the Deputy Minister of Health, wore a suit every day and "had this rug — since he wouldn't admit to having a bald head." Khan provided them with a Land Rover. Failing to obtain help from the Ministry or the army, they could now cruise instead of walking the streets, calling on their respective embassies. Simpson called on the British High Commissioner, where he was turned away flatly without even a pretense of support.

Downtown was a single block that housed the administrative quarter. The American Embassy was the tallest building in the block, cantilevered over the street, with chinaberry and orange-blossoming frangipani growing in the shade. Francis met with a more sympathetic reaction there. In a series of meetings that stretched over days, a channel of communication was set up with the CDC. On one of these occasions, as the dejected pair were leaving the embassy, Francis suddenly bolted. Simpson watched in amazement as the young fair-haired American dodged through the lanes of traffic to reach a gorgeous black woman standing in the middle of the road. Miriam Baseiuni had been Francis's secretary when last in the Sudan. He hugged her wildly . . . "and there I kissed her in the middle of the road." It was a welcome relief to see one friendly face.

The Sudanese Ministry of Health assigned its own virologist, Babakir el Tahir, to travel south with the team. El Tahir had useful prior knowledge of Maridi. In subsequent days the collection of frustrated doctors swelled, as the group was joined by David Smith and the entomologist Barney Highton, before finally, a helpful German Embassy persuaded one of its regular supply planes to drop them off in Juba. This was carrying "essential" supplies, which, according to Simpson, comprised sauerkraut and crates of beer, a couple of which he managed to commandeer before they were ejected by the nervous flight crew with engines still running.

Juba, the capital of the south, is a small town of single-story whitewashed terraces, with an arcaded market along the main street. It is perched on a hill overlooking the Mountain Nile, which is very beautiful as it flows out of Uganda through a mountain pass. Here the soil changes from desert sand to bloodred laterite mud, with grass and plenty of trees. The people are ethnic Africans, a composite of several tribes, some with specific teeth knocked out, others with tribal tattoos or with stretched ears or lips. The majority are Christians or follow animist religions. Here the team was met by the tall Italian Dr. P. L. Giocometti, the assistant WHO representative, already familiar to David Smith, who would subsequently manage radio communications and logistics.

No provision had been made for the expedition. After another two days of frustration, Giocometti gave the doctors his own Land

Rover and driver, and the Juba Ministry of Health added a couple of extra tires for the Land Rover, together with two fifty-gallon drums of fuel. Barney Highton needed two trucks to accommodate his animal traps. He and David Smith made arrangements to follow on a few days later.

Simpson, Francis, and el Tahir set out in the Land Rover, with Giocometti's man driving. They left Paul Brès to coordinate the investigation from Juba. Last-minute delays meant it was already three in the afternoon. Francis would remember taking his place in the right rear seat. He had his arm out of the window. Juba was as scorching as Khartoum, but much more humid. "As I looked down on a tsetse fly lying on my arm ready to chew on me, I thought, this is a hell of a beginning."

The laterite road seemed to wind on and on without end. The Land Rover skirted piles of elephant dung. Baboons ran across the road, or skirted the verges, screaming. As night fell, the road had dwindled into a muddy track, softened by recent downpours. Where they would cruise at twenty to forty miles an hour on the baked laterite, now the going was down to ten to fifteen miles an hour, through tall dense woodland, the lair of leopards. A little before midnight, the driver became exhausted and Francis took the wheel. He soon became impatient and left the transmission in four-wheel drive, crashing through the mud holes. Simpson, who had fallen asleep, crashed his head on the side of the Land Rover, injuring his neck. The arduous drive continued until two in the morning, when at last Maridi loomed into the lurching headlights.

Maridi is situated in a hilly area, the road widening as you approach it, so that in the headlights the first impression is that of a major city. But for Simpson it was a city shrouded in eerie silence. They drove on, a few government offices looming to one side, then the hospital. . . .

It was perched atop a small hill just off the road, with extensive grounds. Said Simpson, "As we drove into the grounds, it remained absolutely silent, no road lights, no movements, nobody." El Tahir knew where to find the health offices, close to the hospital entrance.

Like an apparition, covered in laterite dust, he hammered loudly on the door of the hut and woke up a sleepy man, who proved to be the night watchman. Inside the hut was a sign, with brown letters painted carefully against an ivory background: it read WELCOME TO MARIDI. At four in the morning, they found an English missionary family, who gave them a place to sleep.

At first light, the three exhausted doctors fell out of bed, resolved to go back to the hospital. Now, arriving at the entrance, driving through crimson dirt studded with palm trees, they could make out the corrugated iron roofs of the main ward buildings on the hill above them, scattered among flower beds, ornamental palms, and mangoes. Directly opposite the entrance, peering above a circular picket fence, were two thatched structures. They drove past the curious enclosures and up the hill.

They found three pavilions built out of cinder block, each with encircling verandah. To the experienced eyes of Don Francis, it was, by African rural standards, a large hospital. They knew it was the regional center for training nurses, employing a doctor and no less than 154 staff, including trainees. David Simpson climbed out of the vehicle, stood sweating heavily in the full tropical heat and humidity, turning round in a circle in the grass and palm trees. There was nobody to be seen, not a doctor or a nurse or a patient, not even a curious gardener. These buildings were empty, utterly abandoned. As they entered one of the pavilions and gazed down the desolate rows of bare iron-framed bedsteads, their footsteps and voices echoed with an eerie hollowness. There were unmade beds, personal belongings scattered randomly about the floor.

In an office adjacent to the main hospital buildings they found two Sudanese doctors who had arrived several weeks earlier and who had been expecting them. Pacifico Lolik and William Renzi Tembura had been sent on ahead by the Ministry of Health in Juba. It was they who had organized the building of an isolation facility — the two thatched huts, surrounded by the picket fence. The Sudanese doctors brought the newcomers up to date with what had happened here.

Samir's arrival into the hospital had triggered an explosive epidemic within the hospital. Fifty percent of the ward staff were already dead, including the sole doctor at the hospital. The highest rates of infection and death were in those staff most intimately concerned with day-to-day care of patients, the junior nurses and the doctor's medical assistants. All 6 of the medical assistants had contracted the fever, and 5 had died; 39 of the 95 student nurses, 22 of whom had died; and 14 of the 53 more senior nurses, of whom 5 had died. The tally of death burned through office staff and even the ward cleaners.[1] The survivors had deserted the hospital and fled into the bush.

The plague was still raging out in the surrounding town and villages, and Simpson was determined to get all infected cases back into the isolation pavilions. While a scattering of ambulances went out to find people suffering from the epidemic, the team sought out some of the nurses who had contracted the infection and survived. These nurses, nearly always male, were still willing to help others, not only through nursing but also by donating their blood as a source of immune serum. The missionaries loaned the doctors a house vacated by one of their colleagues. The living room became their laboratory. There was no electricity, but they managed to generate some light from low-voltage bulbs attached to car batteries. They rigged up a gas ring under a pressure cooker so they could sterilize instruments. A table in the center of the floor became their working bench for handling serum, blood, and human tissue, all potentially contaminated with a lethal virus. Simpson had brought his centrifuge, but the lack of electricity made it useless. They were forced to allow the blood to settle under gravity and then, taking what precautions they could, cream the plasma off the top by suction. "It was all a bit horrific," said Simpson.

From now on there would be very strict segregation of patients. The bodies of those who died were buried by the hospital staff without allowing relatives any contact whatsoever. For the villagers, who had no conception of a contagious virus, it was a brutal contradiction of every custom of grief and caring. There was virtually no fresh food: the cordon sanitaire had seen to that. So while the villagers ate their own produce, the doctors subsisted on bread and biscuits from army rations dating from the Vietnam War. The mission gardens supplied them with fresh grapefruits for their breakfast.

Francis began to organize the epidemiology. Simpson was collecting sera, with the object of sending it back to Porton for analysis. The recovered nurses meanwhile were helping Drs. Lolik and Tembura to find new cases. About four days after they had arrived in Maridi, the first batch of patients had arrived in the isolation facility, ready for their inspection.

The isolation huts, segregated for men and women, were each a hundred feet long by fifteen feet wide. A local policeman was posted on guard at the entrance through the picket fence to prevent the entry of relatives. The doctors reached the enclosure at noon. The heat was oppressive. Simpson was wearing RAF pajamas under his surgical suit and a pair of Wellington boots. He had brought two respirators, which were British made, from the virus laboratories at Porton Down. They were designed to cover all of the face, with separate holes for each eye and a thick black hose that came down over the nose. Now he took a piece of sticking plaster and wrote Don's name on it, then stuck it onto one of the respirators. Within seconds the men were so hot, sweat was running down their faces and bodies underneath the protective clothing. Over the surgical gown, they wore plastic aprons that reached from their necks to their ankles. "To those patients," said Francis, "we must have looked like creatures from outer space."

He says he will be forever haunted by the sight that confronted them. There was no door — they entered through an opening in the wall. It was very dark inside. They passed by a little dispensary, then through into the ward. There were fifteen patients lying on sweat-soaked pallets over wire-spring beds. They appeared "emaciated and wild-looking, their eyes staring in otherwise expressionless faces." Simpson stopped by the bed of a young man, a student nurse. His dehydration was so severe he had lost forty-two pounds in weight over the preceding five days. His family would barely recognize him now. The flesh of his face had sunken between the bones of his skull, cheeks, and jaw. Yet Simpson was impressed by the clarity of his mind inside that expressionless mask.

This would be the pattern as they examined patient after patient, their bodies emaciated by dehydration, their faces rigid, their eyes dazed and staring, their minds curiously anxious and clear. Blood was oozing from needle puncture sites, from any part of the skin that had been penetrated, in some cases from the nose, gums, the bowels, and in some women from the womb. Some had suffered conjunctivitis, so their eyes were sore and infected, a few had scleral hemorrhages, so the whites of their eyes were scarlet. From time to time, somebody would vomit altered blood or pass the black tarry liquid stools.

Those close to death were in a stupor from dehydration. Even those who appeared to be recovering were so emaciated they were virtually skeletal. In those early visits to interview survivors, Francis would observe startling examples of the severity of the illness. "I started interviewing people who had recovered. They were incredible. They had suffered a total desquamation. They had shed the skin of their hands and feet, the hair from their heads, even their nails."

Simpson took his specimens of blood and the doctors set up intravenous infusions to alleviate the distressing dehydration. It was tragic that they could not administer serum from recovered cases at this stage because they lacked the facilities to guarantee the serum was virus free. They did what they could in terms of general support and treatment, then asked the male nurse, himself a recovered victim, to let them know when people died. Apart from the limited autopsy on Myriam by Dr. Courteille, there had been no formal autopsies that might provide detailed information on the pathology of the virus attack on the internal organs. The shocked doctors returned to their mission house and talked in awe about what they had seen.

Over the following two mornings, they received word of deaths within the isolation facility. At midday, over the following two days, they returned to the isolation pavilion to perform autopsies. Simpson performed the first. They carried the body of a young woman out of the darkened room, with its odor of death, under the very eyes of her family, who had gathered at the gate wailing that they wanted to take her body home for the rites of funeral. The operation was carried out on the bare dirt at the back of the isolation hut. The following day it was Francis's turn, this time upon the body of a man. "I had no place to lay my instruments, so they stuck out of a bucket of formalin like a porcupine. I had to kneel down to do it on the grass. Then it started to rain. If I pricked myself through my gloves or my rain-soaked gown, I was as good as dead. I opened the guy up — just made a cut and his abdominal cavity oozed this red serous fluid. His liver was like a purple water balloon filled with blood. As soon as I cut it, it was bulging — the normal tissues seemed to have melted away."

2

While working in Burkina Faso, Joel Breman had heard dreadful rumors of political corruption and chaos in Zaire. He saved his worst stories for the flight to Kinshasa. Now that they were finally on the last plane and he had a whisky in his hand, he turned to Johnson and began his yarn spinning. Breman would remember the situation with a chuckle. They were sitting side by side in the center section of a 747, Johnson to Breman's right. Suddenly, a man on Breman's left said, "Hey! Are you guys from the CDC?" The voice was unquestionably American.

By an extraordinary stroke of luck, they were sitting next to Dr. William Close. Close had recently gone home on a short leave after spending most of his professional life working in Zaire. The medical director of the Mama Yemo Hospital in Kinshasa, he had heard about the epidemic, and was heading back to see if he could help. At Breman's suggestion, they played musical chairs, so Close could sit between them. He talked about his experiences in Zaire, giving Johnson an abundance of background information.

Johnson, Breman, and Close arrived in Kinshasa early in the morning of October 18. After a quick get-together at the American Embassy, they made their way to the morning conference in the FOMETRO offices, conducted in French and chaired by Dr. Ngwété, the Minister of Health.

At this time, Kinshasa was enveloped in confusion. Pattyn, also present at the meeting, would remember the atmosphere pervading the city. "Air Zaire didn't go to Bumba anymore. If the local pilots were not already frightened out of their wits, the local papers assisted with their version of aerosol spread. There were tales that the birds were dropping dead out of the sky above Yambuku."

Johnson, through the French interpretation of Joel Breman, asked with his colleagues around the table if they could define what was known in terms of hard fact. He had listened to what Ruppol and Raffier had to say about their hurried previous visit to Bumba and Yambuku, but their stories left a great many questions unanswered. Breman remembers vividly how Ngwété, sitting at the head of the table, turned to gaze at the Americans. "Now," he iterated calmly, "the team is ready to go to the epidemic area." With everybody's eyes upon them, Breman and Johnson realized that he meant them.

In Pattyn's amused recollection, the minister was already talking to his aide, considering how to arrange the flight. By now the only possible travel into the quarantined Bumba was courtesy of the extremely reluctant army. Johnson was running his hand through his hair. "Well, just hold on awhile," he protested. "We don't know what is really going on here." Johnson's equipment, including a fluorescence microscope, had not yet been unpacked from the airport. It was decided they would have to divide the team, Breman making a preliminary trip the following day, while Johnson sorted out a base for operations locally before moving to Yambuku somewhat later. That same day they visited the Ngaliema hospital, where they called in to see the sick Mayinga.

The unfortunate nurse was desperately ill, having failed to respond to the Marburg serum. Lacking the solid rock of faith that had sustained Myriam and Edmonda, she was so terrified of dying that she suffered a torment of anxiety and agitation. Out of compassion, Dr. Isaacson prescribed Valium to calm her.

Patricia Webb's serological confirmation that the virus was different to Marburg explained why the Marburg serum had not helped Mayinga. She died in the early hours of October 20 — and Margaretha Isaacson had already made plans to deal with the situation. Soon after her death, her body, like those of Myriam and Edmonda, was wrapped in cotton sheets impregnated with phenol. These were wrapped in heavy-duty plastic bags and sealed into heavy wooden coffins. Contact with the bodies by relatives and friends was prohibited, the coffins released only for on-the-spot funeral services and immediate burial.

Panic was rife in the city, induced by Mayinga's numerous contacts. Among these was a young girl of fourteen who had eaten from the same plate and a young man who shared a bottle of soft drink with her on the first day of her symptoms. From the international airport in Kinshasa, you could reach anywhere in the world within twenty-four hours, and so the real matrix of fear extended globally. If the lethal virus broke out in the streets of Kinshasa, it would precipitate an international quarantine of the country. The consequences for Zaire's already tottering economy would be devastating. A huge containment exercise radiated out into the city.

The minister, Ngwété, organized four teams of doctors, accompanied by health inspectors and drivers. Their function was to trace contacts and bring them back for quarantine in the Ngaliema hospital. As news of the containment exercise leaked abroad, it alarmed every expert in the know, in particular the monitoring World Health Organization in Geneva.[2] Mobutu had to be dissuaded from throwing a military cordon around every area of the city that had been visited by Mayinga. Meanwhile the Zairian army was on full alert, touring the streets in a volatile state. Thirty-seven primary contacts had already been traced, with 274 secondary contacts at 44 different addresses. A considerable number of both primary and secondary contacts could not be traced, either because their identity was unknown, or their address was unknown or incorrectly supplied.[3] Pavilion 5 was already overcrowded, so Dr. Isaacson set up a second containment area in the old surgical ward, Pavilion 2. Mayinga's contacts were ferried here, willing or unwilling, with the result that Pavilion 2 was soon overcrowded with terrified Zairians, hemmed in by troops armed with rifles and even a machine gun. Dr. Isaacson did her best to defuse the situation by reducing the period of isolation to a minimum.

On the morning of October 21, and despite all of these precautions, there was a mass breakout. It was triggered by the sight of Mayinga's body being taken out of Pavilion 5 for her funeral. The armed soldiers, as terrified as the people they were guarding, did nothing at all to prevent it. Bill Close was sitting in with Margaretha Isaacson at the morning meeting at FOMETRO when this message came through. Close was impressed by "this tiny woman, as tenacious as a limpet," who rushed back out to the hospital and, her voice ringing with indignation, bullied the panicking people back into the isolation facility.

On October 20, Joel Breman left Bumba, where his team had been ferried by army helicopter the previous day, to take the sixty-mile journey to Yambuku. The vehicles were in bad shape and the road soon petered out into a potholed mud track. The landscape seemed eerie. They crept through tunnels so shrouded by forest they drove by day with full headlights, their nostrils full of the pungency of the undergrowth, their ears filled with the screech of monkeys, the medley of bird calls, and the buzzing of flies and insects. They ground to a halt at every village along the way, their progress blocked by makeshift roadblocks — sawhorses with planks straddled across them or tree trunks thrown across two old petroleum cans. They would explain at every stage that they were doctors, here to help control the epidemic. They would inquire if there were any cases in the local villages. The Africans emerging to meet them were for the most part terrified, full of questions, sometimes evasively mute, always resentful of the quarantine and its economic deprivation.

Breman's first impression, as they entered the mission, was more eerie still. "As we drove into Yambuku off the road, the first thing we passed was the church, and then we drove down the avenues of mango trees and royal palms, through the mission grounds themselves. It was no more than late afternoon, yet the place was completely deserted."

There were no patients left in the hospital. One or two Zairian nurses filtered cautiously out of the houses on the periphery, so the doctors could see that not everybody was dead. Then they headed toward the convent building, splendid in its festoons of red and purple bougainvillea, within gardens that seemed incongruously beautiful and orderly. The surviving three nuns peered at them from the guest house, which had been cordoned off with surgical gauze tied to sticks. They had been isolated in this nightmare for ten days. Breman would vividly remember the figures materializing, as if from a nightmare, "wandering out in their heroic stance" to have a look at them. The young Belgian, Peter Piot, was the first to step over the forbidding gauze, ignoring the nuns' warnings about danger and death. Reassuring them in Flemish, he shook each of their hands in turn, Marcella, Geno, the timid Mariette, and the hesitant Père Claes, the only remaining priest, who, with his long tobacco-stained white beard, stood back behind the women.

Breman, shrewd in his assessment of people, gauged Sister Geno to be the sharpest intellectually, the most "action oriented." She addressed them in good French, describing what had happened and debating animatedly what should now be done. Sister Marcella helped to organize the arrangements. The nuns prepared a simple meal. The doctors would be put up in the school dormitory. Marcella retrieved her paperwork so the doctors could inspect it. Breman was impressed by the records they had kept of the epidemic. The nuns, together with Père Claes, had driven about the villages and had compiled a list of more than two hundred cases. The doctors began that same day with an inspection of the hospital, confirming that there were no patients left. The organization had fallen apart, there was not even a dispensary functioning. The nuns and priest had taken all of the mattresses from the iron-spring bedsteads, burning them in a huge pyre in the gardens, then swept out the wards and swabbed them with disinfectant.

Later that morning, when they heard over the radio about the death of Mayinga in Kinshasa, Pierre Sureau wept openly. That same day the investigation proper began. The doctors sat down with the nuns and worked through their lists, searching for the likely index case and the mode of transmission. In the dispensary, Breman inspected the simple ledger where nuns had kept records of every patient, a single line each. Soon they were examining their first patient: Breman took a photograph of Pierre Sureau examining him.

Over the two or three days that followed, they gathered together four Land Rovers, Breman organizing his surveillance teams into four, one doctor and one of the nuns or the priest with each team. The plan was a simple one: to define the extent of the continuing infection. The vehicles with their teams headed roughly east, west, north, and south. In the villages, they kept their inquiry to two very basic questions: are there any people still suffering from the fever? Are there any people here who have had the fever and recovered? They soon confirmed that the epidemic was subsiding, though there were still active, or suspected cases, in eight different villages.

There were a great many investigations that had to be shelved for a later, more definitive, investigation, including the search for the real index case and for the reservoir of the virus. Even so, those figures compiled by the nuns clearly showed that the hospital had been the amplification zone and Breman was curious to understand how.

On the second day after his arrival, as he was conducting his first epidemiological exploration in the company of Marcella, he had what he would later call his epiphany. "I was out going from village to village. People would start to explain the sequence of events to me. They would tell me about this mobile clinic — an antenatal clinic. I wasn't expecting this at all. People would say, 'You know this woman came down with the disease after going to the hospital for her prenatal visit.'" Breman would jot a tiny note into his green-backed notebook. Gradually a shocking notion began to surface.

He had arrived at a small village. There was only himself and Marcella. He was preparing to take a blood sample from a man whose wife had died during the epidemic. "It is my habit to be a little more inquisitive." He chatted to this man about the events that preceded his wife's illness. "Did she go into the hospital before she got sick?"

"Oh, yes," the man replied. "It was for her prenatal visit."

"Tell me what happened during this visit."

Breman was gazing at the man as they sat opposite one another, in the shade of the thatched eaves of a hut.

"She got an injection."

Breman turned to speak to Marcella, who was standing next to him. Did all of the antenatal cases receive injections? She did not know. Breman returned to the mission hospital that day and spoke to Sister Geno. Geno told him that the women attending the antenatal clinic, just like everybody attending outpatients, had no faith in vitamin tablets. They demanded injections. All of the women attending the antenatal clinics were given injections. In the dispensary, Joel found the evidence to confirm his epiphany: there were just four or six syringes issued each day. The needles and syringes were reused for patient after patient without intervening sterilization. They were only sterilized once in the autoclave, at the end of the day.

Breman understood now why the hospital had become the epicenter of the epidemic. Although there were undoubtedly other forms of spread, particularly from intimate person-to-person contact, the main cause for the explosive spread of the epidemic within the mission hospital was the repeated use of contaminated syringes and needles.

3

In the Sudan, one of the two trucks ferrying Barney Highton's equipment and the protective clothing overturned on the road between Juba and Maridi, destroying a very large number of irreplaceable animal traps. They could obtain no fresh dry ice to preserve the specimens, and the ice they had brought down with them was slowly boiling away. They tried to arrange a regular and dependable transport to ferry the specimens out to Porton Down. "We tried to charter a plane from Nairobi," said Francis, "but they said they would burn the airplane."

Communication between Maridi and Juba was only possible through radio messages on the ACROSS radio network. From there, messages had to be further relayed by the WHO representative to Khartoum and Nairobi. Delays were inevitable and answers frequently took an extremely long time. Paul Brès struggled to help the teams from his base in Juba until November 2, then returned to Geneva. In his confidential report to WHO, he included a section entirely devoted to difficulties in transport and communications. "In spite of efforts from the Ministry of Health and the WHO Representative, neither of these two means of transport materialised."[4] The team's eventual report would conclude that of the 46 days spent in the field, Simpson and Francis were able to work for only 22 days, Highton 11 days, and Smith just 9, these figures "a sad reflection of the poor logistic support received by the team."[5]

There remained the nightmare of transport between Juba and Maridi; and within the plague zone itself, between Maridi and Nzara. For the scientists in the field, the implication was stark. Once arrived in the plague zone, they had been quarantined within it, sealed fast within the cordon sanitaire. In Paul Brès's initial objectives for the Sudan expedition, item four was "Plans for Medical Evacuation." "The crucial point will be the evacuation of a sick member of the team on a light plane from Maridi to Juba."[6] Before setting out, Don

Francis had received similar guarantees from John Bryant at the CDC. "If you get sick, we'll medivac you out to Frankfurt, to the military hospital." Now, once arrived, he would go down to the telegraph office once a day, awaiting Bryant's reply to his series of telegrams, demanding further assistance and the continuation of that guarantee. "I got this telegram back from my boss which read, 'If you get sick, we'll send the hospital to you.'"

Francis began his epidemiological survey by collecting together whatever scraps of information he could from talking to the missionaries, the Sudanese doctors, and by degrees the local population. David Smith joined the others at Maridi on November 6, arriving with Barney Highton, who had to work with whatever animal traps and protective clothing that had finally made it through by road. Only when they had collected together information on every possible victim, recovered or dead, could they determine where the virus had originated, the intimate history of its initial mushrooming and horrifying mortality rate.

At first they assumed the worst: that the epidemic was spread by inhalation. Unlike the situation in Yambuku, the epidemic was still fulminating in the farms and villages, hidden in the labyrinth of paths and channels that weaved through savannah grass that was often higher than the searchers' heads. The real poverty of the villagers was pathetic and touching. Inside the circular huts, mud-walled with thatched roofs, they would find little more than a bed, with its sick or dying victim, a table, and a couple of chairs. Every case had a very clear source, whether sexual contact, or more commonly within the hospital or during the traditional funeral within a family. The doctors were obliged, at this stage, to wear the respirators, with their full-face masks and black side tubing and valves. But it soon became apparent that the infectivity between people was simply not contagious enough to confirm aerosol transmission, and with a sigh of relief they abandoned this precaution.

As Joel Breman was simultaneously concluding in Yambuku, Francis soon realized that the hospital in Maridi had amplified the spread of the plague. It was the focus for all of the subsequent ripples of spread into the local community.

In David Smith's opinion there was evidence here also of infection through contaminated syringes and needles on the wards. But there was also powerful evidence for person-to-person spread. The charts Francis was now compiling told a lucid if tragic story. It had been the vocational care of the young nurses and medical assistants that had spread the pestilence. Where the antenatal link in Yambuku had infected a lot of younger women, here the majority of the victims were men, reflecting the gender ratios of hospital staff. The epidemic had burned within these buildings, jumping from patient to nurse, from nurse to patient, throughout July, August, and September, killing both the caretakers and the cared for without discrimination. In the words of Francis: "The hospital had been involved in a great western-type training center for nurse midwives and nursing assistants. These young nurses just loved these patients, cared for them so ably, and that was why so many of them died."

In Maridi as a whole, 213 victims were traced, 93 of these staff or patients infected in the hospital, while 105 had been infected in their homes. Here the intimate contacts between family members and particularly the washing, fondling, and kissing involved in funeral practices had been crucial to the spread of the disease. Only five were classed as "unknown" contacts. The most likely route of person-to-person spread had been through contact with body fluids, with their high content of virus, and particularly blood.[7]

It was from this epidemic focus in Maridi that Britain would have its first close call with Ebola. At this time, the teenage daughter of English missionaries was returning to her boarding school in England and she shared the journey with a sick doctor in the confined space of the monthly mail plane, a tiny Piper Cub four-seater. The following day she had arrived in London. In the words of Don Francis: "Here you've got these poor people, probably never heard of London, just living off the earth. With the exception of their only possessions, a few plastic or metal pots, their way of life must be close to the Bronze Age.

Now suddenly in the middle of nowhere you have this outbreak. Within twenty-four hours this girl leaves the epidemic zone and arrives in London. If she had gotten sick, she would have had an unknown fever, would have gone into the hospital and specimens would have been drawn on her and sent to every lab in that hospital. . . ." The message was frightening.

The young woman did not become ill. Britain was lucky on that warm September day in 1976.

Barney Highton made his way to Yambio by truck on November 5 and David Smith followed him three days later. Smith took up the epidemiological tracing from the point where they had left it during his earlier visit with Cenydd Jones. On November 16, they were joined by Simpson and Francis, who had managed to charter a regular plane, a Cessna, courtesy of the American Embassy. Back-tracing all cases confirmed Cenydd Jones's initial findings: the plague had begun with the illness and death of Yusia, the quiet storekeeper who worked in an office next to the cloth store. The cotton factory, built by the British during the protectorate, now became the focus of intensive study. It was here that Highton decided he would focus all of his efforts to discover the natural source of the virus.

As Highton began his animal and insect collecting, Smith organized the blood sampling of factory employees, before moving out into the town and district to search for cases and relatives. Here the distances involved were much greater than in Maridi. They were forced to enlist local assistance and drive long distances on the winding dirt tracks and roads. They discovered John's Jazz Bar and its importance in the explosive spread of the virus locally. Smith visited the family compound where Yusia had died, and he took a photograph of the picturesque grave. Only love could have constructed that sad little mound, like a pixie's house in the bloodred dirt, protected by its manicured parasol of thatch with a little bobbin on top. He met Yasona, who had survived the illness, and he took blood for serology from him and his family.

No active cases were found in Nzara. The outbreak had ended spontaneously in late October, after infecting a total of sixty-seven people. Here the tiny hospital had played no part in the epidemic. All the cases had arisen from person-to-person spread in the community, and forty-eight of the infected cases, twenty-seven fatal, were traced back to Paul or his girlfriends.

The last case in Maridi was admitted to the isolation facility on November 25. The epidemic in Sudan was now over. They had plotted the circles of spread of the virus to no less than eleven waves of contagion, four of which had occurred after their first arrival in Maridi.

From the two autopsies, the only such examinations performed by either expedition, they concluded that death was the result of overwhelming tissue destruction by the virus. Karl Johnson would also emphasize the role played by shock from massive fluid losses, complicated in many cases by hemorrhage. Intravascular clotting had played some part, but with the paucity of laboratory investigation nobody could be sure to what degree. Simpson and his team had also noticed something interesting about the behavior of the virus: with time, with the later generations of spread, it had seemed to become less aggressive. "We had the impression it was losing its pathogenicity with time." The focus now turned to Highton's search for the origin in nature of the Ebola virus.

The factory was a large establishment, with spinning and weaving workshops spread out over several buildings. To get to the office with its adjacent storeroom you had to pass through workrooms throbbing with looms and carding machines, the air alive with a fine snowstorm of cotton fiber, before arriving at a much smaller room, perhaps four times the size of a regular office. Yusia's desk was still here, with its neat old-fashioned files and ledgers. Highton's first discovery was the huge population of rats. Every room and workspace was infested with them. But there was an even greater surprise when he stood in Yusia's office and gazed up at the ceiling. There were bats of several different species roosting there. The floor was spattered with their droppings.

Every morning Highton would fly up to Nzara, where he trapped everything that crawled, burrowed, bit, or flew. Not only did they collect bats from above Yusia's office, they captured them on the surface of his desk. "We collected rats and bats, birds and lizards, recruiting the local children to bring in anything they could catch." The eventual tally included 7 shrews, 178 bats from seven different species, 309 rodents of eight different species, 4 lizards, and a single toad. The huge volume of sacrificed animals and insects was brought back on the charter to the makeshift lab in Maridi, where the team would sit out on the screened verandah, don suits and respirators, and dissect the carcasses. The blood and internal organs were then dispatched to Porton Down for further analysis.

The volume of biological investigation was such that even a year later Ernie Bowen's group would still be struggling through a morass of viral screening. Still the work was vital — in time it would assume the greatest importance of all, and the greatest mystery.

Ecologically speaking, Nzara was only recently torn from rain forest, a so-called transition zone of derived Guinea savannah. An expert at the British Museum was called in to help identify the bats. These, he explained, were, on the whole, not savannah bats at all but "forest dwellers." It was a conclusion that had more universal application than was realized. Nzara and its hinterland were providing "enlarging habitats for mammals adapting to human environmental changes and making new links between sylvatic and domestic cycles."[8]

Nevertheless, despite every attempt at viral isolation at Porton, no virus was ever found in the large volume of dead animals and insects.

4

In Zaire, with the panic now resolved in the capital, the first expedition's return had given Karl Johnson a graphic lesson on the lack of any real investigative or handling capacity at Yambuku. They needed more detailed information. They planned a much more comprehensive return, when they would have to take everything they needed out there.

The manpower envisaged ran to hundreds. In the WHO bulletin that would summarize the entire Zairian outbreak, Karl Johnson would describe the circumstances as "those of a small war."[9] The eventual costs would amount to a million dollars. Even with this, a great deal of improvisation would be necessary on site.

The second expedition divided into two functional groups. One, under Karl Johnson, based itself in Yambuku, to become the laboratory for serological screening of blood samples and handling the mass of samples that would be sent back to the CDC for more detailed examination. Guido van der Groen was a notable assistant to Johnson in this work. The second group, which again included Breman, Sureau, and Piot, conducted a formal epidemiological surveillance in the community throughout Bumba.

By the end of November, it was apparent that the plague was truly over. An exhausted Simon van Nieuwenhove had completed a lonely survey of the north of the equatorial region: after traveling 3,200 kilometers in seventeen days, with no evidence at all of the spread of the virus beyond the known plague zone.[10] The epidemiological teams working under Joel Breman's direction had also visited 550 villages and interviewed 34,000 families, extending to approximately a quarter of a million people. The full scale and ferocity of the epidemic could now be accurately defined. In addition to the mission town of Yambuku itself, the Ebola fever had extended to fifty-five of the surrounding villages, infecting a total of 318 people, of whom 280 had died. This lethality, approximately 90 percent, was the worst known in all of human history in relation to an acute epidemic virus. Where people had contracted the virus through injections, as at the antenatal clinics and the hospital dispensary, the lethality appeared to be 100 percent. Twenty-four of the dead were newborns and forty others children under the age of fourteen years.

Although most experts, including Johnson, thought the Zairian epidemic had arisen through spread of the virus from Sudan, nevertheless Max Germain, a zoologist from the Pasteur Institute in Bangui, Central African Empire, arrived to attempt a search for a virus reservoir in

Zaire. He was hampered by lack of certainty with regard to the index patient and his investigation extended no more than twenty kilometers to surrounding villages.[11]

In view of the Marburg experience nine years earlier, the scientists were particularly interested in green monkeys as a potential reservoir. But even the local hunters only managed to shoot six *Cercopithecus* monkeys and a couple of small duikers. Locally recruited children, who were more effective than traps, helped them capture 123 rodents, almost all of them from domestic infestation, including 69 *Mastomys,* 30 black rats, and 8 wild squirrels. Small numbers of other mammals were shot, 7 bats were captured by nets, and organ or serum samples were taken from a dead cow and from 10 pigs.

As in the Sudan, no virus was found from any of the insects collected. No evidence of the virus was found in the 6 monkeys or among the 147 other mammals, including rodents.[12]

The Ebola virus had disappeared into the forest as mysteriously as it had appeared.

5

England had a second close encounter with Ebola, when, on November 5, a scientist at Porton accidentally infected himself while injecting a guinea pig with a syringe loaded with virus. Geoff Platt, the bespectacled and bearded senior technician working with Ernie Bowen at Porton Down, was moved for special treatment to the isolation hospital at Coppetts Wood, on the outskirts of London. Establishment panic forced Porton Down to shut down its virus laboratories and send all of the staff home, while at Coppetts Wood some 160 other patients were hurriedly moved to alternative medical facilities. Platt was himself quarantined in a Trexler tent — a huge cuboidal plastic isolator, maintained at negative pressure with regard to the surrounding room, which in effect becomes a room within a room, isolating the patient in his bed. He was treated with human interferon, a substance produced by our immune cells against viruses, and immune serum, taken from Mabolo's wife, Mbunzu, and from a nurse called Sukato in Yambuku.[13]

Geoff Platts is the only known patient to have acquired the virus by injection and to survive. Though he escaped the hemorrhagic manifestations of the virus, he lost a great deal of weight and during his convalescence he shed his skin and a good deal of his hair. During his illness, virus was isolated throughout from his blood, throat swab, his urine, feces, and semen. By the fourteenth day of his illness, virus was no longer found in the common body secretions but it persisted in his semen for at least sixty-one days. A similar duration of virus in semen had been shown in men with the Marburg virus in Germany in 1967, an ominous perception, though the nature of its warning was not appreciated at that time.

The Ebola virus would never lose its capacity to cause terror. Most worrying of all, although there was no direct evidence of aerosol spread, some of the experts felt they could not rule out a poorly transmissible respiratory contagion in the person-to-person spread out in the villages. Had Ebola spread virulently by aerosol, the consequences would have been a good deal more catastrophic.

Kinshasa, with a population of 2 million people, hardly possessed the medical infrastructure to deal with an explosive epidemic. It was connected by regular air transport to most of the capital cities of Europe, from where an epidemic virus would have spread to America and every developed country. For a time, a simmering panic pervaded the knowledgeable circles within the European capitals as the World Health Organization monitored the situation and people in the know worried about the potential of a global epidemic. Bill Close, then working with the International Commission in Zaire, remembers this well: "We probably came within minutes of instructing that the whole country be put into quarantine — this would have been a catastrophe."

No global epidemic followed. Despite the fact that the virus could spread by direct skin contact, by intravenous injection, by sexual intercourse, and perhaps, however tentative and limited the evidence, by aerosol, chance had determined that it was not sufficiently contagious. Of the secondary cases, people who had not been close to the hospital, only 5 percent of those at risk became infected. Modest barrier nursing precautions were enough to interrupt its spread.

If Ebola fever had broken out and spread from person to person through coughing and sneezing, in the manner of influenza, there would have been no vaccine and no effective treatment.

The world was lucky that year.

ELEVEN

The Riddle of Ebola

The most beautiful thing we can experience
is the mysterious.
It is the source of all true art and science.

ALBERT EINSTEIN
What I Believe

1

If the Ebola virus outbreaks in Zaire and the Sudan triggered a lingering sense of fear in the minds of virologists, they provoked an equal sense of biological mystery. It was hard to believe that the two Ebola outbreaks were not related. The virus that first appeared in Nzara on June 27 must have found a means of tracking 825 kilometers through a mixture of rain forest and dense savannah to emerge in Yambuku on August 26. In late October, Joe McCormick was designated by Karl Johnson to test this hypothesis.

Today, as Professor and Community Health Chairman in the Department of Community Health Sciences at the Aga Khan University in Karachi, Joe McCormick is one of the most experienced tropical diseases experts in the world, but in 1976 he was an ambitious young doctor with a wanderlust. On Saturday, October 30, a big four-engined C-131, drafted into assistance through Bill Close's contacts with President Mobutu, picked him up and dropped him at Bumba. With his relief driver, McCormick began the slow and tortuous journey, lurching through dense forest along rain-softened laterite tracks that, when wet, had the deep red color of venous blood. He talked with the villagers along the way, describing the fever and asking if anybody had seen it. Now and then he would take blood samples from people who had suffered a vaguely similar illness and had recovered.

His journey began north of Bumba, in rain forest that later opened out into savannah and riverine forest. He and his driver slept in missions or in villages and eked out their supplies, supplementing them with whatever food they could lay their hands on. It took the lumbering Land Rover six days to get to the Sudan. "It was a rugged, tough trip and often we couldn't find ferries to carry us across the rivers. On one occasion, it was just a bunch of dug-out canoes we put boards across so we could drive the Land Rover onto them, and the guys were bailing the canoes out as they were rowing us across the river."

At this time, the Sudan team was still in Maridi. Joe spent two days in Nzara. Here he interviewed Yusia's brother, Yasona, together with Yusia's wives, workmates, friends, and neighbors, but he discovered "not a single clue." As he left Nzara, McCormick addressed a letter to his CDC colleague Don Francis. "Don — welcome to Nzara. Here are the first few cases. I can tell you where the index case is buried."[1]

After completing his return journey along a different route, clocking a total of 2,600 kilometers on the combined journeys and after visiting four zonal hospitals, 67 dispensaries and rural hospitals, and an untold number of villages, McCormick was convinced that travel between the two epicenters of the epidemics was extremely difficult. This, together with the lack of cases and negative blood tests, led him to a conclusion that was controversial at the time. Although they were almost simultaneous in timing and caused by what appeared to be the same virus, the Ebola outbreaks in the Sudan and Zaire were not connected. In time, he would confirm this impression through scientific experiment.[2] Remarkably, though they arrived only two months apart, the virus that emerged in Nzara was a different strain from the virus that devastated Yambuku. Today we recognize Ebola Sudan and Ebola Zaire as two distinct species of the Ebola virus — and this provides a tantalizing clue in the enigma of its origins and its behavior.

At a meeting in London in early January of 1977, Pierre Sureau underlined the practical significance of that continuing mystery by warning his colleagues of the future threat of Ebola. "We do not know anything of the 'natural focus' of the disease. . . . No more do we know how the virus maintains itself in its natural focus."[3] Given the local circumstances, human cases of the disease would almost certainly be misdiagnosed as yellow fever. If such a case were to be admitted to a hospital, "amplification" would occur.

The continuing mystery surrounding the origins of Ebola would result in a protracted series of outbreaks, extending over the following two decades. In 1977, in Tandala, Zaire, a single fatal case turned up in a mission hospital. In 1980, another case turned up at a hospital in western Kenya, infecting the treating physician in Nairobi. David Smith headed the investigation, and helped save the life of the physician.[4] Little could be learned of the behavior of the virus from these isolated cases, but the return of Ebola in five larger epidemics would be more revealing.

In 1979, in a chilling echo of its original emergence, Ebola returned to the town of Nzara. At this time, Joe McCormick had returned to the CDC, to take up Karl Johnson's post as head of Special Pathogens. How ironic that his first request from the WHO was to return to Nzara, where a new outbreak had broken out in early August. All the old delays had once again taken place. The Sudanese government had belatedly realized the gravity of the situation, shackling the luckless South once again with a military quarantine.

McCormick returned to the Sudan in mid-September with a young epidemiologist called Roy Baron. After four days of frustration in Khartoum, they arrived in Nzara. "There wasn't any hospital, just what the local people called the dispensary, a little mud hut with a thatched roof." There was hardly any light at all. The only light was a little kerosene lantern. Inside the hut it was eerily dark, hot and very humid. By the light of their lantern, they found thirteen patients lying on mats on the floor. McCormick could not locate the doctor. But he could see that the pattern of illness in these patients was identical to that witnessed by Simpson and Francis during the earlier epidemic. In this lantern-lit squalor, their ears filled with the moans of the dying and their nostrils sickened by the odor of death, they took blood samples from all thirteen patients, which would confirm that they were dealing with a new epidemic of Ebola. It was a second opportunity to look for the source.

Baron traced the epidemic back to an index case who worked at the cotton factory. Tantalizingly, the virus had returned to its precise original haunt. Thirty-four people had subsequently become infected. Despite extensive interviews of the index family and contacts, they could find no unusual exposure. "I remember asking about hunting, food — had he been anywhere unusual, come into contact with insects, with any animals at all — but there was nothing there. We didn't get a single clue." This would typify the secretive and always terrifying subsequent behavior of Ebola. Often, the virus would follow the exact pattern laid down in its first manifestation, yet would remain an utter mystery.

For a decade after its second appearance in Nzara, the virus seemed to hide from the world. There was no further epidemic, not even a fleeting appearance, not a single known casualty. The virus had slipped away into the background. Then, in 1989, and on this occasion with startling unpredictability, it was North America's turn to suffer a close encounter.

2

For decades the United States had been importing monkeys for a number of scientific and pharmaceutical purposes. The trade had not diminished since the Marburg fright: on the contrary, during those intervening twenty-two years, it had expanded significantly. Hazelton Research Products, Inc., a company located in Reston, Virginia, was a quarantine facility for such imported monkeys. On October, 2, 1989, one hundred monkeys were flown from Manila, in the Philippines, through Amsterdam to New York, arriving at Hazelton on October 4. They were caged in quarantine room F. During the weeks that followed, workers at the factory noticed that "the mortality in these monkeys was higher than normal."

The monkeys involved were cynomolgus monkeys, *Macaca fascicularis*, commonly called crab-eating macaques. These are attractive animals, with thick brownish green fur, green eyes, catlike ears, and a luxuriant handlebar moustache on either side of a golden triangle about the nostrils. Hazelton employed a veterinarian called Dan Dalgard, who now examined the dead monkeys to see what was killing them. On November 2, he performed autopsies on four animals, two that had died from the mystery illness and two he killed after observing earlier signs of the illness. Dalgard found evidence of infection with simian hemorrhagic fever (SHF) virus, with big spleens, infected kidneys, and hemorrhages in various internal organs. Lacking the laboratory support that could diagnose a hemorrhagic fever virus, he enlisted the help of the U.S. Army laboratory, USAMRIID, at nearby Fort Detrick.

On November 13, after receiving fresh tissue biopsy samples from Dalgard, the army set up tissue cultures looking for a virus. By November 16 they had confirmed infection with SHF. For a time Dalgard thought they had contained the problem. But occasional deaths continued throughout other rooms in the facility, deaths that no longer fitted the pattern of simian hemorrhagic fever. Dalgard communicated his renewed worries to Peter Jahrling, who, with the assistance of electron microscopist Thomas W. Geisbert, set up new lines of investigation.

On November 27, the cell cultures taken from animals in the original room F sprouted a filovirus. The army scientists needed some infected monkeys for a more careful analysis. Hazelton's directors were reluctant to allow them access to their facility, but Dalgard agreed to meet Colonel C. J. Peters and colleagues in the parking lot of an Amoco filling station some way between the two locations. Dalgard brought whole bodies of bloodied monkeys sealed in plastic bags, and a somewhat nervous Peters had to ferry them back to the army laboratory in the trunk of his car. To their horror, the army scientists would now confirm that the lethal epidemic raging through Hazelton's quarantine facility was Ebola fever. Reston was a commuter suburb of Washington, D.C. Here was the scenario everybody had feared since Marburg: America could be on the brink of a human epidemic caused by the most dangerous of all BSL-4 viruses.

Although USAMRIID had not been involved in the African Ebola outbreaks, the lethal nature of the virus made it inevitable that they would be interested. There had been a series of press and magazine articles suggesting that Soviet laboratories were breeding "Ebola germs" for warfare purposes. On January 31, 1978, an article reported by Reuters claimed that American spy satellites had taken photographs of several establishments near Moscow and in the western Soviet Union that, according to intelligence analysts, were biological research and production centers. The photographs purported to show heavily guarded complexes linked to railway lines carrying tanker wagons. The article described how the Russians were adapting Ebola to make it even more lethal, a truly terrifying prospect.[5]

It was hardly surprising that Fort Detrick already had an Ebola laboratory, run by a civilian scientist, Eugene D. Johnson. Peters and Johnson had for many years been conducting experiments on Ebola in primates, searching for effective vaccines or drugs that might be used in treating the illness. No vaccine or treatment had come of these researches, though an important level of understanding had emerged. The terrible lethality seen in Zaire and the Sudan had been no accident. The virus was armed with a frightful propensity to break down and defeat many of the body's conventional defenses. It was relatively resistant to the antiviral actions of interferon. Even immune serum, the agent in which doctors in 1976 had placed so much faith, had proved ineffective. It seemed that, like AIDS, the virus induced immunosuppression in its victim, so the immune response to its own presence would be nullified. Such properties were unlikely to have evolved accidentally: Peters and Johnson had developed a profound respect for the Ebola virus.

General Phillip Russell, head of medical research for the U.S. armed forces, called Fred Murphy, who was now director of the Viral and Rickettsial Division at the CDC. By morning Murphy had arrived in Maryland, together with Joe McCormick, in his capacity as head of the Special Pathogens Branch. It was agreed that the CDC would oversee civilian containment while USAMRIID would handle the source of infection at the monkey quarantine facility. The army scientists wore civilian clothes as they drove into Reston in unmarked vans, changing into BSL-4 field protection in the parking lot. The tall, silver-haired Jerry Jaax was one of them. He would subsequently describe the feeling as like that of going into war. They anesthetized the animals with an intramuscular injection, bled them, then sacrificed them. The scientists moved on to inspect monkeys in a second quarantine room, room H, where they discovered more sick animals. Testing of these confirmed that the infection had already spread to this room, which baffled the scientists, since room H had no open connection with room F.

It was apparent that a lethal Ebola epidemic was already spreading throughout the facility. The potential threat to people was so overriding that it would be necessary to sacrifice all four hundred remaining animals. While the scientists were in the middle of this slaughter, one of the employees fell sick. The Virginia Department of Public Health was now alarmed at the increasing possibility of a human epidemic.

A detailing of the events at Reston is the basis of Richard Preston's spine-chilling bestseller, *The Hot Zone.* The huge trawl of sacrificed animals were individually bagged, a proportion retained for testing, the remainder boxed, sprayed with bleach, and taken to an incinerator for cremation. The building itself, contaminated with blood, feces, and other secretions, all teeming with virus, was hermetically sealed and sterilized using paraformaldehyde crystals heated in open frying pans, which released a toxic vapor fatal to all life within its walls. Of crucial significance was the potential of the virus to hop species, potentially to infect people. The employee who had taken ill was the caretaker most intimately involved with the caged monkeys. His illness appeared flu-like, and he was seen vomiting on the facility lawn. After admission to a local hospital, where he was kept in routine barrier isolation, antibodies to the Ebola virus appeared in his blood. He developed no symptoms of Ebola hemorrhagic fever; instead he made an uneventful recovery.

Perhaps the most alarming aspect of the Reston epidemic was the suggestion, in its movement from room to room, that Ebola was being transmitted from monkey to monkey by aerosol. Although suspicious that this was the case, Peters and his colleagues had no conclusive proof. The contagion might, for example, have been transmitted through syringes, if not needles, which had been shared during antibiotic and vitamin injections of the monkeys in the facility. Though the suspicions remained, the exact mechanism of transmission "could not be confidently established from investigation within Hazelton itself."[6] After a pause, importation of monkeys began again to the Reston factory. The first two shipments of macaques resumed in January 1990. The next three shipments, all from the same supplier in the Philippines, arrived in February and March. All three contained animals dying from the Ebola virus. Not only did it confirm that Ebola was at large in the Asian rain forest, but it gave the scientists at USAMRIID a further vital opportunity to study the means of monkey-to-monkey transmission. On this occasion, in the words of Peters and his colleagues, "disturbing clinical and pathological observations were made."[7]

Infected monkeys had an unmistakable respiratory pattern of disease: a pneumonia, with severe coughing and copious nasal and pulmonary secretions. When scientists took nasal and throat smears, the secretions teemed with Ebola virus. On electron microscopy the virus was seen budding from the alveolar lining, and copious collections of free viruses were seen in the air sacs. These are the appearances of any upper respiratory infection complicated by pneumonia.[8]

Dan Dalgard, the veterinarian in charge of the facility, had learned a hard lesson during the first epidemic. Before the factory reopened, he took extreme precautions to prevent any similar spread between monkeys and from room to room. Yet the new virus did spread from the room containing the new shipment of monkeys to another room containing previously healthy animals. More alarming still, four of the five staff regularly in contact with monkeys developed antibodies in their blood: incontrovertible evidence that the monkey virus had trafficked across the species barrier. Although one member of staff may have infected himself through a cut finger, the other three had no such defined exposure. It was extremely fortunate that, as in the earlier simian epidemic, no serious human illness emerged.

The dense accumulations of virus in the pulmonary secretions and the relentless spread of the infection within the rooms of the factory "clearly established the ability of this filovirus to spread from monkey to monkey and even from monkey to man by droplets and/or small particle aerosols." In Peters's words: "The seriousness of the efficient spread of a filovirus by such mechanisms cannot be overestimated."[9]

3

Dr. Bernard Le Guenno directs the hemorrhagic fever laboratory at the Pasteur Institute in Paris, which is hidden away in a small building to the left of the original structure where Pasteur himself worked. Le Guenno is a tallish, dark-haired man, lean and gently spoken. In late November 1994, he was faced with a taxing diagnostic problem.

For fifteen years a group of Swiss ethologists had been studying the behavior of a colony of chimpanzees, *Pan troglodytes,* living in the Taï National Park. This is a protected zone of African rain forest in the Ivory Coast. In 1987 they counted a healthy eighty members of the colony, yet by 1995, this number had crashed to a mere thirty-three. There had been two dramatic episodes of decimation, the first in November 1992 and the second in November 1994. Several dead chimpanzees were found with evidence of hemorrhages from their eyes, nostrils, and mouth, but decomposition was too advanced to allow the collection of useful biological samples. On November 16, 1994, the ethologists found the body of a freshly dead animal, and Brigitte, a senior member of the team, performed an autopsy in the field wearing a dilapidated pair of kitchen gloves.

Samples taken from the internal organs were fixed in formalin, precluding any viral extractions, before being passed to Pierre Formenty, a local veterinarian who forwarded them to a veterinary pathologist in Nantes. Blood samples were also taken from three living chimpanzees, two older males and one female, and from members of the ethological team. These were forwarded to Le Guenno for testing.

On November 24, the thirty-four-year-old Brigitte developed a fever, with severe muscle pains. She was admitted to a hospital in Abidjan when her fever was found to be resistant to antimalarial drugs. By now she had diarrhea and a rash. Continuing to deteriorate in spite of all attempts at treatment, she was evacuated to Switzerland five days later. In the meantime, Le Guenno began testing the blood samples. The modern system used to identify specific viruses and to differentiate different strains of the same virus is called the ELIZA test. The capture assay devised by Tom Ksiazek is a good example. The patient's serum is serially diluted and the dilutions are placed in a multiwell plastic plate, which contains standardized quantities of the virus (the antigen). A positive reaction is detected not by a fluorescent label, as in the older tests, but by a color change. Fixed antibody has an enzyme attached to it, which produces the color. The strength of the reaction — the quantity of attached antibody — can therefore be measured using a colorimeter. Few laboratories in the world stock ELIZA reagents for testing unusual plague viruses. In Le Guenno's words, "At that time I had no reagent to perform an ELIZA for Ebola, Lassa, or Marburg viruses."

He did have these reagents for a number of other viruses, so he could accurately screen for Rift Valley fever, Crimean-Congo hemorrhagic fever and the main arboviruses, such as yellow fever, chikungunya, and dengue. All of these tests proved negative. He was also able to screen the sera for Ebola, Lassa, and Marburg viruses using an old-fashioned immunofluorescence test using slides that had long been stored in his department: these tests also proved negative.

Brigitte's blood had been drawn on November 27 during her high fever, when she was first admitted to the hospital in Abidjan. As a result of local power shortages, her serum had been kept alternately at room temperature and 37°F for fourteen days, sufficient to kill most viruses and render testing impossible. Despite this, Le Guenno now attempted viral isolation by inoculation into Vero E6 cells, derived from monkey kidney, and a culture called AP61, derived from mosquito tissue. There was no visible cytopathic effect even after twelve days of culture. He carried on, blindly passing homogenates into further cultures.

At length, in subcultures drawn from the original Vero flasks, he noticed a cytopathic effect: some cells were now becoming refractile and detaching from the flask walls. With some excitement, he tested the damaged cells using the limited immunofluorescence assays available to him, but the test proved negative.

Brigitte was making a slow and difficult recovery in a hospital in Basel. Her Swiss doctors sent on a convalescent sample, which could be expected to contain high levels of antibody to the virus that had threatened her life. Le Guenno now tested this serum against Vero cell cultures he had infected with the unknown virus. Using immunofluorescence, the only screening test available to him, he found that when he added this serum to slides made from the cultures, large virus inclusions lit up brightly within the cells. It was his first indication that the virus infecting Brigitte might be either a bunyavirus or a filovirus. Le Guenno was astonished — "I could not believe this!" But when the infected Vero cells were examined under the electron microscope, they showed the unmistakable serpentine forms of a filovirus.

At this stage, the limitation of the testing agents available to him made a more precise diagnosis difficult. He could take the isolation no further until he managed to obtain specific ELIZA reagents to all known

Ebola and Marburg viruses from the redoubtable Tom Ksiazek at CDC. Now, using a modification of Ksiazek's extraction techniques, he confirmed he was dealing with an Ebola virus that had some cross-reactivity with Ebola Zaire. Further testing suggested the startling possibility of a new strain, Ebola Ivory Coast. This virus had remained alive in Brigitte's blood in spite of poor storage, difficult and protracted transport of her serum specimens, and two cycles of freezing and thawing of her sera. It was an astonishing and worrying example of endurance through extreme adversity.

There were two lessons to be learned from the Ebola outbreak on the Ivory Coast. The viral isolation had been very difficult, even in the hands of an expert, and in one of the few hemorrhagic fever virus laboratories still remaining in the world. How often, one wondered, were Ebola and other rare viruses being misdiagnosed in a Third World setting? One more worry to be added to the intractable delays experienced in responding to both African outbreaks. The second lesson? The virus had been witnessed in the wild, in what was presumed to be its natural habitat, the African rain forest. Were we glimpsing its real role in nature in this? If so, how many other such animal epidemics was it causing in the jungle, dark harvests of lethality that went completely unrecorded by man?

The sera from the other ethologists, like the blood samples from healthy chimpanzees, were all negative to testing.[10] As he processed this startling new addition to the growing library of information on Ebola, little did Bernard Le Guenno realize that within a few months he would be traveling to Zaire to help investigate another terrifying outbreak.

On May 11, 1995, Ebola erupted in Kikwit, a Zairian city of 500,000 people, some 225 miles east of Kinshasa. Its focus on the city's municipal hospital was a poignant confirmation of Sureau's prophecy, as was the fact that initial reports were predictably confused and varied. In a heartbreaking repeat of history, photographs appeared in the newspapers depicting the sealed coffin containing the body of fifty-eight-year-old Sister Dinarosa Belleri, a missionary sister from the Catholic order of the Little Sisters of the Poor, and the fifth Italian nun to die from the Ebola virus, being buried by health workers wearing gowns, gloves, and masks.

Banner headlines covered the day-to-day progress in every national paper. In the *Daily Telegraph* of May 11, it was a "plague worse than AIDS that turned bodies to water." In the *Sunday Times* of May 14, the "disease-busters," headed by Pierre Rollin, had arrived in Kikwit on May 11, to be photographed carrying a portable glove-port cabinet and clad in "chemical warfare suits." Although the virus and the methods needed to contain it had been common knowledge in expert circles since its emergence in 1976, this did not prevent a tidal wave of fear. Troops sealed off Kikwit, where ninety people had died at the time of first diagnosis; meanwhile, a second outbreak was suspected at Musango, a hundred miles away. By May 13, a BBC news dispatch reported, "A wall of roadblocks around the town of Kikwit, where the epidemic started, is thought now to be creating unnecessary panic."

The virus did not appear to be spreading by aerosol. Consequently, the likelihood of its spreading beyond Zaire, or beyond Africa, was small, and authorities within Zaire were constantly reassuring the crowds of arriving media. But this did not prevent a state of international alert. Belgium was monitoring all flights. In Britain, where the Department of Health was circulating port medical authorities and infectious diseases consultants with measures of diagnosis and containment, people were being "strongly advised" by the Foreign Office not to travel or pass through the country. All passengers arriving from Zaire into France and Portugal were being screened for symptoms of the infection. In Britain, as the death toll in Zaire continued to rise and the index case still defied discovery, Roger Highfield, in the *Daily Telegraph,* asked if a "doomsday mutant" was the next surprise to come from the Ebola virus.[11]

As the epidemic continued, involving at least three towns and their hospitals, teams of volunteers wearing surgical gowns, rubber gloves, and masks toured the towns each day, collecting the dead. Local people were so terrified they ignored their traditional practices, abandoning the dead in their huts or by the wayside. Western journalists in Kikwit filmed the grim cargo of a truck arriving at the Kikwit hospital with an old woman, still alive, propped in a wooden chair among the corpses.

At the heart of the epidemic, Jean François Ruppol in Kinshasa and Pierre Rollin and Bernard Le Guenno in Kikwit struggled not only to contain the plague but also to discover the index case. Kikwit was hundreds of miles south of Yambuku. There appeared to be no connection at all between them. Yet the pattern of the new epidemic was the very replica of that seen in 1976. In Kikwit 44 percent of the dead were hospital workers, though this time spread through contaminated syringes and needles was denied. Failure to adopt simple barrier nursing procedures had been a key factor in the deaths among hospital employees.

On May 22, Bernard Le Guenno made the breakthrough in discovering the real patient zero, a middle-aged man whose work took him into the rain forest as a charcoal burner. This man had been infected as long ago as the previous December, when seven members of his household of twelve had also contracted the disease and died from it. An attenuated chain of contagion linked this man with the subsequent amplification in the Kikwit hospital. In 1995, almost twenty years after the first discovery of the Ebola virus, and following the desperate consequences of those earlier delays, it had still taken six months for the authorities to register the epidemic and call for expert international assistance.

By June 22, the epidemic was over. The official WHO total was 316 cases with 244 deaths, making this the most lethal epidemic of Ebola since 1976. Biologists from the CDC arrived in Zaire to search once more for the source of the virus in nature. As yet that search has proved unfruitful. In the words of a bemused Karl Johnson: "Ebola is the classical example of the emerging virus. The world knew nothing about it. All of a sudden it happened, as if it came from the sky."

That unanswerable mystery remains today and it perpetuates its own continuing threat. It means that until we discover the virus reservoir, and with it a new understanding of the natural cycle of Ebola, we have no idea when and where the virus will appear again. Whatever the reservoir, whether bat, primate, rodent, or other animal, and whether there is an arthropod vector or not, there are strong indications of its origins in or around the African and Asian rain forests. In itself this is an important conclusion, a piece in a slowly assembling jigsaw puzzle of understanding. Even as the biologists have been searching for the natural host of the virus, a complementary search has been taking place into the molecular biology of the virus. That search is providing another important piece of that extraordinary puzzle — a fascinating revelation through the window on life.

PART TWO

The Window on Life

TWELVE

Origins in Nature

"What a magnificent view
one can take of the world."
The vast sweep of law controlled
the climate, the landscape,
the changes in animals and plants,
with everything synchronized
"by certain laws of harmony."

A. DESMOND AND J. MOORE
Darwin[1]

1

A new plague virus such as Marburg or Ebola emerges and kills people. But where exactly does it emerge from? This is perhaps the most fundamental question one can ask of emerging viruses. Plague viruses do not fall out of a blue sky. Equally unlikely is their origin in outer space. The truth is a good deal more mundane, yet all the more fascinating for that: viruses come from the diversity of life on earth, they are an integral component of the complex and interacting web of nature.

Although many people see viruses as the ultimate in predators, in reality viruses are rarely so malevolent. Their purpose in infecting a host is nothing more than the propagation and preservation of their own species. That a virus should need an animal or plant host should not come as a surprise. Viruses, by their very nature, have a unique vulnerability. Outside the host cell they become inert: they cannot reproduce — indeed they lose all attributes of life. No other form of life is so dependent. Yet, like every other form of life, viruses crave the security of an ecological niche — a home — in which they can enjoy a certain security of tenure while still evolving.

In 1935 the first step toward enlightenment came when a scientist called Erich Traub discovered a virus now known as lymphocytic choriomeningitis virus (LCMV). In the words of Joe McCormick, the world of international virology and immunology was suddenly presented with a veritable *chemin d'or* of wonder and discovery.[2] That virus was the first of a family we now call the *Arenaviridae*. Traub, who isolated the virus from mice, was the first to recognize that it produced "a persistent infection in the natural rodent host, with lifelong virus excretion but little consequence to the animal." Yet, when he injected it subcutaneously into guinea pigs, it killed 80 to 90 percent of them.[3]

Of all the strange facets of his discovery, this lack of injury to the host animal from persistent virus infection appears of fundamental importance, yet it has been the least studied over the three generations that followed.

Tens of thousands of people are infected by this virus each year, through contact with infected mice. Thankfully, infection in humans is often either asymptomatic or causes a mild systemic illness, with fever, headache, and muscle aching — though rarely it may cause a nonfatal type of meningitis, and even a possible teratogenicity.[4] Only a minority of house mice are infected, and the public health threat is slight, so the human implications have been little studied. Since the discovery of this virus in mice, however, a number of other arenaviruses have come to light, many a good deal more dangerous to humans, and some that will invade the brain and central nervous system.

In 1962, then just six years after graduation from medical school, Karl Johnson arrived in the Middle America Research Unit (MARU) to study tropical viruses. Based in the Panama Canal Zone, MARU was a satellite of the U.S. National Institutes of Health at a time when science, imbued with a laudable spirit of global benevolence, could still obtain funding from Western endowments to study and contain epidemics in poorer countries. Shortly after his arrival, Johnson was invited to accompany the epidemiologist Ron MacKenzie on a journey into the wilds of Bolivia, where they planned to investigate a mystery epidemic that was killing people in the eastern countryside. The Bolivians called the epidemic *El Typho Negro* — black typhoid.

MacKenzie had already made a series of expeditions to the plague zone, involving hair-raising flights from La Paz and sorties by dugout canoes through alligator-infested rivers. He had pinpointed the epicenter to the neighborhood of a river called Machupo, before returning to America to recruit help. This was how he came to return with Karl Johnson and a biologist, called Merl Kuns.

Setting out in May of 1963, Johnson carried with him a newly designed portable glove box for viral isolations that had been designed with the cooperation of USAMRIID. They were ferried to the Machupo area by an old USAF B-17 bomber that landed in a field next to the town of San Joaquín. Here they found a population of 2,000 people devoid of health facilities, telephone, sewers, electricity, or running water. San Joaquín, lacking even the hospital facilities of Maridi or Yambuku in the depths of the African rain forest, would typify the wilderness origin of most newly emerging viruses during the second half of the twentieth century. The townspeople were Spanish cowboys or local Indians, surviving on their cattle, home-grown rice, corn, and vegetables. Their contact with the outside world was limited to traffic on the Machupo River, which drained into the Amazon basin. There were many similarities to the situation in Zaire during the 1976 Ebola epidemic, the irredeemable poverty, the rural isolation, the recent felling of rain forest to clear land for cultivation. Half the local people had already been infected, and of those, MacKenzie estimated that half had died.

Autopsies showed extensive tissue destruction accompanying a hemorrhagic meningitis, with a heavy blood-staining of the cerebrospinal fluid: a feature that classed the epidemic as one of the hemorrhagic fevers. Tissue samples from a dead child revealed an infective agent small enough to pass through filters that would trap a bacterium. From this they deduced that they were dealing with a virus.

Kuns enlisted a local army of volunteers to trap animals and insects while Johnson set up a laboratory, using his mobile glove box. He began to search for a likely virus.

It was slow and painstaking work, involving the trapping of thousands of animals, sacrificing them and dissecting their bodies for evidence of disease, then culturing tissues and sera for evidence of viral damage in the cell cultures. In early July, MacKenzie and a Panamanian laboratory technician called Muñoz showed signs that they had contracted the virus themselves. The two sick men were flown out on July 4 by a USAF C-130. As soon as his colleagues had left, Johnson realized that he too had early symptoms of the potentially lethal infection. There was no possibility of another evacuation, so he had to make an arduous journey across Bolivia, Peru, and Colombia before he was finally stretchered into the Gorgas Hospital in Panama. There he sweated and burned alongside MacKenzie and Muñoz, all three waiting to see if they would live or die.

They were treated with intensive supportive measures by a physician experienced in dealing with hemorrhagic fevers. While he was in the hospital, Johnson was nursed by his wife, Patricia Webb, who picked up the infection herself from the intimacy of nursing him.[5] All four eventually recovered.

Johnson, Webb, and MacKenzie courageously returned to Bolivia, where they discovered the virus, now called "Machupo," that was causing the epidemic. It was transmitted in the urine of a wild field mouse, *Calomys callosus.* People became infected when they swept floors, raising a dust that contained the dried excretions of the rodent. Johnson and his colleagues also made an interesting deduction about the nature of the infection in the mice. Although they had paid local boys to capture "sick" living animals, the boys found few likely candidates. Like Traub and his earlier discovery, the scientists were surprised to find that in its host mouse infection with the Machupo virus "may often be relatively asymptomatic."[6]

Why should a virus that causes an asymptomatic infection in the local rodent produce a lethal epidemic in people?

Suddenly, we find important parallels with the hantavirus epidemic. It seems that the local population of mice had proliferated over the preceding four years, and this "plague of mice" had coincided with the onset of the epidemic. Taking their cue from this hard-won discovery, researchers from the Rockefeller Foundation and the University of Buenos Aires arrived at a similar conclusion for a different hemorrhagic fever virus only recently discovered in Argentina. This too had been named after a local river, the Junin. Both the Machupo and Junin viruses were now classed in the family of *Arenaviridae. Arena* is the Latin for "sand"; these viruses coat their surface envelope with the granules of host-cell ribosomes, giving them the appearance of being sprinkled with sand when viewed through the electron microscope.[7]

MacKenzie was convinced that the epidemic had started at a time when there was a local scarcity of cats, which could have been caused by DDT poisoning, following malaria eradication. But the biologist, Kuns, remained puzzled. He doubted that the explanation was so simple. It seemed to him that while the epidemic was caused by the virus, and that the virus host was undoubtedly the field mouse, there must be additional, if as yet mysterious, causative factors — perhaps some more subtle biological or ecological implications — they had failed to discover.

2

In 1969 another hemorrhagic fever appeared out of the blue at an American mission in a village called Lassa in eastern Nigeria. A missionary sister carried the infection into the neighboring town of Jos, where it spread as a lethal nosocomial outbreak among the staff in the Bingham Memorial Hospital. Those infected developed a high fever, muscle aching, and prostration. Hemorrhagic blotches peppered their skin, often a predication of delirium or coma or the onset of generalized convulsions. Blood and swabs from the sick and autopsy samples from the dead victims were forwarded to Jordi Casals, a Catalan scientist in charge of the famous arbovirus laboratory at Yale University in New Haven. Casals, working with Robert Shope, had for many years been amassing a collection of emerging viruses. Although he lacked modern BSL-4 isolation facilities, he was an experienced virologist who worked with mask, gown, and gloves. In spite of these precautions, Casals became infected with the agent he was testing. A technician also became infected and died from it, but Casals survived with the help of immune serum, and he returned to the hunt, subsequently isolating the virus that was causing the outbreak.

That virus is now known as Lassa after the little mission where it first emerged.

After recovering from his infection, Casals began a systematic investigation of the virus. Confirming it was a new arenavirus, he noticed that same curious feature as Johnson when dealing with Machupo. Mice infected with the virus became factories for virus amplification, but they showed no symptoms of infection. Meanwhile, exactly a year after its first appearance, Lassa fever erupted once more at the small hospital in Jos, killing patients and staff and even the resident doctor, Jeanette Troup. The malignant nature of Lassa fever was now alarming virologists around the world.

Subsequent investigation revealed that the virus could be transmitted by inhalation of infected particulate matter, notably dust contaminated with urine from mice infected in the laboratory. It was also very efficiently transmitted by blood products and could enter the human body through contamination of damaged skin. It was vital that they discover the manner of spread and natural reservoir of the virus, so a team of investigators, under the direction of Casals, arrived in Nigeria. But though it was easy to demonstrate the nosocomial spread on the hospital's wards, the natural reservoir of the virus proved elusive. Equally worrying was the fact that most of the infected were Africans. It surprised Casals to find that local people were no more immune than the Western missionaries.

At Yale, the death of Casals's laboratory assistant, coupled with what was happening in Africa, worried the authorities. Casals was ordered to pass on all further research on the virus to the Centers for Disease Control in Atlanta.

At this time Tom Monath, a Harvard graduate with an interest in tropical diseases, was running the fledgling Special Pathogens Branch. Though a physician, Tom had retained a fascination with biology that had taken him as an undergraduate on zoological expeditions to all the corners of the world. His thesis had focused on salamanders, and he had studied reptiles and amphibians from the Ethiopian desert to the Amazon jungle. One day at Harvard he attended a lecture by Telford Work, who had discovered one of the first known emerging viruses, a flavivirus called Kyasanur Forest virus, from a sick monkey in the Indian state of Mysore. Spread by ticks from rodents and even bats in the wild, this disease, which is fatal in 3 to 5 percent of cases, is still slowly extending, with thousands of diagnosed human victims. The young Tom Monath was enthralled by Work's story and intrigued with the arboviruses generally because they appeared to live in a matrix of interacting, sometimes fantastic, life cycles, including vertebrates and invertebrates. The Kyasanur Forest virus could even be transmitted within the ovaries to the future eggs of the ixodid tick, whose bite continued the cycle to its next host.

Work had established the arbovirus program at CDC, and so the intrigued Tom Monath now moved to Atlanta.

In 1970 the five eight, wiry, blue-eyed scientist was sent to Nigeria, which was suffering a horrific epidemic of yellow fever. At least 100,000 cases had already been confirmed and the mortality varied from 20 percent to a horrifying 50 percent. In April of 1972, while still in Nigeria, he received a telegram from Atlanta asking him to travel to Liberia. An American nurse had died in Zorzor. The cause of her death had been confirmed as Lassa fever.

Monath had been involved in some of the early research on the virus in America, but this was an exciting opportunity to investigate the virus in its heartland. He arrived in Zorzor to find the hospital and surrounding district stricken with panic. The nurse, five patients, and eight members of the staff already had the fever. Monath quickly established that this was a hospital-based spread. Every one of the infected could be traced to a single index case, a pregnant woman who had been admitted to the hospital with vaginal bleeding. The nurse had become infected during a bloody dilation and curettage, and most of the other patients and staff came from the same obstetrics ward. The epidemiology proved easy enough, but it was a different matter when Monath attempted single-handedly to discover the natural reservoir. "I spent a lot of time collecting animals and dissecting them by candlelight — in retrospect a pretty hairy thing to do." Once again, the search for the natural source of the Lassa virus was unsuccessful.

Following the outbreak, Monath returned to Atlanta and took over running the newly modernized Special Pathogens facility, the first of the purpose-designed "hot labs." That autumn of 1972, Lassa fever broke out again in West Africa, this time in the diamond mining communities around Panguma, in eastern Sierra Leone. From the number of cases, it appeared to be a major outbreak. It was a second chance to investigate the biological mystery of the viral origin in nature.

Flying into Sierra Leone in September, Tom Monath discovered a very different situation from any of the previous outbreaks. Although there were some people who had picked up the virus in the local hospital, most of the infected individuals had picked up their infections out in the community. He mapped out the geography of the epidemic, pinning down the towns where the cases came from and focusing the ecological hunt on case households and surrounding areas. The known ecology of other arenaviruses made him suspect rodents, possibly bats, so he focused on those. A laboratory was set up in the field to process the sacrificed animals. Blood and tissues were taken from more than six hundred animals, preserved in liquid nitrogen, the specimens indexed to the preserved corpses of the bats and rodents so that absolute identification could be confirmed later. Monath did not know until he dissected out the mass of tissues back in Atlanta that this time the field investigations had proven successful.

"We just went through the laborious process of grinding up these huge quantities of mouse gizzards, inoculating Vero cell cultures and then looking for virus by immunofluorescence," he recalled. From the miscellany of captured rodents, just a dozen were found to contain the virus. Yet for Monath, those dozen were very significant findings, the sublimation of all of his hard labor. All dozen positives came from the same species, a small brown rat called *Mastomys natalensis.*

Mastomys is one of the commonest rodents throughout the broad sweep of sub-Saharan Africa, forced to compete with the larger and more aggressive black rat. In an interesting parallel with observations made in South America by Jamie Childs, when investigating the population dynamics of the Machupo-carrying *Calomys callosus,* Monath showed that *Mastomys* was in fierce competition for territory with other species, particularly the black rat. The virus epidemics coincided with times when the rodent population also increased, which in itself implied aggressive success in this evolutionary war. And like the other emerging viruses, once arrived, Lassa fever never went away. Today, among the new viruses, it ranks second only to HIV in the number of people it has infected and killed; with a widening perception of its high infectivity, however, the percentage of human lethality among the infected is now much lower than originally suspected, at 10 percent or less. Like the hantavirus in America, it has become an endemic human disease affecting hundreds of thousands of people yearly in Sierra Leone, Liberia, Nigeria, and Guinea, with lesser infection in many other West African countries.

And like the *Sin nombre* hantavirus, there are continuing mysteries. Why, for example, has the virus not as yet extended to the full geographic range of the rodent host. There are other revealing, while also disquieting, parallels.

3

The Lassa fever virus is not new to nature. There are related viruses in southern Africa, called Mopeia and Mobala — and IPPV in the

Central African Republic. Even among the same species, significantly different substrains of the virus infect the rodents from different countries and different geographic regions of West Africa. In the opinion of Tom Monath, the virus must have been spreading among the rodents for several thousands of years, and possibly a great deal longer.

Mastomys is a species of rat that has a strong association with human society. Anthropologists have found its remains in intimate association with human activity from prehistoric times. In Monath's reflection: "I think it is worth emphasizing that the rodents associated with viral infections really are grassland species and to my mind this is telling us something important in terms of association with human civilization." The implication is that agriculture has been a major factor in the exposure of humans to rodent-borne viruses. It is seen, for example, not only with the South American arenaviruses, but also with the different species of hantaviruses that infect people the world over.

There is another aspect of the Lassa virus and the arenaviruses in general that fascinates Tom Monath. "It all fits the same pattern that these viruses either cause minimum or no discernible pathology in the rodent hosts."

Many other arenaviruses have emerged as the cause of hemorrhagic fevers in South America and Africa, including Guanarito virus in Venezuela and Sabia in Brazil. With the exception of a virus called Tacaribe, which is not known to cause human disease and that is hosted by the *Artibeus* species of bat, all the others are maintained exclusively in rodents. The machinery for immunological tolerance in the host can be ingenious. Machupo and Junin have been shown to infect the young rodent while it is still immunologically naive in the womb. When this happens, the virus does not cause disease later on in the growing animal; the animal's immune system does not perceive the virus as a foreign invader. It seems that LCMV, Machupo, Lassa, Junin, a virus called Latino found in Brazil, and another called Tamiari found in Florida, all have this same facility to establish "persistent tolerant infection."[8] But if the macrobiological explanation is wonderfully simple, the precise molecular biological explanation for this immunological tolerance has proved amazingly complex. For example LCMV and Machupo can even enter the mouse ovaries, where they are transmitted to all subsequent offspring.[9]

And even with that same virus and rodent species, there are inconsistencies that are difficult to explain. For example, the LCM virus, so benign to one strain of mice, may be lethal to others; or a virus that, like Junin, causes no disease to a fetus or infant may be lethal if it infects an adult of the same species. Even more bizarrely, the Machupo virus, so benign in some mice, can infect other adult mice of the same species and prove lethal, or if nonlethal, will destroy the ovarian germ cells, rendering the females permanently sterile.

These do not appear to be random patterns. Tom Monath wonders if they are an expression of genetically different strains of the mouse species — if so it would suggest an extraordinary refinement of selectivity on the part of the virus.

How interesting that in Traub's first paper, at a time when the electron microscope was yet to be invented, he drew attention to this curious lack of pathogenicity of the virus for the rodent. It is clearly a common, though variable, pattern within the arenaviruses and hantaviruses. How strange also that those same great families of viruses should behave in such an aggressive, often fatal, pattern when they attack humans!

There is a very important extrapolation from this understanding to the origins of new or emerging viruses. Stephen S. Morse, at the Rockefeller University in New York, has contributed greatly to our understanding of this through his editorship of two books that deal with emerging viruses and their evolution.[10] For many decades, virologists assumed that new viruses emerged as a result of mutation. This suggested a fundamental unpredictability — we could never anticipate new epidemics. But Morse's view is entirely different: through understanding of the evolutionary origins of new viruses, "the seemingly insoluble problem of viral origins thus reduces to a more manageable question of viral traffic [between species]."[11] Taking Morse's analogy, what we need to discover are the rules of the highway code.

Emerging viruses, while they may be new to humanity, are never new to nature. They enter human consciousness when they hop species to humanity from a natural host, and most of the serious emerging viruses do so in a wild or rural setting. There is accumulating evidence that all or virtually all of the million or more species of animal life on earth have their own individual viruses. This global reservoir of origins is therefore vast, the biome of life itself. And since, in the opinion of Edward O. Wilson, the eminent Harvard naturalist, two thirds of all the species of life on earth live in the rain forests of Africa, Asia, and Central and South America, it is more likely that new viruses will "emerge" from the rain forests.[12]

Because of their essential nature and the genomic landscape they inhabit, viruses are no less dangerous for our growing understanding of their biological needs and behaviors. But they are also uniquely vulnerable. They cannot come to life other than within the cells of their destined host. Consider the uncertainty this creates.

To survive, a virus must constantly adapt and evolve. Every act of infection involves a battle for access, a means of circumnavigating the phagocytic defensive system, followed by the adaptation to the adamantine set-play rules of the genomic landscape itself. Once established, it has to compel the reluctant genetic mechanism to adopt the progenitive program of its own genome, then outwit the attempts of a now responding and extremely resourceful cellular and humoral immune system in order to survive once outside the amplifying cells. It has to fight off myriad such forces bent upon its destruction while discovering a way of leaving its host and finding another. One solution to this, perhaps a very common solution, is for the virus to discover a way of cohabiting with the infected host for a very long time, perhaps for the remainder of the host's life, with its implications of *de facto* immortality through an evolution of vertical transmission from parent to offspring.

The inference of such a relationship between two such dissimilar forms of life is a much greater intimacy of give-and-take than the anachronistic notion of simple predatorial aggression. Such refinements of evolutionary partnership must flourish in the fields and jungles, in the oceans, the soil, and the air, where over the aeons of evolution, viruses have explored a multitude of wonderful, if often mysterious, genomic archipelagoes, establishing relationships with every species, extant or extinct, that has ever evolved on the earth. But what then is the true nature and implication of these relationships?

Why then do emerging viruses, such as Ebola or Machupo, strike with such immoderate aggression on their first appearance as infections in people?

4

Many experts view this as the arbitrary consequence of people straying into the alien cycle of virus and natural host. The lethal attack, according to them, is nothing more than an accident. I doubt, however, that this is the complete explanation. At the same time it is a mystery that must be solved, since new viruses are emerging to infect humanity rather more frequently than most people realize. To investigate why, the logical thing to explore is the relationship such viruses have with their hosts in nature. For it is out of the very intimacy of such an ecology, the genomic responses to the natural laws that govern that interesting relationship, that the virus, with all of its lethality, emerges.

Consider the vulnerability of a virus that pursues a determinedly aggressive infective cycle. Its entire evolutionary program is directed toward a war against the immune system of its host. If the host evolves or adapts an effective barrier to any stage of the viral life cycle, the virus will become extinct. Such, in a more subtle sense, was the case with human smallpox. It is a unique part of our human defenses that we complement natural immunity with weaponry devised by our reason. Such weaponry — the vaccine against smallpox — resulted in the extinction of the human smallpox virus.

Yet, if on the other hand the virus wins this war — if it totally annihilates the immune defenses of its host, as for the moment does HIV-1 — all of its victims die. Victory is Pyrrhic, since over time this also condemns the virus to extinction. Admittedly, there are experts, notably Robert May, who have demonstrated, through mathematical projection of unbridled natural selection, how an unrelentingly aggressive behavior can be the evolutionary outcome for an emerging virus.[13] A worst-case scenario drawn from such a grim evolution would be very frightening, and I shall return to discuss it later. How likely is such a terrifying situation in practice? Remember the perceptive question of the late Bernard Fields, after his colleagues had asserted both the plenitude of viruses in nature and their capacity to mutate thousands of times more rapidly than the genomes of their hosts. "Why then," demanded Fields, "hasn't this enormous diversity wiped out the earth?"

Were such an unrestrained aggression commonplace, then Fields's wise question would long have been answered: for virus would have competed with virus for increasing rapacity and virulence. The threat of such behavior would have been truly awesome. Life on earth, in any meaningful way, would be long extinguished. There is a simple explanation why such an apocalyptic evolution did not happen. It would have proved a perilous long-term plan for viruses too, and so it is one that is rarely seen even in microcosm in nature — though, admittedly, it is not impossible. What we actually observe is a striking difference between the aggressive behavior of a virus in a totally novel host, such as humans, when compared to its seemingly benign behavior in its long-established host.

Terry Yates's colleagues used to kid him by asking, "Whatever are you going to do with your collection of dead rats?" Terry would defend himself with innate good humor: "One day, we will be able to read the information from their chromosomes. When that day arrives, these will prove to be a genetic gold mine." Terry Yates is a systematist. A slim and hyperactive Kentuckian who speaks passionately, articulately, and with a singularly innovative intelligence, he is the Professor of Mammals at the University of New Mexico. He is also the director of the University of New Mexico Museum of Southwestern Biology.

He does not focus, as most mammalogists, on a single species or even a genus. Instead, his interest lies in evolutionary genetics: in the patterns and processes that give rise to the diversity of animal species, and in particular how this expresses itself in phylogeny — the branching of mammalian evolutionary history. "By observing these patterns, by using various genetic parameters, we are going to be able to compute the genetic history of life."[14]

The museum, which houses his office, has a cavernous atrium with walls that are lined with the heads of exotic animals and a floor space crammed with cupboards. The floor cupboards are lined with wooden drawers that, when drawn, marshal row after row of small carcasses. Here a drawer is laid out with bats, their membranous wings extended to full stretch. This one drawer contains about thirty bats, and there are fourteen drawers in one of the smallest cupboards. There are roughly 4,500 species of mammals on earth and of these a thousand are bats. Another two thousand are rodents. Other drawers display the neatly arranged columns of tiny rodents, from whiskers to tautly stretched tails. Although it appears a museum of death, it is in fact a library of life. In those myriad rows of cupboards, in the thousands of flat drawers, is a collection that is more than a museum — it is a carefully tabulated and calibrated exploration of mammalian evolution.

There are certain facts that can be learned from the animals' skins, others from their bones or their internal organs. Here, in this amphitheater of cupboards, every carcass is carefully filed and tagged. Every stretched skin is accompanied by a little pot of bones. There is a numeric reference to a deep freeze where the internal organs are preserved at minus 70°F. Each specimen has a museum catalog number that refers you to a database. In the database you will find all kinds of information on this one animal: its species, approximate age and gender, its weight, the date it was captured, the place, and the biologist who collected it. You could, if you wished, delve a good deal deeper. Why did the biologist capture the animal? What research project was he working on? Where was it captured? What were the local geographic and topographical conditions, the weather, even the parasites that were found to infect the internal organs. This allows you to take this data from the master file and put it into a global information system. The chromosomes of the animal are stored separately. Other specimens, racked in the cupboards or distributed about the lateral and horizontal surfaces, date back to the late 1800s, collected by naturalists when the railroads were being forged through the West.

It is a remarkable establishment, not only because it contains the largest and most diverse collection of rodents and bats in the world, but for the prophetic brilliance that lies behind it. Terry Yates has been fascinated during his working life by the diversity of life in two major biomes: the upper Sonoran desert of New Mexico and the contrasting rain forests of South America. In both ecologies, his biological interest is the cataloging of mammals; and like ecologists and systematists everywhere, he cannot help but be horrified by the helter-skelter attrition of species that is currently taking place across the world. Today, thanks to the polymerase chain reaction (PCR), his systematic vision of the genomic library of life has matured to remarkable fruition.

To Charles Darwin, nature could appear "abortive, squandering, profligate."[15] Yet Darwin was lucidly aware of the subtleties and immense time scales inherent in the normal accretion of evolutionary change. In this refined perception of the forces of evolution, experts dissect out the intricate step-by-step progressions that lead to speciation and diversification. Such matters are of great importance to biological systematists like Professor Yates. There is one such subtle mechanism of progress that is particularly relevant to viruses: it is called coevolution.

Coevolution was first proposed by Ehrlich and Raven in 1964 to explain the parallel evolution of butterflies and their host plants. Since then, this concept has been a source of growing wonder to biologists, including virologists.[16] No strategy could be more important or fundamental in probing how such very different forms of life may evolve in tandem, each influencing and influenced by the evolutionary development of the other. Here a window is suddenly thrown open onto a wonderful marriage in nature — a partnership in which the definition of predator and prey blurs, until it seems to metamorphose to something altogether different.

To best address it in scientific terms, we need a conjunction of the knowledge from two very different biological imperatives. In 1993 there was such a meeting of minds when Terry Yates, with his systematist thinking in relation to mammals, joined forces with C. J. Peters, with his profound experience of viruses.

Early in 1993, a man stumbled into a hospital in the tropical town of Trinidad in Bolivia and died. The local doctors suspected Bolivian hemorrhagic fever, caused by the rodent-borne Machupo virus. They sent samples of the man's blood and postmortem tissues to C. J. Peters at the CDC, and he confirmed the virus. The Bolivian government, fearing another outbreak, persuaded Peters to travel to South America to track the origins. But they also needed a biologist who could investigate the ecology of the rodent reservoir. Since 1984 Terry Yates had been conducting a major biodiversity study on the mammals of Bolivia, in which he had discovered an entirely new genus of canopy-living rodents. In March 1993, he accepted the call. "Well, somebody who has a long-term project going on in Bolivia doesn't want to turn down a direct request from the president!"

In La Paz, Yates and Peters met up for the first time. They took a single-engined plane to Trinidad, which is close to where the Machupo virus was first isolated, in the neighborhood of San Joaquín. The dead man had lived on a ranch in a floodplain, and the two scientists had to wade through the flooded savannah on horseback, with water up to the horses' bellies. In Peters's memory, it had an almost fairy-tale beauty, an enormous shimmering water garden of lilies and floating flowers, the sky shot with vibrant splashes of color as flocks of macaws scattered overhead. The farm stood on a raised hillock, surrounded by palm forest on the edge of a small river. They found three small children and nobody else. Suddenly they heard a gunshot. Then they glimpsed a couple in the distance, approaching in a dugout canoe. The man was returning home, with a brace of duck and a string of piranha that he had caught in the flooded river.

In this exotic wilderness, without electricity or plumbing, the two scientists tracked the virus that had killed the index case. The man had worked on this farm, helping the farmer now approaching in his canoe and contracting the virus here during the course of his work. The experience had terrified the farmer, isolated in this wilderness with his wife and three children, with a man bleeding and in shock from this contagious illness.

Peters and Yates began a systematic investigation, taking blood samples from the local population and trapping rodents. They shared the perils of the field, in particular a hair-raising return journey during an airport strike, when a pilot called "Happy Tom Trigo" attempted to fly them out on a single-engined Piper through a mountain pass 19,000 feet in the Andes. The plane struggled to break through the storm clouds only to be forced back when they were all fainting from lack of oxygen. In such circumstances the two scientists became friends. As they sorted out what turned out to be a minor outbreak, each became interested in the other's work.

When the hantavirus epidemic broke out in New Mexico, Peters made contact with Terry Yates, then back at work in the Bolivian rain forests. That chance meeting in the Bolivian wilderness had paved the way for a unique cooperative opportunity, in which they could look beyond the epidemic itself into the enthralling crucible of evolution. With Terry Yates's intervention, those fascinating discoveries made during the PCR typing of the virus, the varying genomic sequences from one locality to another, assumed a new light of understanding. For these were the hard evidence of the past evolutionary track of the *Sin nombre* virus.

When I invited Yates to explain to me what a virus meant to him, he reflected, "An even more pertinent question to me is, 'What is a species as far as a virus goes?' Are there species of viruses analogous to mammalian species? It seems to me that the question 'What is a species when it comes to viruses' is best understood in the context. Are they evolving, and are they coevolving with their reservoir organism?"

The advance of molecular science had caught up with his vision. He uses a methodology called cladistics, which is based on a phylogenetic, or evolutionary development, analysis of different lineages of viruses. In Yates's opinion, viruses have their own evolutionary trajectory and their own historical fate, so that by tracing the precise history of that evolution and studying the branching patterns of that lineage, we can define their precise evolution. It is important to realize that we are talking about enormous stretches of time in this branching progression.

Using such techniques, he has been successful in extracting entirely new species of hantavirus from his frozen tissue collection. "And when we look across the board at the hantaviruses, our evolutionary analysis thus far has shown a very tight correlation between the evolutionary tree that illustrates the evolutionary history of the host and the evolutionary tree that illustrates the history of the viruses." The virus and the rodent evolve in a perfect tandem, a genomic change in one accompanied by a matching genomic change in its partner.

Using this phylogenetic analysis, you can take a viral sequence from the hantavirus of the deer mouse and compare it to other species of virus that coevolve with other species of mice. From the viral track, you can extrapolate the evolution of every species of mouse. It is extraordinary to realize that this same phylogenetic analysis applies to every rodent on earth. In Terry's opinion, it links even further back in evolution. For example, we might compare the sequences in viruses coevolving with mammals to those that coevolve with marsupials and the more ancient egg-laying mammals. In this way, a study of viral genomes and their branching coevolution is one of the keys to the genomic evolution of life itself. It is this remarkable evolutionary partnership that is referred to as coevolution.

Coevolution is now an established theme in the biology of virus-host relationships, from the complex relationship between arboviruses and their vector mosquitoes to that between the malaria-causing plasmodium and humans. A precise and elegant mathematical theory has been proposed by Anderson and May to encompass it.[17] Extrapolating from this understanding of coevolution between hantaviruses and rodents, it is interesting to return to what we now know about the likely origins of the Ebola virus.

5

To date, there have been four major human outbreaks caused by the Ebola virus and two major outbreaks involving monkeys and chimpanzees. From a geographic perspective, Marburg or Ebola have manifested in epidemic form in five African countries, varying from Kenya in the east to the Ivory Coast in the west. The filoviruses have not only infected people in this concatenation of shock attacks but have also inflicted lethal outbreaks in forest-dwelling primates, African chimpanzees and Philippine monkeys. The inference is therefore plain. The geography of the virus is gigantic but ecologically restricted: it is the great biomes of the African and Asian rain forests.

The epidemics in primates are too lethal for these ravaged species to constitute the natural host. In the wake of every new outbreak, teams of scientists have gone into Africa, and in the Reston example into Asia, to track down alternative possible hosts. In the words of Fred Murphy, they have followed up some "pretty wild notions."

The Reston outbreak in 1989, for example, was traced back to a single Philippine exporter. In 1992, an outbreak of Ebola-related fatalities in cynomolgus monkeys in a facility in Siena, Italy, was traced to that same Philippine exporter. Unfortunately, limited investigations at the site of exportation and at sites of monkey trapping were unrevealing. Though it points very strongly to an Asian origin for the Reston strain of Ebola, genomically the virus is as similar to the African strains as they are to each other. It is an intriguing hint.

Straddling the borderlands of western Kenya and eastern Uganda is a volcanic massif called Mount Elgon, which is riddled with caves. For C. J. Peters, these caves are the Bermuda triangle of the Marburg virus. Two of the isolated victims of Marburg, though they contracted their infections some seven years apart, visited these caves at or about the time they contracted the virus. Yet when teams from USAMRIID traveled to Mount Elgon, they were unable to find the virus on its slopes or in its remarkable caves. The multiple primary infections of the Sudan strain within the cotton factory in Nzara in 1976 and the return of the virus to the same factory three years later points to a bat, a rodent, or an arthropod vector, a conclusion shared by many of the investigators, including Pierre Sureau, Joe McCormick, and David Simpson. Such experts see a possible connection with the many species of bats that infest the Mount Elgon caves. Simpson, who, with a Dutch zoologist, also searched the slopes of Mount Elgon, did manage to isolate a virus from bats in the locality, but it was not Marburg or Ebola.

A team including Guido van der Groen, from the Institute of Tropical Medicine in Antwerp, traveled to the African rain forest, where they found antibodies to the Ebola virus in the blood of pygmies.[18] In an important recent epidemiological survey in the Central African Republic, which included 4,295 serum samples from people in five distinct ecological zones, ranging from dense rain forest to dry grassland, another team led by Eugene D. Johnson, from USAMRIID found an astonishing 21 percent prevalence of antibodies to Ebola virus strains and a 3 percent prevalence to Marburg.[19] While Marburg was much more associated with dry grassland, Ebola was found with equal frequency from rain forest to grassland. In the Asian rain forest, Susan Fisher-Hoch also found antibodies to the Ebola virus in monkeys, though she was unable to show the presence of virus.[20] In the Central African Republic, there was an intriguing higher antibody incidence in young females compared to young males, suggesting a link to subsistence farming activities in or around the family compound. The range extending into savannah, and the link to farming, might indicate a rodent host, and antibodies have been confirmed in guinea pigs kept for food in kitchen huts in Zaire.[21] Whatever the host, whether rodent, primate, bat, or some much more surprising animal, studies such as these confirm that the filoviruses infect primates and people who live in and around the rain forests. The antibody levels to filoviruses in the Johnson survey were higher than the antibody prevalence of Lassa, Rift Valley fever, and Crimean-Congo viruses combined. This, added to the fact that there is evidence for lethal epidemic or endemic Ebola infection in these populations, suggests that infection with less aggressive strains of Ebola may be much more common than previously realized.

Some factors would appear to favor arthropods as carriers. One of these is the persistingly high blood levels of the virus. In the opinion of Don Francis, "I think what nature was trying to tell us was that cotton was the source and cotton can have lots of insects."[22]

The Ivory Coast epidemics in chimpanzees suggest a seasonality, as does the coincidental infections with different strains in Zaire and the Sudan in 1976. Another relevant facet to Ebola is that everywhere it has broken out in lethal epidemic form it has been tracked back to one or few index cases, which suggests a species of arthropod or animal that only rarely comes into contact with humans, perhaps as a result of a limited breeding season. Yet any such potential arthropod would need to have a widespread distribution throughout the huge geography of the attacks.

In time, we are likely to solve the mystery of the Ebola and Marburg reservoirs. But if we ignore for the present the mystery of the host, while accepting that such a natural host must surely exist, the molecular biology of the virus is very revealing. In 1994, C. J. Peters and Anthony Sanchez coauthored a masterly summary of what was then known about the filoviruses as emerging pathogens.[23] Although written before the two most recent epidemics in Africa, this careful study, with its discussion of the genomic biology of the viruses, would help address the mystery.

Using PCR, scientists can now plot out the entire genome of viruses such as Ebola and Marburg. This has revealed 72 percent sequence divergence between the two genera of viruses. Perhaps this should not surprise us. The viruses are very different antigenically, and this explains why the unfortunate Mayinga in Kinshasa failed to respond to the Marburg serum. Even between strains of Ebola, Joe McCormick's conclusion has been amply vindicated by genomic typing. The virus that exploded in Nzara in 1976 has major sequence differences from the strain that terrorized Yambuku. The Reston virus is also genomically distinct from the other two strains, and all three are different from the Ivory Coast virus. These genomic divergences between the various subtypes or strains of Ebola result from nucleotide sequence differences as great as 47 percent. From this we can draw two very interesting conclusions. Ebola has been prevalent in nature for a very long time, time enough to have allowed it to evolve such differences between the subtypes and to have extended through the rain forests of two continents. Even more intriguing, if disturbing, there is growing evidence that a great many intermediate links — other strains of Ebola viruses — exist within the broad dark heart of the rain forests.

We can add to this a recent, very intriguing discovery. The Ebola virus that broke out again at Kikwit in 1995 was virtually identical in its nucleotide sequences with the Yambuku strain from nineteen years earlier.[24] It means that over a wide ecology of Zairian forest, this strain of virus has been under no evolutionary pressure to change. This implies that the relationship of the virus with its natural host is also extremely stable. Such a relationship must surely have existed for a very long time. And the variation of the Ebola genome with geography points to the same archipelago of dispersion of virus and host — in other words, to the fact that, like the hantavirus and the deer mouse, they are coevolutionary partners, with all of the wonder, and mystery, that such a relationship implies.

Can we extrapolate even further? Is coevolution a universal mechanism for survival among viruses in nature? The hard evidence is not yet available to answer this question sufficiently. But I proposed this hypothesis to the eminent arbovirologist Robert Shope during an interview when I visited him at Yale University. While acknowledging that he was "sticking his neck out" a little, he replied, "I would say it is probably always true — that all viruses have a coevolutionary relationship."

Dr. Shope knew that he was generalizing from a situation of great complexity, necessarily glossing over gray areas of subtle interpretation and even of controversy. Nevertheless, the accumulating evidence does point to coevolution as being very pervasive. This means that emerging viruses, including the most dangerous plague viruses in history, and in the future, have and will emerge from a coevolutionary relationship with a feral host.

It should be possible, therefore, to extend this understanding — and in doing so to challenge its veracity — to the enigmatic origins of HIV, the viral cause of AIDS. In the words of Stephen Morse, "Even apparently new viruses, like HIV, have usually left tracks." Discovering those tracks in nature might answer the questions that have aroused such passion, controversy, and even misinformation since AIDS first made its terrifying appearance.

THIRTEEN

The Stuff of Nightmares

The ravaging epidemic of
acquired immunodeficiency syndrome
has shocked the world.
It is still not comprehended widely
that it is a natural, almost predictable,
phenomenon.

JOSHUA LEDERBERG[1]

1

The words of Dr. Lederberg, Nobel laureate and president of the Rockefeller University, beg some vital questions. Why did AIDS happen? Should we have anticipated the AIDS epidemic? Could a catastrophe such as AIDS be a warning of an even greater global danger?

It is the inevitable consequence of natural selection that every class of life will cast its net as wide as it can, in an attempt to occupy every ecological niche available to it. Viruses are no exception to this, though it takes a certain reflection to perceive the true nature of the ecological ocean into which their net is cast. This is, of course, the sum of all genomes, the entire biosphere of life on earth.

Following the structural elucidation of DNA there was a short pause, as if biological science had so surprised itself, its hand upon the key to such profundity, it needed to take a breath of mixed reflection and wonderment. What difference would understanding of the elegant double helix of life's template really make? Then, little by little, as an escalating and accelerating series of further discoveries opened up the new universe of molecular biology to novel experiment, as each individual genomic wonder materialized, like stars sparkling to light in the dawn of creation, the implications were literally iconoclastic. The landscape of the genome, that for aeons had been explored by the smallest form of life on earth, had become the tool and plaything of the human imagination. In time — provided our own ingenuity does not conspire to destroy us — we will have the godlike potential of altering our own book of life.

The key series of discoveries began modestly enough, through two young and enterprising sources. Even before Watson and Crick's discovery of the DNA double helix, an American plant geneticist, Barbara McClintock, walking her fields of experimental maize at Cold Spring Harbor, New York, realized something that would subsequently earn her the Nobel Prize. What she noticed was that a mutation in the maize readily changed the color of the cobs. Even more surprisingly, this mutation reversed itself frequently and readily — too readily for another chance mutation to be the explanation. The logical deduction was that there was a mobile element within the genes of the maize, now called a transposon or "jumping gene." This curious element could jump from one chromosome to another. From here, one could make but one further deduction: a gene that could jump from chromosome to chromosome could also jump from cell to cell.[2]

In the summer of 1946, a scientific meeting in New York was attended by a number of distinguished contemporary geneticists and molecular biologists. Mutation was the dominating theme.[3] For half a century after the discovery of bacteria, people had assumed they could only inherit their genetic material during reproduction by binary fission. Bacteria, and such simple life forms, were by implication very limited in their capacity for change — unlike higher life forms, which could benefit from the juggling of genes that took place during sexual mating.

But some people were beginning to query this. They wondered if mature bacteria could in some way swap genetic material. But hard evidence was lacking. Toward the end of an exhausting day, a heavyset young graduate student, barely twenty-one years old, stood up and presented a surprising paper. It was a short dissertation squeezed into the close of day by tacit indulgence of the organizers. Calmly, with the lucidity that would become his hallmark, he described an experiment he had recently performed in which a bacterium, *E coli* K12, did indeed share genetic information in the proposed way. The graduate student was Joshua Lederberg, and his experiment would radically alter the way people regarded microbes such as bacteria and viruses, bringing him the Nobel Prize some twelve years later.[4]

A much more revolutionary message would derive from Lederberg's experiment, a startling message indeed: the genome of individual life was not the stable structure people had hitherto imagined.

Until this time, change in the genome of simple life forms such as bacteria and viruses was thought to arise only from random mutation. Suddenly there was the potential for much larger change. McClintock's transposons are now believed to be essential ingredients in evolution. In his subsequent researches, Lederberg would extend the scope and implications of his own youthful discovery, deriving applications that extended far beyond the K12 coliform bacteria. Bacteria and viruses could communicate using genetic carriers, which Lederberg termed "plasmids": these plasmids, together with a growing number of other microscopic systems of communication, such as hybridization and conjugation, would be found to bridge genomes, from the most minuscule, the viruses, to the most elevated of all, humanity itself.

During the 1960s, a young virologist working at the University of Wisconsin made another unusual but relevant discovery. Howard Temin was interested in how viruses could cause cancers in animals. One such cancer in chickens, involving muscle or connective tissue, was caused by the Rous sarcoma virus. Temin was surprised to find that when he added a drug that blocked DNA activity, the virus was inactivated. The Rous sarcoma virus was not a DNA virus but an RNA virus — so the drug should have had no effect. It seemed to Temin that the virus must somehow incorporate a DNA cycle within the infected cells. As with many discoveries that appear to contradict the established scientific canon, Temin was ridiculed by his colleagues.

In 1970, and simultaneously with David Baltimore of the Massachusetts Institute of Technology, Temin discovered the physical incarnation of his earlier prediction, and one of many leaps in understanding that derived from Lederberg's pioneering discovery: a new class of viruses, the retroviruses, which incorporated precisely such a chemical machinery.

Retroviruses are so-called because they possess a unique cellular enzyme, reverse transcriptase, which uses the viral RNA as a template to make a DNA copy, which is then incorporated into the chromosomes of the infected cell. This is then used as a template for future generations of virus. But the ability of retroviruses does not end there. Certain retroviruses, the endogenous viruses, can lie dormant inside the chromosomes, lost as it were in the book of life of the host until the virus decides it will manifest once more.

To paraphrase Shakespeare, retroviruses are such stuff as nightmares are made of. But they are also the stuff of wonder.

Such hidden viral genes are, for example, capable of causing cancer. Those that cause cancer are believed to lodge in cancer-inducing areas on chromosomes called "oncogenes" — from *onco,* for tumor. Retroviruses are not the only viruses to cause cancer, and most retroviruses do not induce cancer, at least given our present knowledge — though the list of such virally induced cancers, whether through retroviruses or other types of virus, is growing.

One niche available only to retroviruses is fantastic. This is the genomic landscape of the germ cells, the seed of future progeny for every species. Endogenous retroviruses have the capacity to seek out the germ cells, the eggs in the ovaries and the sperm in the testes, where they dovetail into the chromosomes and lie dormant. This means that they are passed from parents to children, and on down through all subsequent generations.

It is quite a shock to realize that there are thousands of such retroviral sequences within the human genome. Some found throughout the human species are also found in old world primates, though they are missing from new world primates: suggesting that pandemics of retrovirus infection took place millions of years ago in our common primate past, but after the separation of the early primate ancestors into the new world. Nobody knows what function such viral pieces in our chromosomes serve. The prevailing wisdom is that they are all or mostly silent, suppressed by the body's genomic defenses. Though some, linked to oncogenes, appear to function like time bombs, erupting into life episodically, and unpredictably, to cause cancers in future generations.

In one other respect, there is no doubt at all of the scope and implications of these sinister passengers. Retroviruses, like viruses in general, have evolutionary and biological agendas of their own. A virologist working at the University of California, San Francisco — Harold Varmus — believes that they may have played some part in the evolution of life on earth.[5]

The arrival of AIDS changed the pace of everything, thrusting a savage urgency into such virological and molecular biological research throughout the world. The disease fell upon humanity with brutal suddenness, shattering the illusion that we had infections beaten. Suddenly, it seemed, millions of human beings were dying, and dying horribly. The virological researches that had been the backroom intellectual pursuits of theoretical scientists now entered center stage, a major theme in the needs of society and in the fiscal considerations of governments.

Suddenly everybody became very interested in the potential for human disease of these strange and remarkable entities, the retroviruses.

2

In 1976, a Japanese researcher at Kyoto University, Dr. Kiyoshi Takatsuki, was studying a bizarre form of leukemia that caused a cancer of lymphocytic cells in the blood. The cancerous cells had very curious nuclei, so dramatically convoluted in on themselves they are now called "flower cells."[6] Takatsuki noticed that almost all of the afflicted patients came from Kyushu, a large island to the southwest of Japan. Realizing he was observing an unknown disease, he traveled to Kyushu and found that his colleagues in the local hospitals were treating many people with this bizarre leukemia. At this time nobody knew the cause of this lethal new disease, but they decided they would call it "Adult T-Cell Leukemia," or ATL. They published their discovery of the newly discovered disease and explained their findings to date to alert colleagues throughout the world.[7] Over subsequent years Takatsuki discovered that the ATL leukemia could take many forms, some very rapidly progressive, others smoldering or lingering. One of his colleagues, Dr. Isao Miyoshi, managed to grow T lymphocytes from an ATL patient in cultures (the RT-1 cell line). But the cause of the disease remained a mystery.

In 1980, a Japanese virologist, Yorio Hinuma, left Kyushu to work at Kyoto University. In conversation, in 1980, a pathologist, Professor Masao Hanaoka, talked to him about the clustering of ATL cases in Kyushu. Hinuma began to investigate this and found that an antibody in the sera of patients with ATL lit up some antigen in the T lymphocyte cell line earlier established by Miyoshi. Hinuma now believed that people suffering from ATL were infected with a virus, perhaps a retrovirus, but he could not prove it.

A year earlier, in 1979, a vital new step in the elucidation of this mystery was made by the American scientist Robert Gallo at the National Cancer Institute, or NCI. Gallo and his team, notably a postgraduate student, Bernie Polesz, were examining the blood of a twenty-eight-year-old African-American who lived in Alabama. This man was suffering from a lymphoid cancer of his skin, known as mycosis fungoides. To their astonishment, they isolated a retrovirus from the diseased T lymphocytes, which they labeled the Human T-Cell Leukemia virus, or HTLV.[8] This was the first retrovirus discovered to cause disease in humans. The NCI team found an oncogene in the genome of the retrovirus recovered from the lymphoma cells.

In Japan, Isao Miyoshi, who was said to have "golden fingers" in growing cells, had established a second T lymphocyte cell line culture (RT-2). These cells were found to harbor many virus-like particles on electron microscopy. In March 1981, Gallo met the Japanese scientists at a small research meeting in Kyoto. Listening to Hinuma present some of the Japanese data, Gallo requested some sera from ATL patients to test it for his newly discovered retrovirus. In May 1981, the Japanese researchers were joined by Dr. Mitsuaki Yoshida, working at the Cancer Institute in Tokyo. Yoshida had a good deal of experience working with chicken retroviruses. He was provided with the RT-1 and RT-2 cell lines by Hinuma and within a few weeks had established that these produced a good deal of reverse transcriptase and "its activity was associated with particles having the density of retroviruses." By September, Yoshida was so convinced he had a retrovirus, he named it the Adult T-Cell Leukemia virus, or ATLV. These findings were published, with Hinuma and Miyoshi as coauthors, by Yoshida in 1982.[9]

Gallo meanwhile had proved that the virus causing leukemia in the Japanese islanders was exactly the same virus he had found in the skin lymphoma of the man from Alabama. The two research groups had discovered the same retrovirus.

HTLV infects a subset of lymphocytes with an antigen on its surface membrane, called the CD4 antigen. Soon, a second retrovirus was discovered in America, dubbed HTLV-2, which caused "hairy cell" leukemia.[10] A ripple of excitement paralleled the acceleration in the pace of research.

At the Pasteur Institute in Paris, a French virologist, Luc Montagnier, was very interested in retroviruses. Montagnier had started his career primarily interested in the causative mechanisms of human cancer: from his perspective, viruses were an ideal research tool to help in understanding these mechanisms. He had benefited from some early training in Britain, with the Gitane-smoking Kingsley Sanders at Carshalton, near London, followed by a spell in Ian MacPherson's laboratory in

Glasgow. In spite of his bemusement with the British habits of tea breaks and Saturday nights' beer and whisky drinking in the pubs, the young Montagnier acquired a vital dexterity in keeping cell lines, seeded with virus, alive in jelly-like cultures of agar.[11] In 1972, back in Paris, he took charge of the newly created department of virology at the Pasteur Institute.

In 1975 he was joined by another virologist, Jean-Claude Chermann, together with his assistant, Françoise Sinoussi (later Barré-Sinoussi). At this time, Montagnier's main line of research was avian retroviruses, while Chermann was working with murine retroviruses, a little closer to human retroviruses. By 1977, all three scientists were looking very hard for human retroviruses.

In attempting to work with human cells, and particularly in culturing human cells derived from the immune system, the intrusion of human immunity made it almost impossible to culture viruses in them. Interferons, produced by the human cells, were playing havoc with all of his attempts. Montagnier became aware that a colleague, Ion Gresser at Villejuif, kept two sheep that had been immunized with interferon so that the sheep now produced antibodies to it. Such antibodies might block the interference in his cultures. Gresser generously agreed to provide Montagnier with some sheep serum. Using this ready supply of anti-interferon they could block antiviral activity and increase the retrovirus production six- or even tenfold.

In 1979, Robert Gallo arrived in France, where he presented his findings on the HTLV virus to a conference in Villejuif. The new virus discovery was treated with skepticism, but Montagnier was intrigued by a comment Gallo made toward the end of his presentation, when he announced that in his laboratory Doris Morgan and Frank Ruscetti had discovered a chemical — T-cell growth factor (TCGF) — which would keep lymphocytes multiplying for very long times in cell cultures. These lymphocytes provided an ideal growth medium for retroviruses. Montagnier told Gallo about his discovery with anti-interferon. It was the beginning of a collaboration between them.

Françoise Sinoussi had spent some time at the NIH in Washington, working on a project unconnected with HTLV research. Although she had never worked with Gallo, she had learned the elegant techniques used in assaying mouse retroviruses. Now she brought this additional expertise to the Pasteur group, who were still searching for human retroviruses that might cause cancer. From time to time they would receive some material from human biopsies, or blood from leukemic patients. Montagnier would perform the virology and Sinoussi would look for reverse transcriptase activity.

In the autumn of 1982 Paul Prenet, Scientific Director of the Pasteur Institute, dropped by Montagnier's office to talk about a very different subject, but it was something that was worrying him. Prenet had charge of the Institut Pasteur Production, a company that developed the commercial applications of some of the institute's own discoveries. At that time he was working on the production of a hepatitis vaccine. And he had encountered a major problem. The vaccine was being harvested from serum derived from blood donors, and they could not collect sufficient serum from France alone. They needed to buy some from America. The previous year the quantity of serum imported in this way amounted to 2,500 liters. Whereas in France donors gave blood out of altruism, in America a substantial percentage of people sold their blood. It was a practice that encouraged the wrong types of donors, in particular drug addicts, who were in desperate need of money. Robert Netter, Director of the National Laboratory of Health, was worried that the anti-hepatitis plasma the Pasteur was preparing for human use might be contaminated with retroviruses.

Montagnier was glad to help. This new line of research fitted exactly with the interests of his small group. From now on every donor serum would be screened for reverse transcriptase, an enzyme that was only found in retroviruses.

In the autumn of that same year, Prenet, aware that he needed an experienced immunologist, was about to advertise the job when Jacques Leibowitch arrived to take up exactly such a post at the Raymond Poincaré Hospital, just across the road. AIDS was beginning to worry and intrigue people. Leibowitch warned Prenet of the escalating alarm in America about the potential contamination of blood supplies with the agent, as yet unknown, that caused AIDS. Only recently, a CDC panel, including Don Francis and Jim Curran, had warned a gathering of American blood transfusion authorities that 90 percent of severe hemophiliacs — those requiring very frequent factor VIII transfusions — were either sick with AIDS or already dead from it. Leibowitch warned Prenet that in his opinion the cause of AIDS was almost certainly contaminating the blood transfusion supplies in America. This was alarming information.

3

Although the cause of AIDS was unknown, the frequency of the disease among homosexuals and intravenous drug addicts pointed strongly to an infectious agent. Homosexuals and intravenous addicts were long known to be vulnerable to other venereal infections, not because of their sexual preference but because of their promiscuity. One notorious index case, a Canadian airline steward living in California, admitted to more than three hundred sexual partners in a single year. This behavior, whether homosexual or heterosexual, would have sparked a catalog of venereal transmissions.

The growing frequency of infection in hemophiliacs was another obvious clue. Yet the factor VIII concentrates were passed through filters aimed at clearing them of bacteria, fungi, and protozoa: only a virus would pass those filters. Then in 1982, in a further step toward enlightenment, Max Essex, a Harvard-based virologist, showed that infection with HTLV, a known human retrovirus, damaged the immune system in a way that resembled AIDS. It was looking more and more likely that AIDS was caused by a retrovirus.

In the autumn of 1982, following his alarming discussions with Liebowitch, Paul Prenet returned to ask Montagnier if he was close to detecting the virus of HTLV in the donor sera. Montagnier told him they had not been able to find the virus as such but they had found a clue. Tests for reverse transcriptase on the blood had produced some positives. Prenet left the meeting more worried than ever. But Montagnier was a little more sanguine: they had shown that their technique worked. He was confident they had the tools to hunt for a retrovirus — if a retrovirus this eventually proved to be.

At this time, the only retrovirus known to cause human disease was still the Human T-Cell Leukemia virus (HTLV) that had been discovered by Robert Gallo. This virus infected the CD4 subset of human lymphocytes, which seemed also to be the target for whatever agent was causing AIDS. Gallo was now convinced that AIDS was caused by another HTLV virus.

Luc Montagnier is short, muscular, and handsome, with a full head of gray hair. He has an old-world Gallic dignity and charm, a modesty that cloaks the sharp sparkle of great intellect. On the wall of the room where I interviewed him was a photograph of the more mature Louis Pasteur — labeled "D'apres Nadar, June 8, 1886." Pasteur was wearing a gray coat and a bow tie, and sitting in an oak chair with leather upholstery, his pince-nez on a string around his neck. In our discussion, I brought up Pasteur's inspired observation: *Dans les champs de l'observation, l'hasard ne favorice que les esprits préparés* — in the field of scientific experiment, chance favors the prepared mind. "The prepared hands also!" he replied softly. It was the astute observation of a practical scientist.

Science is not only a question of intellectual adventure and determinism: discovery is by its very nature a practical endeavor and all those years of intensive practical experience would now pay dividends.

Soon after his conversation with Prenet, Montagnier received a phone call from a doctor who had previously been his student — Françoise Brun-Vézinet — who was running the virology laboratory at the Claude Bernard Hospital. She told him she might be able to lay her hands on some material from AIDS patients. It was a pleasant sunny day in December 1982. He would remember her exact words: "We might have something in early 1983 — are you ready to take it?"

"Yes" was his simple reply.

Françoise Brun-Vézinet was working closely with Willy Rozenbaum, one of the small circle of clinicians interested in the growing AIDS problem. A dispute between Rozenbaum and the administration of the Claude Bernard Hospital, who frowned upon his clinical association with homosexuals, had caused him to move to the Pitié-Salpetrière Hospital, but the relationship with Brun-Vézinet had survived this. On January 3, 1983, Rozenbaum told Brun-Vézinet he was about to take a lymph gland biopsy from a homosexual with generalized lymph gland swelling, or lymphadenopathy.

This was the opportunity they had discussed at their meetings. They knew that whatever caused AIDS targeted lymphocytes, particularly the CD4 subset. Very often an early symptom of a viral infection, glandular fever for example, was a generalized swelling and tenderness in the lymph glands. Rozenbaum had already seen some young homosexual patients who had first presented with lymphadenopathy, then developed AIDS. Perhaps the lymphadenopathy was an early sign of AIDS? It seemed better to start at this stage, since, if they found a retrovirus, it was more likely to be the cause than a consequence of the immune depression. Brun-Vézinet arrived at the Pitié-Salpetrière on the day of the biopsy, where Rozenbaum allowed her half of the removed lymph gland, a piece of tissue about half a cubic centimeter. Careful to keep any potential virus alive by storing it only in saline, she rushed the biopsy across the city to the Pasteur Institute.

Luc Montagnier was not in his laboratory. Although Brun-Vézinet had called him first thing to tell him the tissue would be arriving, it was the first day of the new term and he was busy at a lecture. She placed the samples in his fridge. At the end of the afternoon, at about 5:30, he walked back through the grounds to the laboratory. It had fallen dark an hour earlier so that, on entering the deserted laboratory, he had to turn on the light to see what he was doing. He went straight to the fridge, where he found the small plastic bottle, with a note attached. He would subsequently keep the note for its historic interest. It read "Lymph node biopsy from Mr. 'Bru'" — for his initials. "This man has lymphadenopathy. He is homosexual — he may go on to develop AIDS."

Montagnier knew that cells and viruses would not live for very long in saline, so there was no deferring this until morning. "I was a little afraid, you know. I didn't know exactly what I was working with. It could be a very dangerous virus — I could catch it."

Even if the laboratory staff had been present, he would have insisted on handling the live specimen himself in the laminar flow cabinet he had purchased just two years earlier. Donning surgical gown and gloves, he lifted the bottle through the eight-inch working aperture, removed the tiny piece of tissue, and laid it on a sterile dish. He felt it through his gloves. The lymph gland had a hard, rubbery consistency. Carefully, with scissors, he divided it in two, freezing one of the two halves for DNA and RNA studies. The other half he now prepared for tissue culture.

He began by mincing the sample with scissors, then putting it into a small homogenizer. He was careful to homogenize only to a certain level, not sufficient to kill the cells. After several washes, followed by centrifugation, he had a good suspension of lymphocytes from Bru's lymph gland. He then mounted them in two small culture flasks, twenty-five centimeters square, one each for B-cell and T-cell lymphocytes. Then he added a chemical derived from the staphylococcus germ, which stimulates the growth and division of lymphocytes. This, in its turn, encourages any latent virus to activate and manifest its presence. He placed the two flasks, labeled with Bru's name, in the incubating room, which was maintained at body temperature.

Every day after that, he would go to inspect the cultures, using an inverted microscope to scrutinize them through the transparent bottoms of the flasks. This magnifies the cells by a factor of about four hundred. Those of living lymphocytes, unstained with any dyes, resemble tiny rounded amoebae, as transparent as a clear jelly, small compared to most body cells, with huge nuclei taking up a good deal of the total cell volume. He noticed that the cells were entering the first phase satisfactorily: they were expanding. On January 6, he added interleukin-2 and some serum containing anti-interferon, derived from Gresser's sheep. The waiting continued.

In the two flasks the lymphocytes had now entered the second stage of growth: they were multiplying. Every three days he would remove a sample of the supernatant and pass it to Françoise Sinoussi, who would screen for the telltale reverse transcriptase. Sinoussi would wait until she had two or three samples together. Then she would set up the experiment to look for the enzyme that would confirm that Bru's lymphocytes were infected with a retrovirus.

At about the fifteenth day, Montagnier noticed something that could be significant. When the cells had begun to multiply, they had changed shape. First they were round, then they became a little more irregular. This was entirely normal. But on day fifteen the cells had become round again. That was not expected — in fact it was somewhat puzzling. Rounding of the cells in this manner was a known cytopathic effect in cultures from a different kind of cell, called a fibroblast, but it was not known for lymphocytes. Montagnier was now suspicious. Subsequently, he would never quite understand this change, since the quantity of virus was not sufficient to cause such an effect: it would remain unanswered.

At this stage, Sinoussi told him she was obtaining some small activity in the reverse transcriptase testing.

The test she was performing involved the preparation of a cocktail of solutions: dilutions of salts, the addition of some nucleotides that could only be incorporated into DNA through reverse transcriptase activity, and a radioactive label attached to the key nucleotides. She would then extract any DNA present and test for radioactivity. Every day so far it had proved negative. On day fifteen, the supernatant yielded 7,000 counts compared to a background level of 2,000: a result too low to be significant. While it might imply a retrovirus, there were other potential red herrings. The cultures were allowed to continue.

Six days later, on January 23, the count rose a further 15,000. "So by that time, we believed we had a retrovirus. But what retrovirus would it prove to be? HTLV-I, as discovered by Gallo? A new virus entirely? We did not know." Montagnier informed the clinicians, Willy Rozenbaum and Françoise Brun-Vézinet, but cautioned them against excessive optimism. Still the excitement mounted.

They needed to further characterize this virus. But now, suddenly, they ran into a potential catastrophe. The lymphocytes were dying. For the group this was as unexpected as it was heartbreaking. Robert Gallo had already shown that infection of lymphocytes with HTLV-I virus conferred "immortality" on the infected cells. By this he meant that the cells went on dividing and dividing without ever dying, constantly producing more virus. Why then were their cultures dying? Perhaps they had indeed found an HTLV virus, but a variant that killed instead of endowing immortality? On the other hand, maybe they were dealing with a very different virus — a virus that killed lymphocytes rather than making them immortal?

If the virus was related to HTLV, they could try to grow it in normal T lymphocytes. Montagnier called his friend André Eyquem, the director of the blood transfusion service of the Institute, asking if he could provide him with a little fresh blood from one of his donors. He would set up new cultures of uninfected lymphocytes from a healthy donor and see if he could infect them with Bru's virus. The sample arrived, part of a blood donation taken from a Spanish man, only transiently in Paris. Montagnier mixed the fresh lymphocytes with those still alive from Bru and began again. More tense days of waiting and testing. There was no change in the cells after their multiplication, no cytopathic effect. The little group of scientists continued, testing aliquots for reverse transcriptase activity. Suddenly the count rose again. It was working. They were propagating the virus. But they suspected it would soon kill these cells also. Could they keep on reculturing in this way? Would the virus grow in anybody's T-cells? What if it would not? Montagnier called André Eyquem. Could he keep supplying them with small samples from the Spanish donor? No good — the donor had returned home. There was no forwarding address.

Montagnier tried contacting Gallo by phone. He needed reagents that would enable him to assess if this was a variant of Gallo's HTLV.

Gallo was not available to talk to him. Instead, Montagnier wrote him a letter, describing what they were doing and their findings. To his credit, Gallo responded promptly, enclosing antibody to HTLV and also cells infected with the virus. Jean-Claude Chermann prepared some new and crucial experiments, assessing their lymphocytes infected with Bru's virus against Gallo's antibodies to HTLV. Within days they had the results. Bru's serum appeared to react against the cells containing HTLV. It seemed that Bru had come into contact with an HTLV virus. But the cells infected with HTLV were in a poor condition and the reaction might be spurious.

Montagnier asked a colleague, Sophie Chamaret, to check it. Her findings suggested something altogether different — something fascinating. Bru's serum did not contain antibodies to the internal protein of HTLV. Jean-Claude Chermann and his collaborator, Marie-Thérèse Nugeyre, confirmed this, using a different technique. It was a strange, exhilarating finding. In the course of their evolution, the family tree of the retroviruses had diverged, with certain proteins becoming progressively more different among the diverging strains. Yet the internal protein was one of the most conserved throughout the family. It seemed possible that the reaction they were seeing indicated that their virus was from the same family of viruses as HTLV — it was a retrovirus but belonged to a very different group of retroviruses. A new virus for a new disease? "So this was the beginning of some very intense excitement."

Little by little, they were coming to believe that they were in fact dealing with a new retrovirus — and that this virus could be the cause of AIDS.

They needed to see their virus, to get whatever clues they might from its physical appearance. On January 27, believing he now had sufficient information that pointed toward a virus, Luc Montagnier descended the stairs to the basement of the virology building, to speak to Charles Dauguet. In Montagnier's memory, "I gave him the problem — 'We have found some reverse transcriptase activity. There must be a retrovirus present. You must — *you must* — see that retrovirus in the lymphocytes.'"

5

Stocky and of medium height, with silvery white hair, Charles Dauguet wears gold-rimmed spectacles. Although of retirement age, he has a youthfully rounded face, and a neat strap of short-trimmed beard that clings to the arc of his jaw. On meeting him you are overwhelmed by his infectious enthusiasm for his subject. He remembers the conversation with Professor Montagnier exactly, how it was evening when Montagnier arrived in his laboratory and how he warned him that the testing could be dangerous. The Pasteur Institute had no GAMMACEL to kill the virus. If he did not wish to do it, nobody would question it. Dauguet accepted immediately.

He had, after all, spent the previous twenty years of his life working with every dangerous virus that had found its way through the portals of the Pasteur Institute. The walls of his laboratory are lined with the magnified portraits of the menacing bullet face of rabies, the icosahedrons of polio and flu, the symmetrical perfection of the rotavirus, its envelope studded with spikes, like a sea mine.

Montagnier returned with his two precious preparations: a piece of the lymph gland biopsy taken from Bru and one of his culture bottles containing live infected lymphocytes. Claudine, Dauguet's wife, a small, dark woman, who worked by his side with the same passionate intensity, helped him to prepare them for mounting. They began with the suspension of cells from the lymph gland biopsy. Taking those same precautions as Montagnier three weeks earlier, the suspension of cells was removed and spun down in a protected ultracentrifuge, to yield a tiny button of cells, laid down on the surface of the plastic. The supernatant was discarded to be sterilized. A polymer was now poured over the button of cells, congealing the cells into a tiny waxy plug. Dauguet carried it through to the microtome, a guillotine in perfect miniature, with a diamond blade.

The diamond blade, smaller by far than the diamond on an engagement ring, is loaded into an ultracut machine, where the blade is kept stationary and the material is passed over it, sliced off in very thin layers, five hundred angstroms thick. These slide down a tiny slope to be floated off onto the surface of water, from which the sections, finer than the gossamer wings of a fly, are picked off directly for examination under the microscope.

Dauguet lifted one of the sections from the water and laid it on the surface of a copper grid. The grid was perforated with 120 tiny "windows" filled in with carbon but perfectly transparent to an electron beam. From the preparation room, the tiny receptacle was carried into the darkroom adjacent, dominated by the high cylindrical tower of the electron microscope. The light was turned down in the room so that the only window was the screen. In the green actinic aura that bathed his face he gazed into the submicroscopic landscape in which viruses live, the cellular geography that encloses the human genome. Grid after grid was inserted onto the stage above head level, the stage slid in sideways, about halfway up the tower of the microscope, using something a little like a poker with a black handle, then a ninety-degree twist to get the samples to end up horizontal inside the barrel.

Dauguet peered through a glazed aperture, like a lopsided pentagon, scanning with a delicate flick of the wrist through islands of cells, the intricacies of their cytoplasmic and nuclear structures, and the intervening sea, littered with extracellular debris. There was a clicking as he turned the ratchet-like focusing knobs. Now he had stepped up the magnification so that one lymphocyte would fill a third of the screen. Next to a magnification of 50,000, perhaps 100,000. The lymphocytes expanded so that just a fraction of one filled the entire screen.

Gazing for hours into that amazing geography is exhausting to the eyes. Charles Dauguet spent the next seven days peering into that screen, searching for the virus. He would begin as soon as he arrived in the morning, and with breaks to allow his blurring eyesight to recover, return to continue the search until late at night. When Saturday and Sunday arrived, the solitary routine did not alter. The historic day was February 4 — how could he forget the moment? At 5:45 P.M. For the first time, dark shapes were springing into view. They looked like pollen grains in magnification, adherent to the organic edges of flowers. Without question these were viruses.

On the high magnification, each appeared as a ring about half a centimeter in diameter. Now viruses were appearing in other fields of view: he could see one actually budding from the membrane of a lymphocyte. In another view, he could make out viruses between the indentations on the cell membranes, clusters of twenty or more in a single field. Overall, they were hardly plentiful: no more than two percent of the cells appeared to be infected. Yet this virus looked unusual. Dauguet had never seen a virus that looked quite like this one before in his twenty-eight years of experience. *"Je suis très excité. C'est extraordinaire — le premier moment de ma vie."*

Those first pictures of the virus were quite unlike the HTLV virus anticipated by Robert Gallo. Both Montagnier and Chermann were astonished by the density of the core of the virus. It stood out, much denser than the body of the virus, almost black, darker than anything they had ever seen before. In its shape it resembled nothing more than a sleek and featureless coffin.

So the virus, like its immunology, looked entirely novel. A new disease, a new virus: so the small group of researchers still hardly dared to think. But what was it? Was it a retrovirus at all? The tests were repeated. There could be no doubt about it. The reverse transcriptase activity was always confirmatory. A few days after he had first visualized the virus, Charles Dauguet used a slightly different fixing technique to obtain pictures that showed immature virions budding from the membranes of infected lymphocytes. These appearances looked more typical of a retrovirus. They were identical to what they had seen previously with murine and avian retroviruses. The three scientists, Chermann, Barré-Sinoussi, and Montagnier, named their virus LAV, for "lymphadenopathy-associated virus."

It would be several months later, in June that same year and over lunch at the institute cafeteria, when a colleague called Oswald Edlinger told Montagnier how he remembered seeing a virus that looked like the one now christened LAV. It caused a condition called infectious anemia in horses. Montagnier hurried from the institute cafeteria to find out more about horse viruses in the library. The only reference he could find led him to some old veterinary journals, hidden away in the attic of the library building. Here, under the very roof, braving the cobwebs and the accumulated dust, he laid his hands on pictures of the equine viruses. There could be no doubt about it: the virus that caused infectious anemia bore a close resemblance to LAV. Even more intriguing, it was also a retrovirus.

Today we realize that the family *Retroviridae* has several genera, including oncoviruses, which include HTLV, spumaviruses — also called foamy viruses, the disease association of which is unknown — and lentiviruses, the latter associated with extremely long incubation periods and protracted patterns of disease. The equine anemia virus is a lentivirus.

Could it be that LAV was also a lentivirus? If so, this would be a totally different genus from HTLV — it would explain its very different serological reactions, while maintaining the distant family link through the positive surface antigen reaction.

Montagnier contacted his colleagues at the Veterinary College of Maisons-Alfort. Could they possibly send him some immune serum to the equine lentivirus? "To my great surprise, we got a reaction with that serum that was much closer to LAV than that of Gallo." The serum from a horse infected with the equine lentivirus formed an immune precipitate with the core protein of LAV. They were not the same virus, but they showed distinct affinities. As testing continued, it would be the only affinity Montagnier's group could find between the virus they had discovered and the entire family of the retroviruses. "It made me feel that this virus was indeed a member of the genus of lentiviruses."

Rarely in a scientist's career does he or she make a discovery of great importance. But when it happens, there is a phase, soon after the discovery, when the related tumblers begin to fall into place. For the group continuing to work patiently at the Pasteur Institute, those tumblers were now falling. They began to isolate the virus from other AIDS patients. Soon the virus would also be isolated, though under a different name, HTLV-3, by the renowned Robert Gallo and his team at NCI. Today the virus is known as the human immunodeficiency virus, or HIV-1.

They had found the AIDS virus.

FOURTEEN

The Widening Gyre

There was no excuse, in this country
and in this time, for the spread of
a deadly new epidemic. . . . The bitter truth was
that AIDS did not just happen
to America — it was allowed to happen. . . .

RANDY SHILTS
And the Band Played On[1]

1

On June 16, 1983, Drs. Jerome E. Groopman and Michael S. Gottlieb opened an AIDS news article in *Nature* by comparing the American media reaction to the exploding AIDS epidemic to the poet Yeats's apocalyptic vision of the coming of the Antichrist: "as an enlarging maelstrom, a widening gyre, inexorably dragging the world into destruction."[2] Yet, given the statistics in their own article, the media response was all too understandable. There had been an exponential rise in the incidence of AIDS in America, with "four to five new cases daily." It will seem ironic in retrospect that the two doctors were shocked at the forecast of 20,000 victims over the following two years.

AIDS was a startling vision: an emerging virus, contracted in the main by sex among the young, a subtle and sinister infection that kills the very cells that might fight it. This destruction of the immunity of the victim allows all manner of secondary horrors to invade, so people drown in amoebae; MAC germs from tap water invade their bowels and enter every crevice of the internal organs; strange viruses, such as the cytomegalovirus, hop a ride. Their skins are showered with purple-bluish cancers. Bizarre forms of leukemia course their bloodstreams. Every surface of the body, both skin and internal, is prey to a myriad of disfiguring, debilitating, and tormenting afflictions, so that the sufferer is compelled to take a nightmarish cocktail of relatively toxic drugs just to try to keep down some of these secondary manifestations.

AIDS is almost always lethal, at least for the moment. It arrives stealthily, almost silently: the illness caused by the first entry of the virus into a new victim is often unrecognized, causing no more than a mild rash or fever — in 50 percent of cases, no symptoms at all. The virus masks its own presence, often hiding for years in those infected lymphocytes and nobody is any the wiser. Yet, while hiding, it can be transmitted to other victims, through blood, through other bodily fluids, particularly through sexual intercourse, heterosexual or homosexual. It is devoid of prejudice, embracing all races, any age from newborns to the elderly, and both genders.

Consider how this virus mutates inside an infected person so that, some two years after first infection, the original strain of virus is no longer recognizable among the hundreds of mutating progeny. It is a vision of hell worse than anything dreamed up in Dante's fantasy.

When Groopman and Gottleib published their article, the two Los Angeles–based doctors derided the current press coverage because it tended toward panic. But the calm and considered approach essential to scientists was the very opposite of what was now needed. Society needed galvanizing: what was needed was a good deal more media alarm.

As yet nobody, not even the most pessimistic of the doctors at the eye of the storm, could visualize the true magnitude of the impending tragedy. This excellent scientific news article dissected what little was known: already there was a flicker even of professional panic as they acknowledged that they might be noticing just the tip of the iceberg. "Unknown numbers of men with prolonged unexplained lymphadenopathy and abnormalities of immune function are not reported to the CDC." What if there were ten times as many people out there, suffering the lymphadenopathy and early failure in their immune function, itself a possible prodrome of eventual immune collapse and full-blown AIDS?

Even more frightening were the sporadic reports from key census groups. In San Francisco, one report estimated that 50 to 80 percent of asymptomatic homosexual males and 20 to 30 percent of hemophiliacs who needed factor VIII concentrate were showing some evidence of "T-cell imbalance."[3]

Faced, then, with the likely death of at least 20,000 U.S. citizens, and a crisis of mounting global importance, just what steps were the government taking to control the crisis? Such dreadful prophecy hardly dented the wall of antipathy, territory protectionism, homophobe antipathy, and mischievous politicizing of issues that now dogged the AIDS response in America. On a national scale, this continued to subvert any attempts to expand the modest investigative budget. AIDS was inextricably bound up with screaming press headlines and finger-pointing on all sides. A vocal minority of homosexuals also played their part, closing their ears to the advice they were receiving and refusing to cooperate with the suggested prevention measures. No educational programs were introduced for drug addicts — despite the fact this was the main portal of infection for babies and when simple measures such as the introduction of free disposable needles and syringes would have saved lives. Randy Shilts would not be alone in condemning this complacency. In Stephen Morse's words, "AIDS was once an emerging disease . . . it too could have been stopped at this precise stage."[4] But it was not stopped — and by and large, where America led, the world followed.

In Britain, the distinguished John Maddox, editor of *Nature,* was so caught up by the general incredulity that he concluded that there was no real emergency.[5] In France, in August 1983, Luc Montagnier addressed three national authorities, including the director general of the French National Centre for Scientific Research, the Director General of Health, and the director of the French Ministry of Research, informing them that he had confirmed AIDS in a young French hemophiliac. He told them it was caused by the virus LAV and that the patient had probably acquired it through blood factor concentrates. "This virus," he warned, "is extremely dangerous to mankind." He went so far as to warn the responsible authorities of the urgent need to develop a means of diagnosis and screening of blood to detect the virus. But if Montagnier was hoping that his warnings would elicit a vigorous response, he was disappointed. The relevant directors and ministry sent him polite letters and the promise of a hundred thousand francs for further researches. This modest sum would not build the BSL-3 facility he needed. To the now alarmed Montagnier, it seemed that "only a Charles de Gaulle would have understood and taken the decisions that were needed."[6]

So it went, from country to country, the shackling of the imaginative by unimaginative minds, the squeeze on funding from the accountancy mentality of Thatcherism and Reaganism, the taint of homophobe prejudice dogging the struggles of the few doctors and knowledgeable journalists who were dedicated to fighting the spread of the disease.

It was apparent that this disease would never be treated with the respect it demanded. When a virus kills more than the number of casualties during all of the Second World War, all the resources of the developed world must surely be made available to fight it. Instead, the common opinion saw fit to blame the sufferers themselves. Little wonder that few of its victims wished to be identified: few could afford to be identified. With insurance companies demanding intimate details on their preliminary questionnaires, with HIV-positive children being turned away from schools, with people being cast out from the workplace, with, at its most extreme, evangelical Bible-thumpers claiming it was God's vengeance and the lunatic fringe going so far as to propose the genocide of all victims, it was hardly surprising that the sufferers themselves withdrew their cooperation.[7] Where it had been routine in Britain to assess all pregnant women for syphilis, without ever bothering to ask formal permission, with AIDS it would be altogether different. From the beginning of this epidemic it was forbidden to routinely check anybody for AIDS, no matter how relevant. The result was that for years most developed countries had little idea of the true prevalence of antibody-positive cases in their communities.

Throughout the early years of the epidemic, Japan and Russia implacably denied their citizens were infected. In South Africa, apartheid-supporting extremists saw it as an opportunity to denigrate the independent black African nations. So it went around the world, as the ever widening wave of death found a ready ally in the sinister side of human nature and society.

Yet it would be in Africa, with its tragic people suffering from unimaginable poverty, famine, war, and despotism, that the metaphor of the Yeats poem would find its most literal expression: where "the ceremony of innocence would be drowned" and "mere anarchy be loosed upon the world." While in America and Europe AIDS was already extending beyond homosexual men to bisexuals, hemophiliacs, intravenous drug addicts, Haitians, and children, in Africa the spread would, from the start, be overwhelmingly heterosexual. And where it was heterosexual, the risk to children was greatly augmented.

The scientific struggle against AIDS belatedly took on the dimensions of a third world war. The magnitude of the human response, measured in billions of dollars each year, spawned myriad secondary projects of research and discovery, pushing back boundary after boundary, in virology, in molecular biology, in the sociology of homelessness, in the morality of caring. In the words of John M. Coffin: "No group of viruses has received so much attention from scientists in recent years as retroviruses." Yet even today, almost two decades after the disease was first recognized, though much has been achieved in alleviating the worst of its horrors through antiviral drugs, predicting the secondary infections, and treating the cancers, no treatment can cure AIDS. It has also, so far, defeated every attempt at a vaccine.

So what has been the conclusion of this vast industry? What have we learned about the AIDS virus and the mechanisms of its devastating effects in people?

2

The scientific counterattack expanded exponentially as the death toll spiraled to millions. Massive epidemiological studies proliferated in most nations around the world. The prevalence of the infection was easily measured, provided you could persuade populations to agree to testing — not such an easy measure. Redfield and Gallo confirmed its transmission by heterosexual intercourse. This was no longer a disease of homosexuals, drug addicts, and the occasional "unfortunate" hemophiliac. Promiscuity and prostitution were just as effective in spreading it within heterosexual populations as were bathhouses in promoting transmission among male homosexuals, though there were mysteries, such as the much higher heterosexual attack rate in Africa and Southeast Asia, that demanded close study of the virus itself to explain them.

New laboratories came into being, often at the expense of those dedicated to other conditions, to study the cellular biology, the molecular biology, antiviral chemotherapy, and vaccine development. The higher heterosexual infectivity in Africa and Asia might be linked to high background incidence of other infectious illnesses, which amplified the infectivity of the AIDS virus, whether by damaging the sexually exposed mucosa or by stimulating infected lymphocytes to pour out virus.

The viral genome was now dissected apart, every nucleotide calibrated and monitored. And how astonishingly ambiguous that exercise proved — as the precise biological chemistry of its behavior was teased and plotted, its various proteins explored, from the gp120 of the mushroom-shaped spikes to the gp41 protein that anchored them to the virus envelope to the p24/25 that enclosed the coffin core, the reverse transcriptase, the core itself, those two identical strands of RNA, so tiny in comparison to their human DNA equivalent, yet so terrible in their capacity to destroy people!

Gallo's group was the first to demonstrate the single most frightening aspect of the virus: its awesome mutational capacity. It mutates so quickly inside every infected patient that the dominant strain of virus within a single individual actually changes during the course of his or her infection.

In a sense, every sufferer evolves his or her own strain of virus, and within each sufferer the strain is not a single viral genome but a swarm of thousands of related genomes, all furiously mutating, metamorphosing, driven by a genomic intelligence the likes of which had never been imagined before. Self-regulating, it could speed its own production up or slow it right down at will, overwhelming the failing immune system with novelty from week to week, even from day to day. In 1988, in a joint paper that introduced a single-subject edition of *Scientific American,* Gallo and Montagnier made a revealing statement. Following a paragraph that summarized the remarkable proliferation of scientific knowledge about the virus, they agreed, "Yet in some respects the virus has outpaced science." It had also outpaced the billions of dollars worried governments were now throwing at it. Prospects for an early vaccine were quickly dashed.

The two leading scientists in the field were hardly likely to exaggerate. So what then is it about this minuscule life form that makes it so seemingly impregnable?

How terribly ironic that the AIDS virus is very fragile! It dies within seconds of exposure to air. If free virus enters a new human body it is often inactivated through contact with blood enzymes; free virus, therefore, is much less infectious to humans than virus contained in the cells of an infected person. This is why transmission by contaminated needles or, in the case of hemophiliacs, by contaminated intravenous infusion of clotting factor concentrates, is much more successful than sexual intercourse with an infected partner.

When virus-containing cells from an infected person (the donor) enter the blood or tissues of a new host, the donor cells, usually lymphocytes, are programmed to release the virus continuously. They are likely to impact with host cells in a direct cell-to-cell confrontation: this activates the donor cell, triggering a graft-versus-host response, which causes it to amplify and release large numbers of virus. And as soon as virus leaves the donor cell, it is sitting right on the surface of the new host cell, where it can easily obtain access, protected from inactivation by plasma enzymes. This explains why infection by the sexual route is much more effective if there is preexisting disease or injury: the damaged mucosa of the vagina, or of the male urethra, allows cells from the donor to get past the first-line barrier of the new host. This is also the mechanism of infection of infants. The general consensus is that two thirds of infected infants contract the virus during childbirth, the additional third becoming infected through placental passage prior to birth or through maternal milk in breast-feeding.

After first infection, the virus goes through a predictable sequence of behavior, replicating in huge numbers in the blood. This is the stage when roughly 50 percent of infected people develop a mild fever or a rash. If the pattern were to follow the usual pattern of virus infections — if AIDS had not proved so lethal — a mild infection of this nature, even if pandemic, would have spread around the world without our being unduly aware of it. If the example of roseola is anything to go by, it would have taken years, very likely decades, before virologists searched hard enough to isolate the virus.[8]

Within a few weeks of infection, the viral load falls in the blood. Stimulated antibodies and immune cells, which have taken days to weeks to develop, are now sufficiently tuned to attack the alien invader. Normally these would remove all trace of virus from the body, retaining a permanent memory of its antigenic structure so that no renewed invasion will ever succeed. At this stage, every person is producing high levels of antibodies to HIV. But then, by degrees, the real nature of the AIDS-human battle becomes less and less predictable.

The AIDS virus is never actually cleared. After variable periods of time, from about a year after intravenous infection to as long as twelve years after sexual infection, the virus reappears in large numbers. The infection has now evolved from HIV-antibody-positive to the aggressive stage of full-blown AIDS. A wholesale destruction of the body's immunity now results in a proliferation of bizarre infections and the appearance of cancers, such as Kaposi's sarcoma, leukemias, and lymphomas.[9]

The genome of HIV is 100,000 times smaller than that of a human cell, a mere 9,749 nucleotides — a letters in the book of life analogy — fashioned into nine or so genes, or words. How could such a tiny repository of genomic intelligence so outwit the combined onslaught of human immunity and the frantic efforts of every branch of science?

Since the discovery of the virus, every letter, every comma and parenthesis of that minuscule book of life has been translated, probed, assessed under the ratiocinative focus of molecular biology. Such scrutiny has been full of surprises. HIV governs its life cycle in astonishing and unforeseen ways. Indeed the study of viral behavior may hold the key not only to the control of AIDS but also to a clearer understanding of how cells regulate their own growth and activity.

That cycle begins during the first moment of infection. Here, as large numbers of viruses are released into the blood of the new host, the HIV surface structures bind to the membrane protein of any host cell that contains the antigen, CD4. This is preeminently found on the subset of T lymphocytes called helper cells and on up to 40 percent of the amoeboid immune cells, called macrophages. For reasons as yet undetermined the virus can also infect a small variety of cells that do not have the CD4 antigen, including certain cells in the bowel wall and the support matrix within the brain, called glial cells. This may account for the prominence of diarrhea, weight loss, and nervous system disturbance that are common features of AIDS. The CD4 antigen fits the stereochemistry of the mushroom-shaped spikes on the viral surface like one piece of a jigsaw puzzle fits another. Once slotted into the cell membrane, the virus penetrates the membrane and injects its coffin-like core.

The core contains two identical strands of RNA as well as the structural proteins and enzymes that begin the process of taking over the cell. Still within the cytoplasm, the viral enzyme reverse transcriptase converts the viral RNA to DNA, then copies it to make a double-stranded helix, while an additional enzyme destroys the remaining viral RNA. The virus-coded DNA now enters the nucleus, where it is spliced by another virus enzyme into the very chromosomes of the human genome. The virus is now a permanent resident. From now on, the viral-coded DNA is replicated every time the cell divides. In the clandestine interstices of the nucleus, the virus is hidden away from the body's immune defenses, lying dormant in this way during those years of clinical silence.

Firmly entangled in the cell's machinery of life, the virus will decide when and how it will reactivate. To replicate, all it needs to do is to instruct the genomic chemistry of the human cell to use the viral DNA as template for more virus. Proteins are formed that will become the virus's outer and core membranes. These migrate to the surface of the infected cell and attach to the membrane. All by themselves, they aggregate until they form a complete virus capsule, producing a telltale bulge in the human cell membrane. Meanwhile other virus enzymes, all coded within its released genome, fashion the coffin-shaped core itself and the virus RNA is packaged within it. Other viral enzymes manufacture the spikes that are swept up by the growing virus envelope, and begin to arrange themselves like the protruding triggers of a floating mine. There are further steps, including the deconversion of virally coded DNA back into RNA, all of which the virus masterminds. But only some of this RNA becomes the genome for new viruses. Other of the RNA carries the intelligence for some truly remarkable behavior.

At the Dana-Farber and National Cancer Institutes, William A. Haseltine and Florrie Wong-Staal registered with astonishment what appeared to be an elaborate set of management controls over every aspect of virus production.[10] There were genes that switched production of virus off or on, others that hugely accelerated the viral assembly line, jacking up production a thousand-fold when the infected cells encountered a stimulatory antigen. There were still others that had the opposite effect, that switched off this positive feedback burst of replication. The more people studied the virus, the more strategically calculating its behavior appeared. It was now apparent that the long interval during which the virus lay dormant owed itself at least in part to a programmed homeostasis, a strategy arising from subtleties of interaction of these governing mechanisms. In other words, the virus coded itself for long-term survival, during which it could reproduce itself endlessly without necessarily killing huge numbers of host cells.

Other scientists, such as Redfield and Burke at the Walter Reed Army Institute of Research in Washington, D.C., found evidence for a slow but progressive conflict with the immune system of the host, in which the virus maintained a persistent and slowly escalating level of infection. They proposed a new model for the infection, based upon a war of attrition in which the virus gradually overcame the body's defenses, multiplying and strengthening as the body weakened.[11]

A paradox for the scientists was the fact that even in fully developed AIDS, only a minority of the CD4 T-helper cells are infected with the virus, yet as the condition progresses to full-blown AIDS, it slate-wipes these cells. Why should infection with the virus, which infected so few, still result in the death of vast numbers of T cells?

There are three likely explanations for this. During prolific multiplication, as virus particles bud from the cell membrane, the virus envelope protein tends to tear holes in the lymphocyte membrane. An imbalance of osmotic pressures causes the punctured cell to swell up and die. A second mechanism arises where large quantities of virus envelope protein (called gp120) attach to the CD4 sites on surrounding healthy cells, by degrees mopping up more and more normal cells and coalescing into multinucleated giants, condemned to a moribund existence. Finally, even a trace of the virus capsular protein on infected, or uninfected, lymphocytes leads the body's other immune cells, such as the killer T cells, to destroy the cells. It is the aggregate action of all three of these processes that accounts for the massive destruction of CD4 lymphocytes. And this is the key event in the illness.

In order to fight off foreign invaders, the human body needs a strong, highly orchestrated response in which the various divisions of its immune army are working together. The loss of these cells has a disastrous effect on the body's ability to fight off other infection, and the presence of the virus, with its tumor-causing potential, in the chromosomes of its victim's cells further gives rise to a panoply of different cancers, leukemias, and lymphomas. This disastrous concatenation explains the clinical syndrome of AIDS.

3

By 1985, thanks to the joint researches of Gallo and Montagnier, there was a serological test that could be used to screen for the virus. Blood transfusion services — provided they were prepared to spend the money — could screen out risk donors. In developing countries, where they lacked the money, for far too long they simply ignored it.

By July 4, 1988, and quite distinct from the one million plus antibody positives, 66,464 adults and children in America had been registered as full-blown AIDS sufferers, more than half of whom had already died. More than 80 percent of those diagnosed for three years or more were among the dead, confirming that dread in the hearts of the experts: infection with HIV-1 was virtually a death sentence. While the risk of spread from a single heterosexual contact was relatively low, transmission by injection was a different story. Transfusion of a single unit of HIV-contaminated blood resulted in an infection rate of between 89 and 100 percent, and the subsequent manifestations of the disease were ferociously aggressive.

By January 1986, the epidemic had already been confirmed in thirty-five countries, extending to every populated continent, as far north as

Greenland and as far south as Chile. More than five new cases a day were being diagnosed in Brazil, where São Paulo and Rio had the same "gay" reputation, with bathhouses and promiscuity, as San Francisco. Canada, which had registered 350 cases, had set a forthright precedent in deciding to make the reporting of AIDS compulsory, a measure not of prejudice but in order to accumulate reliable national statistics that might then become a platform for appropriate aid and action. In Hong Kong, where homosexuality could carry a life sentence, the shaken authorities had been forced to accept anonymous help lines. In Japan, where businessmen had long enjoyed the sexual license of Thailand, and where the authorities had considered the dozen or so registered cases solely the result of contaminated blood transfusion, serological screening was suggesting that thousands might be carrying a latent time bomb. In January 1987, the death of a Tokyo prostitute, the country's first woman patient and the first to contract AIDS heterosexually, sparked a panic, with Tokyo's AIDS information line swamped by 150,000 calls within its first forty-eight hours of operation.[12]

Meanwhile, despite its more limited means of spread, the virus traveled effortlessly across national and continental boundaries, across religious and cultural divides. And wherever it traveled it would carry that same fury in its wake, of blame and counter-blame, social stigma, discrimination, fear, and anger. Pakistan was forbidding people who had lived abroad in recent years to donate blood. And with the hysteria of the first wave of infection came the superstitions, the misinformation, all fueling the growing pyre of anger and prejudice: it could be transmitted by mosquito bite, by personal contact, by kissing or cuddling, or by sharing a lavatory, glass, or towel with an infected individual. When the virus was found in saliva and tears, there was an explosion of panic: surely this meant it could be transmitted through a cough or a sneeze. Fortunately, this view proved to be mistaken. The virus was transmitted neither casually nor by aerosol.

Educational programs fell back onto scare tactics: coffins, menacing backstreets, minor-scale music, with the sad innuendoes of loneliness and death. In Africa, the wholesale losses of the populations most affected, the more educated middle classes, the teachers, politicians, even the medical care workers themselves, threatened the very fabric of society.

So pervasive were the politics of paranoia, hysteria, and blame that Dr. Jonathan Mann, founder of the World Health Organization's Global Program on AIDS, would group them under his generic title "The Third Epidemic." In a speech to the United Nations on October 20, 1987, he warned, "The third epidemic of social, cultural, economic and political reaction to AIDS . . . is as central to the global challenge as AIDS itself." AIDS would scourge society as much as it would the physical and mental well-being of individuals: an acid test for the love within families, the strength and durability of friendship, the courage and frailty of the caring vocations, and the temper of justice. Dentists and doctors would, across courts of law, face patients they had themselves carelessly or inadvertently infected.

As the population studies proliferated, the information on virus behavior expanded with them, often more quickly. And a cotraveler along every avenue of expansion was the spiraling cost. It was not cheap to set up long-term nursing and community-based support for vast numbers of young, previously fit people with a progressive and intractable illness, an illness peppered with very complex relapses, strange infections that were very expensive to treat and even more expensive to prevent, the surgery of the cancers, the costly leukemias, the hospices, the medical care — not to mention the huge and burgeoning costs of the epidemiological, public health, virological, molecular biological, pharmaceutical, and a myriad other aspects.

The entity that had been forecast by the prescient few — their views utterly ignored by governments, health-funding bodies, the public at large, and most notably and tragically by virtually all of their qualified medical colleagues — had become a true pandemic virus, emerging from a chrysalis of ignorance and cynical disbelief.

4

In October 1985, Luc Montagnier was analyzing blood samples brought to his laboratory by a visiting investigator from Portugal. Many of the samples were from immigrants originating in Guinea-Bissau, a former Portuguese colony in West Africa. Among them were people who were definitely suffering from AIDS. It baffled him that their blood tests were HIV antibody negative. Despite the negative serology, a virus was definitely found to be infecting the lymphocytes in the blood samples. The virus was isolated. DNA probes were prepared. If this was truly the HIV virus, they would bind to it. The probes did not bind very well. What Montagnier's group had discovered was a second AIDS virus. It was labeled HIV-2, to distinguish it from the original virus, which was now relabeled HIV-1.

Another complex twist added to the burgeoning global epidemiology: which virus were they dealing with? Was it merely two viruses, or were there in reality others?

In time it would become clear that HIV-1 was associated with America, Europe, central Africa, Asia, and Australasia; HIV-2 was almost always associated with Africa, particularly certain countries in West Africa.

By January 1, 1988, 129 countries had reported at least their first case of AIDS to the World Health Organization. The majority of these were a consequence of earlier unrecognized spread of the virus in the 1970s. There were major differences in true incidence from country to country, and for the different AIDS-linked viruses. Even as late as January 1988, 70 percent of known cases of full-blown AIDS were being registered in the United States, where there was 90 percent compliance of reporting. But WHO epidemiologists knew by now that there was gross underreporting elsewhere, even in developed countries, and a massive ignorance in Africa, where limited access of large sections of the population to health care facilities and the general lack of epidemiologic and diagnostic facilities had led to an underestimation of the size of the problem.

Screening of stored blood now yielded much earlier evidence for the presence of the virus in Africa.[13] But what did this mean? AIDS had actually been diagnosed in Africa after the epidemic had begun in America, but now there was gathering indication that the epidemic had been growing there for some time without being diagnosed. There was clear evidence, both clinical and from blood screening, that a low background level of the disease, that had been maintained for decades, had undergone an explosion within urban areas during the late 1970s. This suggested that a major African epidemic of HIV-1 began at much the same time as it had in America and Europe.

In contrast to the situation in big cities, such as Kinshasa, where a tenfold increase in HIV-1 antibodies was documented in pregnant women between 1970 and 1980, the percentage of people who turned up positive on blood screening in rural Zairian surveys appeared stable and persistently low. Many such studies pointed to changing human behavior playing as vital a role in Africa as it had in America and Europe. Social changes with interruption to rural lifestyles, in particular the movement of young men into the big cities, had played an important part. Another vital factor was once again promiscuity. But it would follow a very different pattern from that initially seen in America and Western Europe.

Traditional African culture has a low incidence of homosexuality, but multiple heterosexual contacts were a common finding, particularly in the young men frequenting prostitutes in the big cities and ports. Where AIDS in the West was predominantly a disease of young males, AIDS in Africa was about equally distributed between men and women. A further source of infection was contaminated needles and syringes, not only through intravenous drug addiction, which was rarer than in America and Europe, but through carelessness and economy in poorly equipped clinics and hospitals involved in vaccinations and drug therapy. Their third and most tragic spread of all was in children.

The sum effect was an overwhelming spread of the virus throughout the 1970s, eighties, and nineties that made Africa the tragic epicenter of the pandemic, with a staggering burden of 11 million cumulative infections.

Even in 1995, the World Health Organization was reporting a relentless global increase of HIV-1 infection and cases of full-blown AIDS. The statistics were harrowing. No less than 190 countries were now reporting AIDS cases. "Since HIV-1 began its spread in North America, Europe and sub-Saharan Africa, over 19.5 million men, women and children are estimated to have been infected. . . . To date it is estimated that close to 1.5 million children have been infected with HIV-1 through mother to child transmission. These children rapidly develop AIDS and usually die before the age of 5 years."[14] But by now this dreadful pattern was showing a distinct dynamic development.

Prevention efforts were winning in the developed countries, where the new infection rates were stabilizing or even decreasing. But the trends were very different in the developing world, "where the health care infrastructure is already overwhelmed by other causes of morbidity, and where around 700,000 and close to half a million new cases of AIDS are projected to occur annually by the year 2,000 in Africa and Asia, respectively." Although the long-term dimensions of the epidemic cannot be forecast with confidence, by the turn of the century the WHO anticipates that between 30 and 40 million men, women, and children will be infected with HIV-1 alone.

For decades the people in so many sub-Saharan African countries have survived the most dystopic conditions in all of their history, racked by famine, war, epidemic disease, and the absence of a consistent and just government. In ways that have scarcely been acknowledged by people in the more affluent nations, in their recourse to spirituality, their stoicism in the face of adversity, the people of Africa have lit a beacon of hope for all of humanity.

And it is to Africa too that we must turn to discover the real origins in nature of the AIDS pandemic.

FIFTEEN

The Biological Origins of AIDS

Everything that happens,
happens as it should,
and if you observe carefully,
you will find this to be so.

MARCUS AURELIUS ANTONINUS
Meditations

1

In the growing awareness of cross-species trafficking of viruses, the potential of viruses to hop from primate to human was a growing source of interest and anxiety. Experience with some of the most lethal human viruses, such as yellow fever, Marburg, Ebola, and smallpox, pointed to monkeys and apes as the only other species to share the infections. In 1982, two years after Robert Gallo discovered the first human retrovirus, HTLV-1, the Japanese researcher Isao Miyoshi, of Kochi University, discovered a related virus in the Japanese macaque, designated the simian T-lymphotrophic virus, or STLV. Like the human virus, the simian virus was capable of immortalizing laboratory cell lines, a characteristic of cells transformed to the cancerous state. Miyoshi wondered if monkeys were also the source of retrovirus infections in humans.[1]

Soon other researchers discovered a number of related simian viruses, in new world monkeys in Asia and in old world monkeys and chimpanzees in Africa. It was clear, both from the vast geographic dispersal and the fact that up to 40 percent of the animals in certain species were infected, that T-cell lymphotropic viruses had been infecting these primates for a very long time, possibly hundreds of thousands, even millions of years. Not only did STLV possess similar biological properties to its human equivalents, its genome had much in common — called genetic homology — with HTLV-I. The viral proteins were remarkably similar: antibodies to the human virus would cross-react with the monkey virus, as would antibodies to the monkey virus cross-react with the human viruses. It surprised people that the STLV found in wild monkeys had a powerful genetic homology with the virus causing human disease, notably that seen in African green monkeys: no less than 95 percent of their genetic sequences were shared.

The broader inference was not lost on scientists interested in retroviruses: they focused their efforts on nonhuman primates. One scientist who would become a world leader in the search for a primate source of AIDS was Max Essex, then professor of virology at the Harvard School of Public Health and now Mary Woodward Lasker Professor of Health Sciences.

In 1982, that same year Yoshida became interested in the cross-species traffic of ATLV viruses, Ron Hunt, a mammalogist at the New England Primate Center, noticed that certain monkeys in his laboratory were suffering the ravages of an immunosuppression that was very similar to what was then being described in human AIDS. Other primatologists soon came to notice this strange affliction, which was only seen in Asian macaques. The monkeys suffered a wasting illness, with invasions by opportunistic infections, and a number were dying from a lymphomatous cancer of the lymphatic tissue. At this time the clinical syndrome of human AIDS had been recognized, but the viral cause had yet to be discovered. Hunt began to investigate this further, and he phoned Essex to talk about it.

Max Essex is of medium height, gray-haired, and gifted with a fine scientific intellect. He is also possessed of the quintessential prerequisite for the creative scientist: a profound curiosity for the labyrinthine mysteries of nature, coupled with the unflagging probity and endurance needed to pursue a decades-long quest to its ultimate goal. For some years Essex had been researching a related condition in cats, which resulted from infection with the retrovirus that caused feline leukemia. This was at a time when there was gathering evidence from research being performed by Robert C. Gallo that human AIDS was caused by a retrovirus. During their conversation, Hunt invited Essex to travel out to the Primate Center and talk it over with a group of his researchers.

When Essex met Hunt and his colleagues, the Harvard scientist offered his laboratory expertise to screen blood samples from their monkeys for retroviruses. "If a primate immunodeficiency virus really was causing the immunosuppression seen in these monkeys, the causative virus might share some of its genome, and therefore its surface antigens with the virus causing human AIDS."

By 1984, the initial offer had expanded into a major program of research and collaboration between Essex and Hunt, drawing in Hunt's colleagues Muthiah D. Daniels, Ronald C. Desrosiers, and Norman L. Letvin. They screened large numbers of primates for retroviruses, and soon they were able to confirm that there was indeed a retrovirus infecting the immunocompromised monkeys. They could show this was a retrovirus because it cross-reacted with serum raised against other known retroviruses, and it looked like a retrovirus under the electron microscope. The coresearchers published their evidence, showing that monkeys suffering from lymphoma and some of the ones that were immunosuppressed had a virus that was cross-reactive with — and hence morphologically similar to — the retrovirus that caused HTLV in Japanese people.[2] An extraordinary picture, pieced together through a combination of intellectual brilliance and dogged experiment, was emerging from the close inspection of nature.

It was about this time that Luc Montagnier and his colleagues at the Pasteur Institute isolated the first of the two AIDS viruses, HIV-1. Essex had isolates of HIV-1 in his laboratory by May of 1984. "We then asked the questions: Is this virus related to the monkey virus? Will it cross-react serologically with virus in those monkeys? Are our monkeys infected with a virus exactly like that? Is there a direct link between the human AIDS virus and the virus infecting our cluster of monkeys that were developing immunodeficiency?"

Essex had taken on an able doctoral student, Phyllis Kanki, who now coordinated the collaboration with the primatologists, Desrosiers and Letvin. By degrees they proved that the monkeys afflicted by lymphomas and immunodeficiency-related illnesses were infected with two distinct retroviruses. One of these viruses was clearly HTLV-1. But what was the other? Using serological cross-reaction studies and electron microscopy of the viruses, they detected a second virus that showed some tantalizing immunological cross-reactions with HIV-1. While many of the monkeys were concomitantly infected with both viruses, it became clear that monkeys solely infected with this new virus showed every sign of immunodeficiency. It must be the new virus that was causing the immunodeficiency. Essex decided he would call the new virus STLV-3 — but today we know it as the simian immunodeficiency virus (SIV).

It was an important step along a tortuous and thorny road, but it was only a beginning.

The SIV virus was clearly related to HIV, it infected the same subset of lymphocytes, it looked just like HIV under the electron microscope, and it had similar biochemical and biophysical properties. Antibodies from humans with AIDS cross-reacted with the monkey virus — just as monkey antibodies cross-reacted with the human virus. Essex was intrigued to notice that the human antibodies reacted strongly with the monkey virus core protein while only minimally with the envelope proteins that coated the surface of the monkey virus. He knew that the core proteins were the most conserved during evolution. Rather as Luc Montagnier had discovered in the testing of his LAV (HIV-1) discovery, this suggested that the two viruses were members of a common family or group but were not the same species.

Subsequently, mapping of the nucleotides that make up the genomes of the two viruses confirmed the immunological evidence: the viruses have 50 percent of their nucleotides in common. Moreover, the way in which the genome is organized — the structure and regularity of the component genes — is virtually identical for both viruses. Like humans, Asian macaques, when infected with SIV, suffer an attrition of T4 lymphocytes with eventual immune system failure. Infected monkeys waste away, suffer intractable diarrhea, and their immune responses fail, resulting in lethal opportunistic infections as well as a primary brain infection with the virus itself that causes a dementia similar to that seen in human AIDS. They also develop cancers such as lymphomas. The human and simian immunodeficiency viruses are clearly members of the same group of viruses, the immunodeficiency viruses, but they are different "strains" or "species."

For experts such as Essex, this opened the investigative doors to a close laboratory study of the pathological mechanisms involved in AIDS, even in time an opportunity to test the effects of antiviral drugs and vaccines. But there was an additional fascination. "We wondered whether the geographic distribution of the monkey virus might provide clues to the origin of the human AIDS virus itself."

2

That first simian virus had been found in unnatural circumstances: monkeys imprisoned in the captive conditions of primate laboratories in America. To probe this new puzzle further, researchers needed to examine monkeys in the wild. But when they did so they received a jolt. Applying the same screening tests to macaques in the Asian rain forest, they found not a jot of evidence of the simian immunodeficiency virus. The SIV discovered by Essex and his collaborators was limited to Asian monkeys held in captivity. What could this mean? Had the Asian macaques been exposed to some alternative source of the virus? And if so, what was the original source?

At about this time disturbing reports of rampant AIDS were emerging from central Africa. Beginning in 1982 and 1983 with sporadic scientific observations from Belgian and Dutch doctors, the pace of recognition soon accelerated. By 1985 the reported rates of infection with AIDS in certain urban African centers were so dramatic that some experts wondered if AIDS in central Africa might have predated the epidemic in America and Europe. To Essex, his scientific curiosity now aroused, there was an urgent need to turn his attentions to Africa. "Perhaps the common link to human AIDS virus would be found there?"

He began this new line of research by collecting blood samples from representative African primates, including wild-caught chimpanzees (*Pan troglodytes*), green monkeys (*Cercopithecus aethiops*), baboons (*papio spp.*), and patas monkeys (*Erythrocebus patas*). The samples were analyzed for antibodies to SIV. Almost immediately Essex made the first of two remarkable discoveries. All of the primates tested proved negative with the exception of the African green monkeys, of whom an astonishing 50 percent proved positive. Professor Hayami of Tokyo University discovered the causative virus — which is now called SIV(agm) to distinguish it from Essex's SIV(mac). Since then, Essex and his collaborators have tested thousands of wild green monkeys from sub-Saharan Africa, together with captive monkeys of the same species from research facilities around the world. The initial results were intriguing.

They found incontrovertible evidence of SIV infection in a staggeringly high percentage, between 30 percent and 70 percent of the animals tested. In the course of this mammoth undertaking, Essex made the second of his remarkable discoveries.

The African green monkeys infected with SIV showed no sign of the illness seen in the captive Asian macaques infected with a similar virus. Their T4 lymphocytes, although infected with the virus, were not in the least depleted. On the contrary, they persisted in normal numbers and there was no clinical evidence of immunodeficiency. Supportive of this was the fact that these monkeys, widely infected with the virus, are among the most successful of the African primates in exploiting their rain forest ecology. It was clear that the very high infection rate with a virus that proved lethal to captive Asian macaques "had not been exerting long-term adverse selection pressure on the African species."[3]

At the time of its discovery, this finding was a considerable surprise, more than enough to lead to a proliferation of new studies.

Clearly the African green monkeys had evolved some mechanism that protected them from the pathological effects of the virus. We now know the likely explanation for this strange absence of pathological effect of the virus on its host monkeys: it is exactly what we have seen with hantaviruses and rodents — and similar to what we have seen in arenaviruses. Was Max Essex observing something that was an integral part of the texture of life? A virus of a similar species as HIV-1 had, it seemed, over an unknown period of time, come to an accommodation with the monkey. It seemed altogether likely that Essex was witnessing that same wonderful marriage of widely differing species, the mutual evolution toward a harmonious relationship we recognize as coevolution. The implications were of major importance. Essex turned to the local human population.

It seemed logical to focus on people who were at the highest risk — female prostitutes and male patients attending sexually transmitted diseases clinics. Essex already had research links with West Africa, as part of an ongoing study of the hepatitis B virus. New investigations were now set into motion with those same collaborators, screening blood samples from the two at-risk populations, with antigens from both HIV and for the SIV virus they had found in the local monkeys. In early 1985, with the help of Souleymane M'Boup of the University of Dakar and Francis Barin of the University of Tours, Essex found exactly what he was looking for — antibodies to SIV in the serum of prostitutes from Senegal in West Africa.

About 10 percent of the prostitutes proved positive to both viruses. How interesting it was to observe that the positive sera reacted more with SIV than HIV, the strength of reaction indistinguishable from that in the African green or Asian macaque monkeys infected with SIV. The scientists were sure they were dealing with a new species of human virus.

"People were very clearly infected with a virus that was very closely related to the monkey virus, in fact virtually indistinguishable from it." The virus infecting these people in Senegal was clearly related to HIV-1, but it was equally distinguishable from HIV-1. It seemed to react much more strongly with, and was therefore much more closely related to, the SIV virus they had found in the monkeys in the New England Primate Center.

When Essex first presented his African findings in Belgium in 1985, Luc Montagnier was at the meeting. Soon, Montagnier, together with his colleague François Clavel, confirmed Essex's findings with the discovery of the second human immunodeficiency virus, HIV-2. HIV-2 causes a definite AIDS syndrome in people. But on the whole the disease expression is much milder and the progression more protracted. Essex found that there was an increasing frequency of infection with this virus as the age of the prostitutes increased — which suggested that the virus had been around, infecting people, rather longer than was known for HIV-1. When the nucleotide sequencing of HIV-2 became available, it confirmed the close similarity to SIV.

Today, thanks to scientists like Essex, we know that many of the species of monkeys in West Africa coevolve with an archipelago of SIV viruses, with genomes that vary significantly with geography, in much the same way as the hantaviruses and arenaviruses. The similarities between human and primate immunodeficiency viruses can be very striking. Another research group, Marx and his colleagues, isolated a simian immunodeficiency virus in a sooty mangabey from West Africa that had a very close genomic relationship with HIV-2.[4] Even more remarkably, the SIV that afflicts sooty mangabeys, in exactly the endemic area Essex was studying, shares 85 percent of its genome with the HIV-2 viruses in people in that locality. Yet there are HIV-2 viruses in people in neighboring countries, Ghana for example, where the HIV-2 is significantly different at the nucleotide level from the virus infecting people and mangabey monkeys in Senegal, five hundred miles away. The AIDS viruses vary in their genomic sequences throughout the human diaspora just as they do in monkeys. Two viruses, one in people and the other in monkeys, cannot be almost identical by accident. It is so unlikely, the odds are virtually incalculable.

Essex would come under criticism at this time, because Africa was a convenient scapegoat for the bigotry and opprobrium of the ignorant in the wave of panic accompanying AIDS. But the scientist was courageously impenitent. He emphasized the fact that some of the simian viruses were so similar to HIV-2 that the differences between the human and simian viruses are less than the variation seen between strains of HIV-2 in different groups of humans or between different strains of SIV in different groups of monkeys. He concluded that, from a classification perspective, HIV-2 was simply a strain of SIV that had become endemic in people of West Africa.[5]

These observations would be given interesting confirmation by infections in laboratory personnel working with monkeys. In 1994, the potential for such cross-species infection with SIVs from primates to man was confirmed by Rima Khabbaz, Tom Folks, and coworkers from the Retrovirus Diseases Branch of the CDC, who, in 1994, described at least one, and possibly three, cases of human infection in laboratory workers exposed to the monkey virus.[6]

Essex's interpretation is surely correct: "HIV-2 and SIV are exactly the same virus — it's just that you call them one thing if they are in people and another thing if they are in mangabey monkeys."[7]

Today there is overwhelming evidence that points to an origin for both human immunodeficiency viruses from Africa. This owes nothing to human decadence or racial triumphalism but to proximity to the real Garden of Eden, the most primal of the rain forests, where all of the primates, including humanity, first evolved. But the viruses that cause the two forms of AIDS are different. Although HIV-1 and HIV-2 are lentiviruses, belonging to the taxonomic class of retroviruses, they only share 40 percent of their genetic sequences. These divergences explain the observed differences in the patterns of human infection.

Essex and his collaborators went on to perform exhaustive epidemiological studies throughout fourteen African countries, whereby they discovered an intriguing geographic competition between the two viruses. On the whole, where you find a high incidence of one virus, you find a relatively low incidence of the other. For example, there is a low rate of HIV-1 in West African countries, where HIV-2 is most prevalent, and there is almost no HIV-2 in central African countries where HIV-1 is prevalent. To Essex, this suggested that they were observing an important pattern in the texture of viral-host relationships. And it opened up interesting avenues for future study. Prostitutes in West Africa, initially shown to be infected with HIV-2, were at a substantially lower risk of becoming infected with the much more aggressive HIV-1 than those who were negative for HIV-2 to start with.[8]

3

Another observation of fundamental importance emerged from attempts to infect primates with HIV-1. Despite many attempts to infect a diverse series of primates with the virus, the only success was with chimpanzees. Even more curiously, when chimpanzees were infected with HIV-1, no lethal disease resulted. In fact, the chimpanzees remained perfectly healthy, with no symptoms or signs of disease at all. Essex was well aware that similar retroviruses were known to infect mice, chicken, and cattle, where they often behaved in this benign fashion. Could it be that wild chimpanzees had encountered the virus over a long period of time, long enough to have evolved an immunity to its pathogenic effects?

Every virus in the process of replication makes mistakes in its genomic copying. These are what we mean by mutations. The remarkable thing about RNA viruses — and the AIDS viruses are RNA viruses — is that almost every replication leads to a mutant progeny. So we start with a single viral genome and as soon as it has been amplified to a few thousand, a few million, or a few billion, we have a swarm of related mutants and not a single entity at all. To think of a single virus is misleading, we need to consider the resultant swarm of mutated viruses, all competing with each other.[9] The swarm analogy has implications that go beyond the mutational evolution within a single infected monkey. The "species" swarm is nothing less than the genomic diversity of the entire virus proliferation across the genomic landscape of the entire infected species of monkeys. The majority of mutations produce no natural selection advantage or disadvantage and so are discarded. Occasionally they make the virus more harmful, occasionally less harmful. But the profligate mutations that arise from infection in an entire species of animal make it much more likely that some resultant strain of the virus will adapt, for example to facilitate coevolution toward nonvirulence, just as other mutations might enable a virus to hop species, as, notoriously, from a primate to man.

Infected monkeys must be coming into contact with humans on a very large scale from day to day and from year to year. And through the mechanism of the swarm, it is the sum of all such contacts that creates the effective potential for species trafficking. All it requires is a single successful bridge between one monkey carrying a particularly virulent trafficking virus and a vulnerable human.

When a virus mutates, the change is recognizable in the nucleotide sequence of the affected gene. Comparing the viral sequences over time, it is not difficult to calculate the rate of mutation with time. This rate is fairly steady. On this basis, matching human immunodeficiency virus with simian, scientists can calculate how long it might take for the two AIDS viruses to have emerged from primate to man. Although the evidence for HIV-2 is strong, since the sequences are similar enough to allow for a very recent origin, the major differences between the genomic sequences of the primate SIVs and HIV-1 have been a source of controversy for many experts. Then, in 1990, scientists working at the Centre for International Medical Research in Gabon, West Africa, found serological evidence of an HIV-1–type virus infecting two wild chimpanzees. Admittedly, the chimpanzees were the only positives in fifty screened, and the virus was only successfully extracted from one of them. But sera from the two animals cross-reacted with all of the

HIV-1 proteins, including the specific envelope proteins from the virus surface. Nucleotide typing of the extracted viral genes confirmed a close relationship to HIV-1; though, with only 75 percent concordance with HIV-1, it was not quite close enough to confirm identicality.[10]

In a "News and Views" column in *Nature,* Ronald Desrosiers concluded that the new virus, if not the actual progenitor of human AIDS, might be the missing link between SIVs and HIV-1. On the other hand, Myra McClure, a virologist at the Cancer Institute, in London, though intrigued, expressed reasonable doubt, based on the genomic differences, together with the fact the chimpanzee virus had been isolated from only a single animal.[11]

More recently, and in one of the most up-to-date reviews, Myers, MacInnes, and Myers have endorsed the unrepentant Essex's view, not only for HIV-2, where the evidence is now overwhelming, but also for HIV-1. As our understanding of the HIV viruses has increased, the professional caution, in attributing HIV-1 origins to cross-species trafficking, would be made to look altogether timid by the behavior of the virus itself. Myers and colleagues concluded: "The random sampling and sequencing to date . . . presents a sobering picture of what has to be expected in the future from these emerging viruses."[12]

Retroviruses, like all RNA viruses, have a ferocious capacity for mutation. Viruses isolated from a single infected person, representing over time the swarm that had resulted from mutation in that person's original infecting strain, were showing envelope gene sequence variation of up to 6 percent. By 1985, some five or six years after the AIDS pandemic began, viruses taken from people who had been infected with exactly the same initial strain were already showing envelope gene sequence variations as high as 12 percent. Six years later, investigation in AIDS patients in Florida showed a staggering 19 percent nucleotide variation. The viral genome was evolving in an explosive burst. For comparison, some experts assess the total genetic differences between a human and a chimpanzee at a mere 2 percent. Myers quotes nine separate authoritative sources all now convinced that AIDS has in fact crossed species to man from primates that lived in the African rain forests.

The actual mechanism of crossover is not too difficult to imagine. There is no need to propose bizarre sexual practices, as suggested by prurient imagination, to facilitate such viral traffic. Any blood contact between the two species, for example as a result of a bite or a scratch, will enable the virus to cross. Africans hunt monkeys on a huge scale, both for food and to supply the enormous Western demand for medical research and culture tissues. The capturing techniques are unsophisticated, sometimes brutal, so that biting and scratching of the handlers by monkeys is commonplace. The potential for blood to blood contact exists every time a hunter traps or butchers a wild primate. It seems hardly accidental that humans should have acquired the virus from primates: it was probably inevitable.

In 1986, Joe McCormick, then head of Special Pathogens at the CDC, together with his younger colleague Kevin De Cock, asked themselves whether AIDS had been circulating in Zaire in 1976 even at the time of the Ebola epidemic. Blood samples taken from the population around Yambuku back in 1976 were still retained in the deep freezes at the CDC. When they screened those samples, taken several years before AIDS had ever appeared in America, they found that almost one in a hundred contained antibodies to the AIDS virus. Just as intriguing, when Kevin De Cock returned to Yambuku and took fresh serum samples from those same remote villagers, he found exactly the same prevalence of AIDS antibodies.[13] In McCormick's words, "This suggests that, under the right circumstances, and with the typical behavior you find in these rural African communities, the virus could quietly tick over in the background without causing epidemic disease." What happened in 1976, and the dreadful pandemic that has followed it, was the result of the coming together of an old virus and new forms of human behavior.

This scenario would be given powerful confirmation from the continuing molecular research. Mathematical calculations of the time needed for the chimpanzee virus, SIV(cpz), to evolve to the genome of

HIV-1, assuming a steady rate of nucleotide mutations, worked out at about forty years.

4

We can, therefore, put forward a cogent and reasonable argument for the origin in nature of AIDS. Both viruses originated from African rain forest primates. While the evidence is not yet absolute, it seems most likely that the more malevolent of the two viruses, HIV-1, emerged from a coevolutionary relationship with chimpanzees. The evidence, meanwhile, is conclusive that the less virulent HIV-2 emerged from a coevolutionary relationship with African monkeys, very likely the sooty mangabey.

There is an immediate gain from such an understanding: subsequent research by Essex and his coworkers has confirmed that prostitutes infected with HIV-2 are, at least partially, protected from subsequent infection with the much more virulent HIV-1.[14] In effect, the HIV-2 virus is behaving like a vaccine — with possible implications for the future development of a prevention for human disease.

But there is another gain, perhaps even more vital in the broader perspective. An understanding of such viruses and their interrelationships with their hosts in the wild must surely reveal something of how they cause such dreadful human suffering, and in this way shed light on ways to reduce the risk of other pandemic viruses emerging.

Simian immunodeficiency viruses are believed to be transmitted from monkey to monkey through the sexual route or through biting. Just as in the case of HTLVs, for a virus to have dispersed and adapted to such a wide range of primate species and over such a vast geography, it suggests an evolutionary lineage extending over thousands, if not hundreds of thousands, of years. There is early evidence that the HTLV-2 virus may be coevolving with South American spider monkeys. The simian immunodeficiency viruses must have begun their infections of African primates after the separation of the old and new world rain forests, which, coupled with the fact that immunodeficiency viruses have not been found in the primates of Asia, suggests that simian AIDS viruses first infected African primates less than a million years ago. At this time, even the monkeys must have come into their first painful contact with the virus from an alien natural host.

In this great assembling jigsaw puzzle of understanding we perceive a vital gap: a piece is missing, a new enigma, the solution of which would add an important new ingredient to the composite picture of enlightenment. Such a pattern of behavior, of benignity toward its natural host and lethality to naive new hosts, hardly surprises experienced virologists: on the contrary, it is the pattern they commonly observe and would indeed predict from a wide experience of other viral infections. But why should this be so? Is there a profound commonality here, the level of unyielding necessity that signals a natural law?

If so, the clue must lie in the very ferocious aggressiveness of viruses such as the *Sin nombre* hantavirus, HIV-1, and Ebola toward immunologically naive hosts. In other words, is the terrible destruction of the invaded species really the gratuitous accident that is commonly assumed?

SIXTEEN

The Aggressive Symbiont

Most impediments to scientific understanding
are conceptual locks, not factual lacks. . . .
We know ourselves best and tend to view
other creatures as mirrors of our own constitution
and social arrangements.

STEPHEN JAY GOULD[1]
Bully for Brontosaurus

1

The greatest discovery in the biological sciences during the nineteenth century, natural selection remains a cardinal principle of evolution today. Although for some its message is so all-embracing it devours all rivals, like a "universal acid," natural selection does not in fact explain all aspects of evolution.[2] The primary creative role, since it has to involve genomic change, can only be genetic, for example through processes such as the recombination of parental genes. Without the mixing of genes that comes from sexual procreation, there would be no differences in the offspring, no choice for nature to select from. Other genetic processes that may sometimes play a part include mutation, which arises from coding mistakes at replication, and hybridization, where two species actually swap parts of their genome. These can give rise to sudden changes in evolution that bear little relationship to the incremental changes over long periods of time that were envisaged by Darwin.

Natural selection will still come into play in deciding which of these changes is eventually to survive as stock attributes of future offspring — an editorial role that is also subtly creative in the long term. But those sudden changes, dramatically altering the genomes of the offspring, are powerful forces indeed in simpler forms of life, such as viruses.

The evolution of species can only have been a direct response to the ecological niche in which the species lives, its landscape, climate, access to shelter and nourishment, and the life-or-death competition with other life forms that occupy the same ecology.

For viruses, this ecological niche is of course the genome of any potential host. If we are to understand why plague viruses behave as they do, we must look to their struggle for survival in this progenitive crucible. Never is this of more crucial importance than at the time when the virus is exploring the genomic landscape of a novel host. In this situation, the common assumption is that human epidemics are nothing more than accidents: people have accidentally strayed across the virus-host cycle, and now that terrifying virulence in some way gives the virus a better chance of survival. But is this the complete explanation? Or is there a more subtle, yet equally dangerous, explanation for such viral behavior?

Certainly, the virus needs to fight its way through those host defenses before it can get into the new genome. Competition is very fierce. In the opinion of Joshua Lederberg, there is a huge selective advantage for the more aggressive strain of virus at this stage. The mathematicians May and Anderson have drawn up algebraic models, assuming that unrestrained natural selection controls the behavior of viruses during this initial invasion. It is rather scary that they anticipate, solely from Darwinian principles, that viruses are as likely to evolve toward greater aggression as they are to ameliorate their aggression with time.[3]

But how then do we rationalize this with the observation that, over time, we very frequently — some think invariably — witness an accommodation: viruses that are initially very aggressive exhibit less aggression toward their new host species. They evolve toward the interesting if curious partnership of coevolution.

It is hard to explain all of this behavior in terms of the unrestrained application of natural selection. The coevolutionary relationship appears to contradict the ruthless selfishness of Darwinian thinking. Why then is coevolution such a common pattern in nature?

A certain debilitating complacency has followed on the heels of the theory of natural selection, which is all too often assumed to explain everything. Yet Darwin himself never claimed that his theory explained all of evolution. Perhaps it is time we questioned some of our own assumptions about the true nature of viruses. We need to ask ourselves if there is a more complex executive control over their observed patterns of behavior, a rational explanation for the emergence of new viruses — and particularly for their sudden eruption into human awareness as lethal epidemic diseases.

Viruses, of course, have no brain. And because they cannot think at all in the higher sense that we commonly imply by this term, we assume that their behavior cannot be purposeful. Consequently we tend to dismiss them as nothing more than the naked expression of Darwinian natural selection. One way around this obstacle is to define a broader concept of "intelligence."

But what can we take for our broader definition? The first step is to conceive intelligence in any life form (and perhaps one day in computers too) as the ability to receive important information about its surroundings and then to be able to change its behavior, perhaps its very heredity, so as to respond to that information. Put that way, then there is of course a simpler, more primal form of intelligence — a genuine very basic executive control — that governs the innate behavior of all life on earth. We are all familiar with instinct, which is one expression of this intelligence. And it plays a bigger role in human behavior than we might be prepared to admit, monitoring and dictating much that is vital in our lives. The programming of this intelligence is genomic. It drives such powerful urges as hunger, sexual need, and, ultimately, self-preservation: it is this programming that guarantees the survival of species.

Drugs that alter human perception will profoundly change the web-making activity of spiders. A hypnotic that puts the human mind to sleep does the same for virtually any other mammal. At this primal level, the British biologist Richard Dawkins focuses all such activity at the level of the gene: he has coined a simple if wonderfully derivative term to embrace a great deal of complexity, "the selfish gene."

In Dawkins's hypothesis, even the most altruistic and morally guided actions of humanity are governed by nothing other than these primal forces. It is rather a controversial notion, suggesting that we, with our higher powers of reason, are as obliged as the amoeba to obey the banal dictatorship of the genome — though, in his defense, Dawkins implies a very wide definition of what selfish in this context really means. At the level of the virus, however, there is no controversy. Driven by the purest applications of natural laws, viruses would appear to be the ultimate selfish genes. Yet at this level, viruses too are endowed with an executive control that could be regarded as intelligent: though it must be the simplest, most primal intelligence in all of nature.

To embrace this different concept of intelligence, I have called it "genomic intelligence," because it monitors and governs a genomic pattern of behavior. If we hope to understand viruses — and there is an urgent need that we should — then we need to understand the role of this genomic intelligence as applied to them.

By genomic intelligence, I mean the capacity of the genome to be both receptive and responsive to nature. This capability has evolved since the first glimmerings of life in direct response to all of the interacting pressures of natural law and with the express purpose of survival of the organism. It does take a little imagination, but the effort is richly rewarding. Genomic intelligence involves a fascinating interaction between the genetic template and nature that embraces every form of life on earth, no matter how complex the species. Recognizing that such an executive intelligence really does govern the behavior of viruses gives us a means of understanding them. Their behavior is no longer as unpredictable as they are invisible. They are responsive to, and therefore governed by, the same natural laws that control all other forms of life on earth, only one of which is the canon of natural selection.

From a viral perspective, genomic intelligence is the only possible explanation for coevolution. It is easy to see what the virus gets from the arrangement, but why does the host tolerate it, perhaps over millions of years? Surely pure Darwinian evolution should produce a strain of host species over time that would be resistant to the bothersome presence of the virus?

A number of explanations spring to mind. Perhaps the host is immunologically unable to shake off the viral infection? Even a moment's reflection makes this rather unlikely, given that we are talking about contact across the genetic diversity of an entire species. In keeping with this, as we grow in our understanding of the immunological tolerance necessary in the host to allow coevolution, it seems that the host is not trying desperately hard to shake off the virus. While there is an obvious interaction between virus and host at a complex immunological level, there is also a much deeper inter-reaction, both struggle and intercourse, at a genomic level. And that genomic intercourse is the key.

Much in nature may happen through chance, but very little persists long term through chance. There is an obvious implication: the host accepts the persisting viral infection, now at a relatively or absolutely nonpathogenic level, because it confers some advantage to do so. But what is this advantage?

So let us look hard to nature. Are there other examples of such a relationship between a potential predator and prey, between totally dissimilar species?

Of course there is precisely such a relationship well known to biologists, a relationship between two very different species of life, living together in a very close affinity that confers mutual benefit. It is called symbiosis.

2

It was a German botanist, Albert Bernhard Frank, who first defined the concept of symbiosis to describe the mutual interdependence between two distinct species. He was fascinated by the living partnership of algae and fungi that gave rise to an apparently new species, lichens. A year later Frank's idea was adopted and expanded by another and more influential German botanist, Anton de Bary. It was de Bary's inspiration that the concept had very wide ramifications throughout biology, extending from the mutual assistance between species (mutualism) through a series of relationships to outright parasitism.[4] But in the tidal wave of excitement and controversy that followed Darwin's discovery of natural selection, symbiosis was shunted to a minor track of biological curiosity.

It is easy to see why natural selection took center stage: the idea has a wonderful universal appeal. The tragic outcome of all this has been a tendency for scientists to view natural selection and symbiosis as mutually contradictory. In fact they are by no means immiscible doctrines.

Lynn Margulis, a biologist at Amherst College and one of the leading proponents of symbiosis today, believes that through symbiosis radically different life forms combine to accomplish more than either could separately. In her opinion, symbiosis is one of the most potent creative forces in evolution.[5] For example, the mitochondria that provide animal cells with oxygen-derived energy were once independent bacteria that were engulfed by the primal cells that later evolved into animals. This is almost certainly the explanation for a curious paradox. Mitochondria have their own hereditary DNA, which needs to divide by its own private little binary fission every time the cell itself divides. And plant cells too have their astonishing equivalent, chloroplasts, which provide them with the ability to photosynthesize. Much of the early research on symbiosis was conducted in Russia, and it was two Russian biologists, Andrei Famintsyn and Konstantine Merezhkovskii, who invented the term "symbiogenesis" for this fantastic synthesis of utterly new living organisms from such symbiotic unions.[6] It would be hard to imagine how the step by step gradualism of natural selection could have resulted in this brazenly passionate intercourse of life!

It is hardly surprising that experts like Margulis are convinced that symbiosis is one of the fundamental principles in evolution. How exciting to realize that it is equally fundamental as a principle and mechanism for the coexistence of the diversity of life on earth.

We may think it lovely to observe how bees derive their sustenance from the nectar of flowers. In fact it is a matter of life and death to both flower and bee. The bee depends on nectar for its energy needs and the flower on the bee for pollination. The beautiful lichens that decorate the surface of tombstones are a symbiosis, as are the 13,000 other species of lichens that carpet the most inhospitable regions of the earth, from the searing desert to Arctic wastelands. All are symbiotic unions — "holobionts" — of fungi and algae. The cell bodies of coral, an animal, embrace in their living protoplasm the bodies of algae, which are plants. The algae make sugars and starches through photosynthesis that provide nourishment for the coral, while the coral provides a shelter and supplies of otherwise scarce phosphate and nitrogen essential to the algae. Bacteria fix nitrogen around the roots of leguminous plants; other bacteria fix sulphur or methane in symbiotic partnerships with marine invertebrates.

There appears to be an inexhaustible multiplicity of such examples in nature, where symbiosis is essential to the life cycles of both partners and where the interdependent partners are called symbionts.[7]

All of the great cycles of life on earth are symbiotic in a more holistic sense. So green plants, on land and in water, produce the oxygen that animals, including humans, breathe. Animals, in their turn, produce the carbon dioxide the plants need to manufacture carbohydrates. The humble bacteria that teem in soil recycle elements such as carbon and nitrogen, derived from the complex chemicals and proteins of animals and plants, and enable the new generation of plants to reutilize them in their growth. How extraordinary it is to realize that if these humble microbes that live in the soil were to die off, all of the higher life on earth, including humanity, would become extinct! The harder one looks, the more important the role of symbiosis in all of the great cycles and balances that play such a vital role in life on earth.[8]

Viruses, as we know, are uniquely vulnerable in their life cycles. They cannot live independent lives but are compelled to live in partnership with another life form. Outside the limits of that partnership they enter a long sleep, the most profound hibernation in existence, only awakening to dramatic effect when they enter the genome of the host cell. In such circumstances, symbiosis offers a compelling advantage.

In 1992, Professor Werner Reisser, at the Plant Physiology Institute in Göttingen, Germany, recognized this when he proposed a new definition of symbiosis: the interaction not between different species or life forms but between dissimilar genomes.[9] Yet the concept of symbiosis is not even listed in the index of the otherwise encyclopedic textbook *Fields Virology*. This omission seems part of a more general uncertainty on the part of scientific historians and students of evolution, for whom "symbiosis as a source of evolutionary novelty is virtually never mentioned."[10]

But these concepts have been scientifically verified in many experiments over the years, even the literally iconoclastic symbioneogenesis. One such remarkable experiment was conducted in 1966, when a "D strain" of a protozoan, the *Amoeba proteus*, was deliberately infected with a microbe called x-bacteria, containing two plasmid packages of DNA. At first the amoebae tried to digest the invading bacteria, but the x-bacteria resisted this and almost every infected amoebae died. However, a few amoebae survived the infection, continuing their life cycles with the bacteria still within them. In time, the amoebae accepted a burden of infection where the bacteria amounted to 10 percent of the weight of the amoeba's body. Incredibly, the resulting "xD amoeba" had now become dependent on the contained (endosymbiotic) bacterial genome for its very survival.[11] The amoeba and bacterium had fused into a new life form, a holobiont, just like human cells and their contained mitochondria.

This experiment amply demonstrated a general principle in which "the origin of symbionts is likely to derive from parasite-host relationships and to evolve into mutualistic symbiosis."[12] In other words, the intriguing virus-host relationships we recognize as coevolutionary must have begun as such a parasitic attack upon the host. The myxomatosis example in Australia is the classic example. The virus that killed the Australian rabbits was taken from Brazil, where it coevolved with the local rabbits, causing no illness at all. But when the same virus was deliberately introduced into Australian rabbits by scientists, the result was an epidemic disease that killed 99.8 percent of the rabbits throughout the entire Southwest, an area the size of Western Europe, in just three months. You could not get much closer to extinction than that. Yet, over the following ten years, the tiny population of surviving rabbits bred a new population that was now immunologically tolerant to and coevolving with the virus.[13]

Here we see a very important example of a more general pattern. A virus begins as a lethal attacker, yet, given time, it ameliorates its behavior through coevolution until, after many generations of both virus and host, a totally new modus vivendi emerges — a true symbiosis in place of what was formerly a predator-and-prey relationship. How do we come to understand such a spectacular evolution?

Some authorities would extend symbiosis to cover all parasite-host relationships, no matter how virulent the infection. But what we are considering is "mutualistic" symbiosis, where the relationship is by and large mutually benign and supportive. We can begin, therefore, by teasing apart what we mean by coevolution and mutualistic symbiosis, for they are not quite the same. Coevolution implies a dynamic process of change, where two very different species evolve in parallel, the host in response to its environment and the virus in response to the changing genome of the host. Symbiosis, in this sense, is the state of equilibrium that results from this coevolution over a long time and many generations. The meeting point, when virus and host arrive at a stable and mutually interdependent relationship, is not, however, altogether easy to define. The symbiotic union must necessarily retain some of the dynamics of coevolution, though at a much lower intensity than when the infection first took root. But unlike natural selection and its ready fit to epidemic viral behavior, symbiosis does not seem to conform to the mathematics of epidemics devised by May and Anderson. The aggressive or lethality factor is now zero, or close to it.

3

In the forests of Borneo grow species of rattan cane that have a symbiotic relationship with ants. The ants construct a nest around the cane and protect it from browsing herbivores. If the leaves are tugged, the ants swarm from their nests, first beating out a tattoo, rather like the warning of a rattlesnake. If the attack upon the cane continues, the ants charge out and sting the offending animal about its sensitive mouth parts: in the experience of Sir David Attenborough, they have a vicious bite.[14] In the savannahs that encircle the African rain forests, there are many varieties of acacia trees that rely on a similar symbiotic protection from ants. A browsing herbivore, such as a giraffe, is subjected to a remorseless stinging assault upon its tongue and lips until it ceases to damage the acacia. In return, the tree repays its protector with living quarters and a ready and consistent supply of sugary nectar.

In the South American forests, the relationship is even more marvelous. There are species of acacia here that produce a waxy berry of protein at the ends of their leaves that provides the main sustenance for the growing infants within the ant colony, while the ants not only keep the foliage clear of herbivores and preying species of insects but make hunting forays into the hinterland about the tree, ravaging the growing shoots of potential rivals to the acacia.

This is a very common form of symbiosis. The contribution by the minuscule partner is not one of mutual provision of sustenance but aggression. For this reason I have termed it "aggressive symbiosis," though the aggression is not directed at the symbiotic partner but outward — "exogenously" — at a potential rival of the symbiotic partner. No biologist would question this as a classical example of symbiosis. So what is the difference to the situation where a virus protects its host in a similarly aggressive way? The host benefits from the reduced competition from a rival species, while the virus enjoys protection and the facility of replication within the host. The virus, like the ant in the acacia tree, is the "aggressive symbiont."

A striking example of such aggressive symbiosis is the behavior of herpes-B viruses in their infective relationships with primates. One such virus, *Herpesvirus saimiri,* has a coevolutionary relationship with the squirrel monkey, *Saimiri sciureus,* that lives in the Amazonian rain forest. If a rival species, for example, the marmoset monkey, which shares the same rain forest habitat, comes into contact with the squirrel monkey, the virus hops species to wipe out the entire rival species. Not only does the virus induce cancer, it does so with a voracious rapidity, the marmosets dying from a fulminating cancer of the lymphatic system.[15]

A similar virus, *Herpesvirus ateles,* which coevolves with spider monkeys in the South American jungle, is also virtually 100 percent lethal to exposed monkeys from related but immunologically naive species.

The herpesvirus coevolving with the squirrel or spider monkeys is behaving exactly as one would expect of an exogenously aggressive symbiont in eliminating competition with its host for food and shelter in the local ecology. Maintained in the host species by vertical transmission through oral contact with virus-containing saliva, the virus is shed copiously over the monkey's ecology throughout its life. The analogy with urine shedding of hantavirus is striking. Puzzled by the fact that the virus would not harm its host, scientists tried to force the virus to induce disease. Yet even when the natural host was immunologically suppressed with drugs, the virus could not be induced to harm it. Could anything have more dramatically illustrated the true symbiotic nature of the relationship!

This relationship, between the simplest of all life forms and every other form of life on earth, has an endearing aesthetic appeal. But one should not be deluded by this. The aggression of a viral symbiont is on a different plane from that of an acacia-dwelling ant: this aggression is directed not at the lips or mouth parts of a herbivore but into the very heart of the rival genome. In its expression, we witness the awesome potential that is latent in the viral intrusion into that all-powerful ecological landscape.

Another herpesvirus, called the *Herpesvirus simiae* B, is symbiotic with various species of Asian macaques. There have been at least twenty-two cases where humans who came into contact with this aggressive symbiont contracted a serious infection, twenty developing severe inflammation of the brain, or encephalitis, which resulted in fifteen deaths.[16]

In life the situation is invariably complex and associated factors may increase or decrease the likelihood of such aggressive attack. The Four Corners hantavirus epidemic only began when there was a massive increase in deer mouse population, with a presumed sharp increase in the rodent's demand for food and living space. Evidence from other rodent-borne epidemics is similarly suggestive, from Jamie Childs's experiences with arenaviruses to past outbreaks of the admittedly bacterially caused bubonic plague. Given the uncertain nature of the animal reservoir of Ebola, it is not possible to determine the precise links here, yet the seasonality, the hinterland of the rain forest, and the outbreaks now in three species of primate all suggest a very similar pattern of behavior. This suggests that the known epidemics of Ebola have been triggered by human intrusion into the virus-host symbiotic cycle, perhaps during a reproductive phase or season.

Darwinians will see in symbiosis the expression of natural selection: certainly it is in the interest of the symbionts to support one another, since this passes on an advantage for the survival of the offspring. But it is equally possible to see natural selection as the road to symbiosis. Common sense would suggest that there is no real competition between these two powerful principles: both apply, and simultaneously, though at stages one may predominate over the other.

And with this we are in a position where a synthesis is possible: a hypothesis that integrates both natural selection and symbiosis in the dynamics of the evolution of plague viruses and their behavior toward their victim species.

The genomic intelligence, which exerts an executive control over viral behavior, is influenced by and responsive to both natural laws. When the virus first hops species, the predominant expression of that executive is aggressive symbiosis. At this stage, viral behavior is primarily directed to protecting its symbiotic partner and not toward colonizing the new species. Only later on, as in the example of myxomatosis in the Australian rabbits, will natural selection come to predominate, at the stage when the attacked species have survived the initial viral assault and only if conditions are then amenable to the long-term coevolution, through natural selection, of the virus and its new host. It is a curious reversal of the common perception: where symbiosis is the hawk and natural selection the dove.

A number of interesting deductions arise from this hypothesis. Firstly it implies that the critical time in the emergence of new viruses — the most dangerous time as far as people are concerned — is when aggressive symbiosis predominates in viral behavior. And this, in turn, has important implications. The current pattern of human expansion brings people into frequent contact with other animals, from mammals to insects. Since every species of animal will have one, and often a good deal more than one, species of virus associated with it, there must be a multitude of opportunities for viral traffic daily. The question demanded by the late Bernard N. Fields becomes all the more relevant: Why hasn't this enormous diversity wiped out the earth?

In part, Fields answered his own question. Any virus that has coevolved to become symbiotic with one species will have an uphill struggle to establish itself sufficiently to attack a totally novel intruding species. It has to find a route of trafficking. It needs to penetrate surface defenses, protect itself from immune attack, discover a cell similar enough in the new species to its former host cell to allow its highly evolved mechanisms to penetrate the cell. Then, within the cell, it has to find a means of taking over the genomic machinery for its aggressive purposes. The vast majority of viral attacks must necessarily fail.

Where the host species is very different from humans — if it is a fish, for example — such an attack is very unlikely to succeed. Again, where two species have been in close contact for millennia or more, for example, humans and dogs, any such attacks will have worked through the ecology of both species long in the past. Though, as the *Sin nombre* hantavirus has shown, increased contact between a common feral animal and humans can still provide a key junctional zone. There will be many other such gray areas, where perhaps a new virus is on the move in the animal kingdom or a change in host or human behavior tilts the situation from safety to risk. The greatest danger, however, is all too predictable. It will arise from contacts between people and animal species that have rarely if ever made contact before, what might be termed "new junctional zones."

This behavior of viruses has evolved over billions of years. The genomic ecology of humanity is little different from any of the other primates, and from the perspective of an aggressive virus, it is not too different from most terrestrial mammals. It is hardly accidental that the increased emergence of viruses, and other microbes, affecting people in the second half of this century has paralleled the extent of human invasion of wilderness ecologies such as the rain forests and of massive alteration of savannah and other similar ecologies through farming. Even here junctional zones must be arising very frequently, yet the same considerations apply. Most aggressive attacks from viruses will simply peter out. But the sheer volume of contacts is enough to ensure that sooner or later a virus will emerge with the formula for success. Genomically speaking, the closer the host species is to humans, the more likely the virus will possess that perilous formula.

And there is another, vital consideration to all this. If the hypothesis of aggressive symbiosis is correct, then the "accident" theory of viral attack on people, for example in the African Ebola outbreaks, is misleading. The attacking virus is programmed to injure and kill, even if in doing so this portion of the swarm sacrifices itself in an evolutionary cul-de-sac. The symbiotic relationship is well served by this sacrifice. This is as important as it is a radically different perception.

The Ebola and hantavirus outbreaks fit the first step in this integrated scenario. They are precisely what one would expect from aggressive symbiosis on the parts of the viruses. And if they have made humanity more wary of encroaching on the African rain forest, as such attacks may well have succeeded in doing in our hunter-gatherer past, they could be seen as biological successes. But Homo sapiens is no longer primarily a hunter-gatherer. We have radically moved the goalposts, with huge urban populations and a mobility that would have been impossible for animals in nature. And it is this that makes it possible, more rarely of course, but of immense human significance, for a combination of successful viral strategies, extreme infectiousness, and human behavior to produce the right circumstances for an epidemic in the new human host while the virus is still behaving as an aggressive symbiont.

What then are the circumstances that make it possible for such aggression to become epidemic?

Occasionally, the attacking virus discovers an avenue of contagion that enormously extends its penetration and infectiousness. This can arise from spread through a universal medium, such as food or water, through carriage by a common biting insect, or most dangerous of all, through spread in a highly efficient way from one member of the victim species to another. In human terms, the most extreme manifestation of such circumstances will take place when the aggressive symbiont, now termed an emerging virus, discovers the means of mounting a global attack. This is the most threatening scenario of all, an epidemic that sweeps across continents, referred to as a "pandemic."

Such an event took place in the late 1970s to produce the worst human catastrophe of the twentieth century when two immunodeficiency viruses, HIV-1 and HIV-2, long established as aggressive symbionts with their evolutionary partners, primates in the African rain forest, attacked an intruding rival species of primate — humanity — and discovered the venereal route as an efficient means of spread from person to person. The result was the AIDS pandemic.

4

In the early 1970s, while researching the Spanish conquest of Mexico, the historian William H. McNeill came across a curious observation. It was well known that Hernán Cortés, with fewer than six hundred men, conquered the Aztec empire, whose subjects numbered millions. "How," McNeill asked himself, "could such a tiny handful prevail?" A casual remark in one of the accounts of Cortés's conquest suggested the answer. When Cortés first attacked Mexico City his army was in fact defeated. But on the very night he and his men were in full retreat, with many dead, an epidemic of smallpox was already raging in the city. In McNeill's words, "The paralyzing effect of a lethal epidemic goes far to explain why the Aztecs did not pursue the defeated and demoralised Spaniards, giving them time and opportunity to rest and regroup . . . and so achieve their eventual victory."[17] McNeill became so intrigued he searched for other historical parallels. The result was a book, *Plagues and Peoples,* which is regarded as a classic on this poorly researched subject.

The conclusions of McNeill's book would startle many of his readers, not least historians. For he put forward the hypothesis that plague microbes, bacterial and protozoal as well as viral, have played important roles in the history of human civilization, and very particularly in the aggressive wars and conquests of one group of people over another. Prolonged interactions between human host and infectious organisms, particularly such interactions over very many generations, created a pattern of mutual adaptation that allowed both to survive. "At every level of organisation — molecular, cellular, organismic, and social — one confronts equilibrium patterns. Within such equilibria, any alteration from 'outside' tends to provoke compensatory changes throughout the system so as to minimise overall upheaval." For McNeill's "equilibrium" I would substitute "symbiosis" and for his compensatory changes, I would substitute "aggressive symbiosis."

The smallpox that decimated the Aztecs was in the process of symbiotic adaptation, through coevolution, to Europeans, including the Spanish. Where the herpesviruses and SIVs crossed species, the smallpox trafficked from conquistador to Aztec, across immunologically different human populations. The result was nevertheless a typically aggressive symbiotic attack upon the Aztecs, which profoundly altered American history. While there is some evidence that the Native Americans may have been unduly vulnerable in originating from a relatively small population base — this might mean they lacked a wide enough genetic variation — even a genetically diverse human population that was also naive to smallpox would have been devastated. Many other examples can be found for the role of aggressive symbiosis in human history, particularly in a pioneer-versus-native confrontational situation.

Measles is a viral disease that probably emerged from a virus such as rinderpest in cattle at about the time of the Roman empire. Like smallpox, it coincided with the rise of human populations in the historic centers of Rome, China, India, and the Middle East, where its emergence into immunologically naive populations brought massive epidemics with a harrowing mortality.[18] Where today it has coevolved to the extent that it is relatively benign in its infection of Western children, in recorded outbreaks in novel human populations, such as the Fiji Islands and the Amazon basin, high mortalities have been recorded, even the extinction of entire tribes.[19]

Other microbes, and particularly viruses, have coevolved with every known species of animals, insects, and even plants. The end result — what McNeill refers to as a "stable, well-adjusted, and presumably very ancient" relationship — is symbiosis. It seems astonishing that even the malarial plasmodia are symbiotic with certain rain forest primates and rodents, and the trypanosomes that cause epidemics of sleeping sickness are symbiotic with the ungulates of the African plains. The advantage to the host of their aggressive symbiosis is fascinating when you place it in its historical context. Malaria and yellow fever — the latter also symbiotic with monkeys — have probably done more to preserve the ecology of the rain forests than any aesthetic or moral scruple of man. And, in McNeill's opinion, "It is mainly because sleeping sickness was and remains so devastating to human populations that the ungulate herds of the African savannah have survived to the present."

Primates are believed to have originated in Africa. The most ancient pools of primate genomes with their symbiotic partners are nowhere in greater abundance than in the African rain forest. This is why African primates support a huge diversity of microflora, including viruses. One only has to compare the fate of the Africans, no more equipped to protect themselves against European attrition and exploitation, with that of the Native Americans. The more temporate regions of America, relatively unguarded by aggressively symbiotic microbes, provided an easy lebensraum for European colonists, while Africa, at least within the hinterland of its great and ancient rain forests, became the White Man's Grave.

So the plague rickettsia of typhus is symbiotic with its carrier tick. The hantaviruses and the aggressively emerging arenaviruses, including the South American hemorrhagic fevers and the African Lassa fever, are symbiotic with their rodents. The seal plague that emerged to destroy large populations of seals off the western European coast in recent years is probably symbiotic with other species of seals.[20] The more one considers the evidence, the more extensive becomes the list — and it extends beyond viruses to bacteria and even protozoa. How incredible to realize that even bubonic plague is symbiotic with certain rodents, cholera with algae, and bilharzia perhaps with humble water snails.

In 1995, when I met David Simpson, the British virologist who led the Sudan Ebola team, I asked his opinion on this hypothesis of aggressive symbiosis. He agreed that most viruses rely on a symbiotic relationship for their very existence.

Joshua Lederberg also supports this hypothesis, including in his warning of the very real threat of new epidemic viruses, the "potentialities for change in the evolution of their symbiotic relations with their hosts." Indeed he takes the inference further. "It is not hard to imagine the sources of resistance to these evolutionary concepts. It is scary to imagine the emergence of new infectious agents as threats to human existence, especially threatening to view pandemic as a recurrent, natural phenomenon."[21]

Why then has the threat of such circumstances increased in the last half century? And just how great is the future risk?

In Simpson's words: "There are very many more viruses at large in nature. Man simply has not yet entered their natural ecosystem. When man does interfere with that ecosystem, he may end up becoming infected. When that happens, new epidemics will occur."[22]

SEVENTEEN

The Tear in Nature's Fabric

The world is always surprising us,
overthrowing beliefs based alike on tradition,
superstition, common sense or science.

RICHARD LEVINS[1]

1

For those readers not familiar with the "Gaia" concept, this was an idea first put forward by J. E. Lovelock in 1979, when he adduced that "the entire range of living matter on earth, from whales to viruses, and from oaks to algae, could be regarded as constituting a single living entity."[2] This fantastic synthesis, named Gaia after the Greek goddess of the earth, was extended to include the entire interactive biosphere, the atmosphere, seas, and soil. The core of the hypothesis was that all of these entities when added together amounted to a whole that was greater than the arithmetic sum of its individual parts. This holistic concept had a life of its own. It could, for example, manipulate the earth's atmosphere to suit its overall needs.

At first acquaintance, the hypothesis seems preposterous — a fantasy from the overly romantic imagination of a romantic naturalist.

But Lovelock's idea is more subtle and science based than this. It grew out of his struggle to define what constitutes life for NASA's space exploration program. What it really hinges on is a rather broader definition of life than would be familiar to most biologists, a definition first designed as a means of searching for its presence on an alien planet, such as Mars. While we may circumvent the more unorthodox if fascinating proposals in the Gaia hypothesis, the notion of a self-regulatory homeostatically balanced earth is entirely acceptable to ecologists. This way of thinking opens our imagination to a broader and very important perspective on the biosphere and the mechanisms by which it maintains the balance of its living and nonliving constituents.

It now seems likely that all aspects of life on earth have their viruses, from the great redwoods of California to the humblest bacteria on the bottoms of the oceans. Although their primary goal may be selfish replication, the more universal effect, the holistic effect, will inevitably be far-reaching. The diversity of life on earth, the very sum of the myriad of species, the matrix of their interactions — since by and large the biomass is a matrix of interactions rather than the product of predatorial battles — is to create an interdependency of life. Viruses, so often thought to be nothing more than parasites, play a much wider role in this interdependency through their symbiotic relationships with other forms of life. Indeed some people might not unreasonably derive from the hypothesis of aggressive symbiosis a "Gaia" self-regulatory effect. It could be argued that viruses have, through the empirics of evolution, become unwitting knights of nature, armed by evolution for furious genomic attack against her transgressors. Although not primarily designed to attack humanity, human exploitation and invasion of every ecological sphere has directed that aggression our way.

No other life form could have been more efficiently honed for such a counterattack. However one views their holistic role, there is no denying that once emerged, in the words of Richard M. Krause, plague viruses "know no country. There are no barriers to prevent their migration across international boundaries or around the twenty-four time zones."[3]

For half a century, the rate of new disease emergence has been increasing.[4] Warnings about the potential consequences of such a situation are not new. Even a savant such as Joshua Lederberg was shocked by the arrival of the Marburg virus, conceding that it came as a total surprise, a reawakening.[5] Then in 1976, with the dramatic emergence of the Ebola virus in Africa, the British virologist Gordon Smith added his voice to the growing chorus of alarm. He had no doubt where the danger zone lay. "The emergence of a previously latent zoonosis is usually due to a change in the ecology of its maintenance cycle or to changes in the ecology of neighbouring areas. . . . The larger the scale of man-made environmental changes and the more they involve areas little frequented by man, the greater must be the probability of emergence of a zoonosis, old or new."[6]

The prime factor behind the emergence of a new plague virus is therefore obvious. The threat to humanity derives in particular from rodents, since they are the most numerous mammals on earth, and from primates, because their genomes are so much closer to our own. Most species of rodents and primates live in the rain forests or their hinterland of grassy savannahs. Millions of years ago, before humans began to disrupt their ecologies, the balance between predator and prey had been stable, or nearly so, for very long periods of time. Until recent times, taking the African rain forest for example, human intrusion was kept in strict check by the rich variety of parasites lying in wait.

On the other hand, the mammals that took their origins there had, over these vast periods of time, developed symbiotic relationships, through coevolution, with very many microbial life forms. Today, for example, every individual monkey, baboon, chimpanzee, and gorilla is carrying at least ten different species of symbiotic viruses. While the situation in rodents, bats, and marsupials is not so well defined, it is likely to be very similar. It is significant in this sense that Ebola, Marburg, and HIV all derived from the African rain forest or its hinterland savannah. When scientists mark the epicenters of origin of newly emerging virus infections on the global map, it is clear that almost all of these have emerged from these formative biomes.[7] And this means that interference with the rain forests, and deforestation in particular, is the most dangerous activity with regard to the emergence of epidemic viruses.

2

Until recently, Robert Shope was the director of the Arbovirus Laboratory at Yale University, in New Haven. In the same way that Terry Yates is a systematist in studying mammals, Shope is a systematist when it comes to viruses. He collects unknown viruses and he classifies their evolutionary lineages and trajectories. His collection currently includes some five hundred different species, each tagged according to time and place of capture, its precise hosts, vectors, and vertebrates, the geography and ecology in which it was found, and its molecular biology and epidemiology.

Arboviruses are much more prevalent around the world than people might realize, causing a range of diseases from plagues such as yellow fever and dengue fever to the more subtle teratogenic effects that have only recently been suspected from a large variety of mosquito-borne Bunyaviruses, such as La Crosse and Jamestown Canyon viruses.[8] If Robert Shope is sent a virus from a sick traveler who has arrived in New York City, he will not only diagnose the causative virus but tell you in what country, rain forest, subtropical savannah, town, or district the person picked up that strain of virus.

In the mid-1950s, immediately following the Korean War, he found himself at Camp Detrick, now known as USAMRIID. When he requested overseas experience, he was sent to Kuala Lumpur, in Malaya, where he worked with Colonel Robert Traub, an entomologist with a special interest in mites. Shope became involved in a number of research projects, but one in particular fascinated him.

This was during the Malayan insurrection, when the British army was parachuting SAS counterinsurgency troops into the jungle. The troops would return carrying infections. Leptospirosis, caused by a spiral bacterium, was a particular problem, but of greatest interest to the budding virologist was the fact that once in the jungle these fit young men seemed to pick up mysterious fevers. These infections did not appear to have a bacterial or protozoal cause. To Shope, it seemed likely they were picking up new viruses. It was apparent that there were a great many unknown viruses in the jungle, some of which — perhaps many — had the potential to infect people.

For the Harvard ecologist Edward O. Wilson, as noted, the rain forests are home for two thirds of the species of life on earth.[9] Most if not all of these species have microbes associated with their ecological and life cycles. Malaria, leptospirosis, and dengue are all to be found in the Malayan rain forest. Shope, who was now in charge of leptospirosis diagnosis for the army, found that he could predict where in the jungle the afflicted soldier had picked up his infection from the serotypical strain of leptospira he was isolating. Moreover, there seemed an interesting link between the infections picked up by the SAS parachuting into the jungle and those afflicting the people living in the nearby city.

At much the same time, Stefan Pattyn was picking up a related intelligence from his experiences in what was then the Belgian Congo. In the early 1950s he found himself working in a hospital in the Katanga province, where one of his colleagues, Dr. Ignace Vincke, the head of the hygiene laboratory, was conducting experiments on the chemotherapy of malaria. Malaria is caused by a protozoal organism called a plasmodium. Today we know that there are very many such organisms in the rain forests of Africa. Before 1950, there was only one experimental species of plasmodium known, *Plasmodium gallinarium,* a species that infected birds, so this was the only malarial parasite that could be used for such experimental purposes. This microbe was so far distanced from human disease that its usefulness was severely limited. Vincke decided he would look for alternative sources of plasmodia in nature.

Katanga was a savannah region, with forested areas along the rivers. Vincke searched the riverine forests and captured animals, staining their blood for the presence of the parasite. In one species of rodent he found what he was looking for — a new plasmodium he called *Plasmodium vergae.* This was a major step forward because the plasmodium was much closer to the human parasite. It could be cultured in the laboratory. New and old drugs could be tested against it. The discovery helped this innovative scientist eradicate malaria in the white settlers living in Katanga. But there was a subtle lesson to be learned from this experience.

Later on, when biologists and microbiologists visited the gallery of riverine trees to assess the precise geography of the newly discovered parasite, they were astonished to find it was very limited. The *Plasmodium vergae* inhabited no more than twenty kilometers in extent in stretches along the narrow green banks of the river. It was a brilliant confirmation of the modus vivendi of those archipelagos of virus-host cycles that had been discovered by genomic mapping. In the words of Pattyn, "There must be many factors, environmental, humidity, light, temperature, that modify both the host life-cycle and that of the microbe."

Pattyn has a simple analogy to explain the human consequence that might arise from this. He compares the microbe-host cycle to a colony of ants that have been living in their isolated anthill in the wilderness for millenia — in the virus analogy, perhaps aeons. Human intrusion into the local ecology causes the anthill to be destroyed. The ants swarm out and attack.

For decades the Rockefeller Foundation had satellite laboratories in many Third World countries, many working on the control of yellow fever. Two of the countries involved in this initiative were Brazil and Colombia. Yellow fever, the most notorious of the arboviruses, is spread by mosquito throughout its own archipelago of symbiotic rain forest monkeys. One of the things the researchers would do is to capture forest mosquitoes and grind up their bodies to look for the yellow fever viruses. While examining these mosquitoes, the scientists began to find other viruses. Worried by the potential of these unknown viruses to infect people, the Rockefeller Foundation established research laboratories in India, Brazil, Trinidad, Nigeria, Uganda, and South Africa. By now Robert Shope wanted to follow up his observations in Malaya. The Rockefeller initiative gave him the opportunity to study viruses in the South American rain forest, and he traveled to Belém, in Brazil, near the mouth of the Amazon. Here, working under the direction of Otis Causey and his wife, Calista, he found many hitherto unknown viruses.

Some Japanese immigrants were actively involved in colonizing newly cleared areas of forest. They would employ foresters brought in from the northeastern part of Brazil looking for work. These people were unfamiliar with the local ecology and had little immunity to the native viruses. Soon after the foresters began felling trees, they began to get sick, each succumbing to some strange fever or another. At much the same time the Pirelli rubber company was also employing foresters to clear rain forest for rubber plantations. Once again, the foresters who felled the trees, those who worked on the front line of destruction of the ecological habitat, were the ones to fall sick. Shope investigated their illnesses and concluded they were, in the main, caused by viruses. But these were not the familiar perils of yellow fever or dengue. These were new illnesses, caused by a group of viruses now classified as "Group C" viruses. They would be given Brazilian names after the villages, rivers, or small streams where they were found, such as Oriboca, Caraparu, and Itaqui.

Each time they found a new virus, Shope and his colleagues would search for the hosts of the viruses and for the route of their trafficking to people. The scientists devised a number of experiments to trap insects and their vertebrate hosts, both in the canopy and on the forest floor. One ingenious device was to look around for people who were lying down in hammocks during the daylight hours. Realizing that these hard-pressed foresters could not afford to lose such time from work, Shope soon confirmed that those found in hammocks during the day were likely to be sick. These men would be stretched out between trees close to where they worked in the forest. So Shope and his colleagues would take blood samples from the sick foresters and collect mosquitoes in the vicinity of the sickbed, using light traps. They set animal traps, for example, baited with baby mice, protected by a hood from the rain and equipped with a fan that sucked the mosquitoes into a collection bag. In a still more illustrative experiment, they employed "human bait" by paying volunteers to enter the forest. Nothing much seemed to happen when such incursions took place by day. But during the hours of darkness it was a very different story. Suddenly many volunteers were coming back from the forest febrile, invaded by viruses.

By the time they left Brazil, Shope and his group had discovered at least fifty new viruses — and even this was a considerable underestimate because they could not identify all of the species. Many were arboviruses, transmitted by scores of different mosquitoes. The entomologists made wholesale discoveries of new species of mosquito, their numbers so prolific that they ran out of ideas for names and allocated numbers — Culex 1, 2, 3, and so on.

The hosts for these viruses were very often rodents or marsupials. On the whole, when people intruded upon the ecological cycle they became sick but they did not develop fatal illnesses. A typical illness would involve malaise, listlessness, fever, and the skin might erupt into a rash. On the other hand, when the viral host was discovered in nature, Shope could discover no illness resulting from the host infection, no matter that the animal's blood teemed with virus. To Shope and his colleagues, it seemed very perplexing — all the more so when they found that if they injected the viruses into white mice, the mice died.

In the forest surrounding Belém, Shope found six groups of C viruses. Each was associated with its own particular ecological cycle, some in the forest canopy, where they were coevolving with marsupial hosts, others on the forest floor, where the host was more likely to be a rodent. Each was associated with its particular insect vector, its own vertebrate host. Like hantaviruses, the viruses belonged to the family of *Bunyaviridae* and they followed a similar pattern of parallel evolution between virus and host. The viruses had a stable relationship with their hosts that implied a coevolution over many thousands of years.[10]

At the same time that Robert Shope was working in Brazil, the government decided it would carve a highway through the Amazonian rain forest that would link Belém, on the coast, with the newly created capital city of Brasília. In 1960 Shope and his colleagues conducted an autopsy on a dead sloth they had come across by the side of the new road. In the sloth's blood they identified a virus. The same virus had been discovered five years earlier near a village called Oropouche, on the neighboring island of Trinidad. At this time, the disease potential of the Oropouche virus was unknown. A year later the virus showed its plague potential when an epidemic broke out in Belém, with a flulike cough, severe muscle aching, and high fever, which infected 11,000 people. The Oropouche virus would subsequently go on to infect more than 200,000 people living in or close to the Amazon between the years 1961 and 1981.[11]

From what natural cycle had this virus emerged? How had it moved all the way to the coastal city where it infected humans? Though scientists struggled hard to find the answers to these questions, it would be nineteen years before it was discovered that the virus was transmitted by a biting midge, which had undergone a population explosion when settlers started to clear the forest to plant cacao trees. The epidemic nature of the virus had only manifested itself when, following its emergence through deforestation, a second human behavioral factor had also intruded into its ecology. The midges bred in the discarded husks lying around the cacao farms in huge waste piles.

A great many strange viruses still lurk in the rain forests of the world, awaiting their first contact with people. While the vast majority will probably cause mild illness, if anything, the diversity of such viruses, coupled with chance opportunity and circumstances related to human intrusion and agricultural and industrial practices, make it inevitable that a minority will behave with considerably more aggression. If one wishes to examine a worst-case scenario of what might happen when an aggressive symbiont goes on a successful attack, we need only look at the impact of yellow fever in the Central and South American rain forests. When the virus, which causes no harm at all to its symbiont host among the African monkeys, first arrived in the Americas, it caused a biological conflagration, borne by vector mosquitoes throughout the length and breadth of the rain forest. Several species of South American monkeys, such as the Ateles and Alouetta, were driven perilously close to extinction.

Deforestation is therefore a dangerous practice. Satellite pictures demonstrate the forests burning and the razed wastelands left behind by logging or burn-and-run farming. According to the United Nations, 5 percent of the rain forests' land surface is torched annually, a loss amounting to 4.75 million acres. Though occupying only 6 percent of the earth's surface, tropical rain forests, as Wilson has made clear, contain the greatest diversity of life on earth. Nobody is quite sure just how many species that embraces because nobody has ever been able, or is ever likely to be able, to count them. In a single patch of Brazilian forest, a few square kilometers in extent, two biologists, Emmel and Austin, identified eight hundred species of butterflies alone. The consensus of opinion among the experts is that the numbers of species that live in the rain forests, including many that are rather more dangerous than butterflies, runs to tens of millions, a great proportion of which remain obscure or completely unknown.

The loss of a major ecological zone such as a rain forest will inevitably have devastating impacts on the diversity of life within it and that impact will ripple out to a vast surrounding hinterland. The consequences and complexity of interactions from such devastation will be labyrinthine.[12]

Take, for example, the consequent loss of birds. The reduction in bird numbers may mean a plague of insects. In Robert Shope's words: "Every time you sample a new species of insect, you find new viruses." In the Amazon, as in the African and Asian rain forests, there are more than a million species of insect alone. Insect breeding patterns are also further dependent upon climate, terrestrial ecology, moisture, and so on, all of which will change. So the repercussions, through a myriad of poorly understood pathways, will prove continental and, in certain circumstances, possibly global. Wherever man intrudes, where Pattyn's "anthills" are knocked to the ground, the disturbed microbes will attempt to emerge.

The notorious clearing of the Brazilian forest for the building of the road through Rondônia has already resulted in between 400,000 and 800,000 additional cases of malaria.[13] In 1989 an epidemic affecting 100 people in Venezuela was traced to the emerging Guanarito hemorrhagic fever virus that broke out in a rural community that had begun to clear the rain forest. Deforestation or intrusion into the rain forest played an important part in the emergence of Ebola and, perhaps, with the hunting of primates, in the emergence of AIDS.

3

Agricultural practices involving intrusion into hitherto wild landscapes or major changes aimed at increasing efficiency — both vital to support the massive increase in human population — are also radically changing the ecological landscape. In the words of Karl Johnson, "Man, in the last forty years, has invaded every ecological niche on an unprecedented scale."[14] Such invasion of wilderness brings miners, farmers, water engineers, and fish farmers into contact with emerging viruses in the same way as foresters involved in jungle clearing. And as forests are felled, wetlands drained, or savannahs ploughed to grow cereals or grass to feed cattle, rodents that thrive on the seeds of these crops and the proximity to human habitation discover new ecologies in which they too can proliferate.

Such a scenario is believed to have played an important part in the emergence of the arenaviruses, such as Lassa fever in Africa and Hantaan in Asia. In Argentina, ploughing of savannahs to grow maize led to a dramatic increase in the population of the mouse *Calomys musculinus,* which was the symbiont of the Junin virus. In Bolivia the spawning of small agricultural settlements on the hinterland of the rain forest led to a similar increase in the field mouse, *Calomys callosis,* the symbiont of the Machupo virus. In India, the breeding of cattle in areas cleared from forests played an important role in the emergence of Kyasanur Forest disease. The building of the Aswân Dam in Egypt in 1970 led to an epidemic of Rift Valley fever, which infected 200,000. This disease, a major infection in sheep and cattle in western and South Africa since the 1930s, is transmitted by mosquitoes. Dam building elevates the water table and encourages the mosquitoes to breed. In 1987 a smaller outbreak took place in Mauritania, following the damming of the Senegal River. Other human activities can provoke the most subtle and unexpected ecological changes that result in the emergence of new diseases.

The emergence of Lyme disease in New England resulted from a complex change in the ecology, where forests that had earlier been cleared for agriculture, eliminating deer and their natural predators, were replanted. The returning deer were without predation to control their numbers. The increase in deer resulted in a disease caused by a bacterium, *Borrelia burgdorferi,* which was transmitted from deer to humans through a biting tick. An even more subtle example is the role played by air-conditioning in the genesis of Legionnaire's disease or the intervention of modern medical practices, with its access to syringes and needles, in the intravenous spread of hepatitis B, the Ebola virus, and AIDS. What, one might inquire, is the future potential for the present situation where organs from primates, invariably carrying six or more symbiotic viruses, are being transplanted into humans who are already afflicted with iatrogenic or disease-provoked immunodeficiency?

In Britain, cattle feed, contaminated with sheep brains infected with the disease scrapie, is thought to have triggered the epidemic of "mad cow disease," or bovine spongiform encephalopathy, BSE. Beginning in the mid-1980s, this fatal disease, which destroys brain tissue, making it resemble an aging sponge, has since infected approximately 170,000 cattle on 33,292 farms. The infectious nature of the disease is truly novel, for current scientific research suggests that it is caused by a nonliving yet contagious protein, called a "prion." A penurious lack of vision appears to have afflicted the British government's response in the late 1980s. In spite of a laudable early diagnosis of the nature of the disease by scientists at the government's own Central Veterinary Laboratory in Weybridge, ministers blandly denied human risk from the disease. The assumption appears to have been the common response of the "Age of Delusion." Scrapie was not known to cross the species barrier from sheep to humans: therefore the contagious agent of BSE would not do so. There appears to have been no clear vision of the important role of change, which has underlain the return of so many dangerous plagues, such as cholera, bubonic plague, tuberculosis, and the wide range of common acute germs now furiously evolving resistance to antibiotics.

In May 1990, the current Agriculture Secretary, John Gummer, went so far as to thrust a hamburger into the hand of his small daughter, Cordelia, in a misguided effort to allay public worries. At the same time, the Meat and Livestock Commission announced a £1 million advertising campaign not to warn people of danger but, completely to the contrary, to bolster sales of red meat.

From the onset, Ministry policy was restricted to the slaughter of affected cattle. Significantly, in the late 1980s, British farmers were offered no more than 50 percent compensation, this at a crucial time at the beginning of the epidemic when a more enlightened policy would have encouraged a greater openness in declaring diseased herds. Instituting a committee to monitor the situation, the authorities were compelled, by degrees, to introduce safer practices in slaughtering and meat handling. Such measures, themselves the consequence of timidity in respect to herd slaughter, were aimed at preventing high-risk tissue, such as brain and offal, getting into the food chain. In programs such as the award-winning television documentary *World in Action*, the limitations of such a policy were ruefully exposed.

In the Irish Republic, where cattle and people were also threatened, the epidemic was nipped in the bud by a more vigorous reaction from the government, which not only ordered the slaughter of entire affected herds (16,485 cattle, where only 124 affected cases had been identified) but also gave full compensation to farmers from the beginning.

Although it is perfectly true that the British government could claim no definite scientific evidence to link mad cow disease with Creutzfeldt-Jakob disease (CJD) in people, even the possibility of such cross-species transmission of this dreadful and apparently 100 percent lethal agent should have galvanized draconian measures from the start. The actual response must seem disturbingly complacent when one considers that the disease is extremely difficult to diagnose, unamenable to antibiotic, antiviral, or vaccine treatment, and, most pertinently of all, is well known to have many years of latency in humans before manifesting with first symptoms. The lack of hard scientific evidence for contagion from cattle to humans was therefore inevitable at this early stage. This also implied that if and when the first cases duly appeared, many more people might already be infected.

Tragically, it was not until early 1996, when some ten younger people were shown to be infected with a new pattern of Creutzfeldt-Jakob disease, different from that of scrapie but very similar to that in BSE-affected cattle, that the Conservative government grudgingly accepted the need for more decisive action. By this stage, seven of these ten were already dead. How many more of the British population are already harboring this lethal agent? Will it be small numbers or a real epidemic, as some predict? Time alone will answer this question, as the trusting public is made to pay for the violation of natural feeding methods, including the feeding of diseased sheep and cattle brains to herbivores as a method of improving meat yield in mass-production farming.[15]

A comprehensive litany of viruses have newly emerged, or have reemerged, as major pathogens during the last half century.[16] In almost every case, ecological factors, including deforestation, agricultural change, and human exploitation, destruction, or intervention, have played a major role in the disease emergence. And the threat of new epidemic infections extends a good deal further than the immediate impact of disease in people. Animals and plants also have their own diaspora of microbes — and the impact of disease in those we depend upon for food could be devastating. It is a fundamental weakness of the mass production of crops and the mass herding of animals that the species being grown or bred tends to what biologists call a "monoculture." Such genetic uniformity makes an entire crop or herd unduly prone to extinction in the event of a lethal virus. With the massive burgeoning of human population, particularly in developing countries, where famine is ever threatening, such a threat to food supplies is particularly menacing.

George Lomonossoff is the senior research scientist in the Department of Virus Research at the John Innes Centre for botanical research in Norwich, England. Lomonossoff's thesis at Cambridge focused on the awesome self-assembly of the tobacco mosaic virus. When I visited him and the John Innes Centre, I encountered a refreshingly innovative perception. Just as with animals, plant viruses form intimate and subtle interactions with their hosts, and plant virologists have been studying how viruses function in such a relationship, how they replicate and interact, how and why such viruses cause plant diseases.

Plant viruses, like the viruses of every other kingdom and phylum of life, are coded by a genome based on DNA or RNA, though most are RNA-based, single-stranded, and positive sense. But the methods of transmission must necessarily be different. Where animal viruses will usually gain entry to their chosen cell through chemical interaction with a specific receptor on the cell membrane, plant cell membranes are encased in an armor plating of cellulose. This forms such an impregnable barrier to conventional viral access that you may pour a highly concentrated solution of tobacco mosaic virus over the pristine leaves of an uninfected plant and nothing will happen. Plant viruses must rely on a vector, commonly a whitefly or leaf hopper, either to inject them through into the sap or to allow them entry through chewed-up cell walls. Fungi and nematodes can similarly gain entry for their carried viruses into the plant's roots. But once inside, the virus has much easier access from cell to cell through intercellular channels, called plasmodesmata.

Plant viruses also live in the landscape of the genome, where, in Lederberg's phrase, they become essentially entangled. They are imbued with all of the power such an ecology confers, yet — revealingly — most plant viruses appear to cause little or no harm to their hosts. It will surprise many plant lovers to learn that those beautiful yellow stripes or leopard spots on their favorite ornamental honeysuckle demark the presence of a coevolving virus.

But viruses in the plant world are also the cause of epidemic diseases that can blight entire harvests. Such viruses include maize streak, African cassava mosaic, tobacco, pea, and cowpea mosaic. In Pakistan, where cotton exports amount to 60 percent of the foreign exchange, the cotton leaf curl virus, transmitted by a whitefly, *Bemisia tobaci*, recently decimated sixty hectares of cotton crop next to the town of Multan. When Ian Bedford, an entomologist working at the John Innes Centre, attempted to reproduce the infection cycle in his laboratory, the infection would not take until he had the inspiration to increase the ambient temperature to 104°F, when the virus proliferated wildly. It was an illustration of the affinity viruses have not only for their host and ecological landscape but even for the precise climatic conditions that prevail in that landscape. Another virus, Rice Tungro, which is widely distributed throughout eastern India and Southeast Asia, causes losses in food yields of the order of $10 billion every year.

The forced growing of food crops outside their natural ecologies and the excessive use of insecticides has resulted in the emergence of a new strain of whitefly, *Bemiscus tobaci.* This superbug, derived from a common pest indigenous to the tropical and subtropical regions of the world, has evolved an amazing fecundity, with the female laying two to three times as many eggs as the normal. It is strongly resistant to most insecticides. Where most indigenous species of whitefly colonize no more than two or three local species of plants, the superbug can extend its pestilential range to more than 500 different species of plants, and can carry 60 different plant viruses in its traveling suitcase.[17] This virus has "a marked tendency to attack plants being grown under forced conditions and out of their natural ecologies." It is currently ravaging crops around the world, from watermelons in the Yemen to cotton in Pakistan, from tomatoes and tobacco in Mexico and the Caribbean basin to greenhouse-forced crops throughout Europe. The superbug is assisted in its travel across continents by the shipping of infested ornamental plants such as poinsettia. In 1991, it ruined winter tomatoes and vegetables in the southern states of America, causing $500 million worth of damage to their harvest. When Ian Bedford visited the area he was startled to observe a white mist rising from broccoli fields — a living cloud of trillions of superbugs.

To understand where such viruses come from, a single example is illustrative. Ageratum is a "weed" species of plant that grows in Singapore. It has a very attractive yellow veining in its leaves. The veins denote the presence of its cohabiting symbiont, a virus that does not harm the plant. If this native "weed" is extirpated and root plants, such as beans, tomatoes, or a cash crop such as tobacco, are transplanted locally, the *Bemiscus tobaci* will suck sap from pockets of surviving ageratum and then move on to ravage the agricultural crops with its aggressively symbiotic viruses. One does not need to be radically green to grasp the implication.

Plant virologists such as George Lomonossoff and his entomological colleague Ian Bedford have mutated the genes for the coat protein in maize streak virus, or changed the DNA interference resistance in cassava to the African cassava mosaic virus, and in similar ways, effectively immunized plants from the ravages of some of these viruses. Lomonossoff has even extended his researches into plant viruses to look for potential in the prevention of human diseases. Such viruses, genetically engineered so they carry key antigens on their surfaces, may one day offer the world a cheap source of vaccines against diseases such as foot-and-mouth in animals and perhaps HIV in humans.[18]

4

There are repercussions with regard to the threat of infection even from our pollution of the very air we breathe. While the effect of wholesale deforestation on atmospheric oxygen remains to be determined, what is certain is that we are manipulating and radically changing the atmosphere at a time when we know next to nothing about the homeostatic mechanisms that maintain it. Increasing levels of carbon dioxide, largely due to the mass burning of fossil fuels, are fueling the greenhouse effect.[19] This additional density of carbon, already 20 percent higher than at any time in previous history, reflects back radiant heat that would normally escape into space from the earth. Coincidentally, it is estimated the average temperature of the earth has increased by a degree Fahrenheit since the middle of the nineteenth century.

One degree might not seem much, but this mean rise is not evenly distributed. It involves greater rises at the poles than at the equator, where small temperature differences will have major implications. For example, a fall in mean temperature by 9 degrees would trigger an ice age and a rise of four degrees would be enough to melt the polar ice caps. Even with the current one-degree rise, glaciers are said to be retreating in certain areas, sea levels rising, sea temperatures rising.

While there is scientific disagreement on the precise cause of the yawning hole in the ozone layer over the Antarctic, many suspect an important role from industrial production of chlorofluorocarbons (CFCs) used in refrigerators, air conditioners, and Styrofoam insulation. These rise above the ozone layer of the earth's atmosphere where UV light can release their chlorine ions to catalyze the destruction of ozone.[20] Skin cancer detection rates are rising in southern hemisphere countries such as Chile and Australia, where health authorities are anxiously monitoring ultraviolet-B and its effects. The thinning of the ozone layer, with its resultant increase of ultraviolet-B radiation, may also have more subtle consequences in relation to the threat of infection. Amminikutty and Kripke, from the University of Texas, have drawn attention to the effects of UV-B in causing a hitherto unrecognized immunosuppression, an effect that, though not yet fully understood, might turn out to be important.[21] Any change that might mitigate or reduce human resistance to infection is a cause for concern.

A rise in temperature and increase in carbon dioxide will also have dramatic effects on biodiversity; and as the hantavirus experience has shown us, change or interference with the cycles of life and the interdependence of species from the great mammals down to the humblest microbes, will alter the patterns of infection and, very likely, trigger the appearance of new infections. The year 1995 was the hottest ever recorded in Britain, and that single hot summer was enough to cause an alteration in the ecology. A superwasp, *Dolichovespula media,* invaded from the European continent and traveled as far north as the Lake District. It was accompanied by exotic butterflies, such as the European Swallowtail and even the North African Large Whites. Ladybird beetles, which had boomed in the glut of aphids during the mild spring, ran out of food in the hot summer and took to biting people. What if they had been carrying a dangerous virus?

What, one also wonders, will be the ecological impact of a progressive warming of two or more degrees Fahrenheit over decades and even centuries?

Paul Epstein, together with his colleagues in the New Diseases Group working in the Harvard School of Public Health, has championed the educational effort focused on the growing danger of climate change and marine and atmospheric pollution. Even a small rise in temperature will have a dramatic effect on the distribution of mosquitoes and ticks. This would allow malaria, dengue, and yellow fever to invade North America as far north as New York, European cities as far north as Berlin, Rome, and Paris — perhaps even Stockholm, London, and Dublin — and major Asian and Australasian centers, at present spared, such as Tokyo, Melbourne, and Auckland. There is already a preliminary report of malaria transmission in Houston, Texas, and malaria projections for the likely spread of both the vivax and falciparum plasmodia cover an enormous new band, stretching across the Eurasian landmass from the British Isles in the west to the Japanese islands in the east.[22]

There is a risk of no less than five vector-borne plagues returning to, or newly emerging in, the United States. These include malaria, yellow fever, dengue, Rift Valley fever, and a variety of arbovirus-induced brain infections, called encephalitides. Meanwhile the four arboviral encephalitides already found in the United States, eastern equine, western equine, St. Louis, and La Crosse, are all expected to increase their range of spread.[23] In the same way, arbovirus diseases, including encephalitides, will prove a growing problem in Australia and throughout Asia.

This evolving situation proved sufficiently worrying for the editors of the *Lancet* to commission a series of eleven articles directed to look at "how anthropogenic damage to the biosphere has potentially important implications for health." The articles, all published in 1993, concluded that "the underlying processes are global in scale and the natural systems affected are part of earth's life-supporting infrastructure."[24] In the opinion of Thomas E. Lovejoy, at the Smithsonian Institution, in Washington, "If [global warming] continues, its effects . . . may well become the dominant force in epidemiology in the future."[25] It is a sobering reflection.

5

The threat extends in an important way to the great oceans that cover two thirds of our earth. In 1989, a group of Norwegian scientists used transmission electron microscopy to demonstrate an abundance of viruses in a variety of marine habitats, including Chesapeake Bay in the United States, the North Atlantic, the Barents Sea, and a variety of Norwegian fjords. What they found was astonishing. Where the concentration of viruses had previously been thought low — the assumption was that viruses were ecologically unimportant in such natural waters — they discovered total virus counts of between 5 and 15 million per milliliter, orders of magnitude greater than any previous estimates.[26] It would appear that people bathing in lakes and seas and the marine life that live there do so in a soup of viruses. What are these huge quantities of viruses doing in this ecology?

Bacteria are the host of many such viruses. Prior to the Norwegian study, ecologists had assumed that bacterial numbers were kept in balance by protozoal grazing. But these scientists cautioned that this was insufficient to keep bacterial numbers in check. In their estimate, as much as one third of the bacteria in these natural waters experienced an attack by viruses every day. Viruses, it seemed, were part of the reason why the waters of lakes and the seashore were reasonably clean. But there were other important implications. Evelyn Sherr, a biologist then working at the University of Georgia, saw great significance in these findings. The tiniest plankton in the seas and lakes — called heterotrophic bacterio-plankton — was also much more abundant than previously realized. "Along with the greater densities of heterotrophic bacteria, the reports of high concentrations of picoalgae, including unicellular bluegreens and eukaryotic algae the size of large bacteria, were electrifying."[27]

It seems likely that the tiniest life forms in the world's oceans carry out most of the carbon fixation. Indeed the greatest fraction of particulate organic matter in the water column of the open ocean amounts to nothing other than metabolically active bacterial cells. The carbon fixated by such tiny life forms does not enter into the commonly perceived pelagic food webs, where the larger phytoplankton are eaten by copepods and so on up to the energy requirements of whales. These tiny life forms are so small they pass through the copepod's sieving system without ingestion. Suddenly we glimpse a vast and complex ecological infrastructure, with intercalations between myriad different virus strains and their minuscule hosts — an entirely new and vital microscopic landscape.

In Sherr's opinion, the focus in aquatic microbiol ecology for the 1990s might well prove to be the evolutionary implications of the

Norwegians' observations. And there are other cogent implications with regard to human and animal health. Algal blooms, red, green, golden, brown, and bioluminescent, cover vast expanses of the oceans, particularly encroaching upon estuaries and inland waters. Satellite pictures have shown a huge increase in recent years in the size of these shelves, from California to Tasmania and from Iceland to Guatemala. In the words of Paul Epstein and colleagues, "This increase is a direct consequence of human activities," both local and global.[28] Toxic phytoplankton blooms are ingested by shellfish, which may then be consumed by people, in whom they cause serious illness. In 1972 an outbreak of paralytic illness in New England was associated with a marine red algal tide. In 1978 toxic algae caused an outbreak of ciguatera fish poisoning in Florida. Today, according to Epstein, there are 200,000 cases of such poisoning globally. The problem is so widespread and worrying that it led, in 1987, the International Oceanographic Commission (IOC) and UN Food and Agricultural Organization to set up a panel to investigate harmful algal blooms.

Some bacteria and viruses have aggressive symbiotic relationships with such toxic algae. The bacteria may be found living within the algal cells, helping them compete against other species through the production of toxins. In Paul Epstein's opinion, certain viruses can also play a role in converting an alga into a toxin producer.

It is clear that the sea plays just as vital a role as the rain forest in the ecological balance of our planet. Quite apart from its role in the human food industry, green algae are major producers of oxygen, and the continental shelves may be vital areas for the regenerative cycles of life. The pollution of coastal waters with sewage and the carriage of fertilizer and industrial waste into tidal estuaries by polluted rivers converts the sea into a tragic universal cesspool. The teeming matrix of viruses and microbes, particularly near our polluted shorelines, is likely to result in significant exchange of genetic material with potentially dangerous microbes. The researches of Christon J. Hurst, who works with the U.S. Environmental Protection Agency, has shown that on land, and in freshwater lakes and pools, viruses with potential for human infection can survive for lengthy periods, given certain temperature and moisture requirements. And once again, the major source of these potential pathogens is human waste and untreated sewage.[29]

Little is really understood of the complex interactions and interrelationships such tiny life forms have with the diversity of species large and small, but it seems inevitable that aggressive symbiosis will be taking place on a vast and intercalated scale. In Sherr's words, "Natural genetic engineering experiments may have been occurring in bacterial populations for aeons."

The antibiotic- and chlorine-resistant cholera germ that ushered in the seventh cholera pandemic is one such very important evolutionary consequence of this. The seal plague that devastated whole populations around the shores of Europe in 1988 is another. This lethal epidemic of harbor seals, *Phoca vitula,* was investigated by Brian Mahy, who confirmed the cause as an emerging morbilliform virus. A similar outbreak among the harbor seals of Lake Baikal in Russia in December 1987, extending into 1988, wiped out 70 percent of the local population. When investigating the source of the virus, Mahy made what he termed "a striking observation." In populations of gray and harp seals, they found evidence of the virus but no demonstrable disease. The harp seal is normally confined to much colder northern waters but "for unknown reasons" it had migrated southward in 1988, bringing its presumed symbiotic virus with it.[30]

Morbilliviruses, a genus within the negative strand RNA-based family of viruses, the *Paramyoviridae,* include some of the most common and dangerous epidemic viruses that afflict people and animals, such as measles, canine distemper, and rinderpest. The message hardly needs underlining. People have an intimacy with the sea. They swim, sail, and eat the fruits of its global waters. Pollution of our seas may well constitute the greatest hidden environmental threat of our age.

Though nature has a vast capacity to withstand such insults, one cannot escape the conclusion that the greatest threats to human health in the early decades of the twenty-first century will include a new proliferation of diseases, directly or indirectly inherited from the twentieth century's profligate abuse of the environment.

Serious international cooperation on such catastrophic threats as global warming and atmospheric ozone depletion has only just begun. On March 28, 1995, the UN opened an eleven-day global conference in Berlin, attended by 1,000 delegates from 100 countries, "to try to reduce pollution and prevent potentially dramatic changes in the earth's climate." Yet it is uncertain if such initiatives will get the support they need at a time when so many other potential catastrophes appear to threaten the biosphere.[31]

EIGHTEEN

Winning the Battle

Infectious disease which antedated
the emergence of humankind
will last as long as humanity itself,
and will surely remain,
as it has been hitherto,
one of the fundamental parameters
and determinants of human history.

WILLIAM H. MCNEILL[1]

1

The message is a vital one, but will the world listen? In 1968, Joshua Lederberg was so shocked by the implications of the Marburg outbreak that he felt obliged to issue a personal word of warning in the *Washington Post.* "Marburg virus is extraordinarily contagious and rapidly lethal in a distressingly high proportion of cases. . . . What might have been an epidemic of world-shaking dimensions was contained by the sheer good luck that it did not spread to man at London airport. . . ."[2] In 1977, Karl Johnson, in discussing the Zairian Ebola outbreak, was equally blunt. "No more dramatic or potentially explosive epidemic of a new acute viral disease has occurred in the world in the past 30 years. . . . No better example comes to mind to illustrate the need for national disease surveillance and the prompt solicitation of international assistance, nor of the need for the development of international resources . . . that can be made available to cope with such emergencies."[3]

That same year an international gathering of the experts involved in the two African epidemics met in Antwerp. In his introductory remarks, C. E. Gordon Smith, of the London School of Hygiene and Tropical Medicine, made it abundantly clear where such dangerous outbreaks took their origins. "In almost every case where an explanation *has* been advanced, these outbreaks have been attributed to intrusion into or interference with previously little frequented areas, because of population pressure and/or agricultural developments." The greater the scale of human degradation of the environment, the more likely that new viruses and other infections would emerge. He concluded with these words: "The Ebola epidemics exposed many of these problems and we have learnt a great deal from them. . . . I hope that when the next serious epidemic occurs, we will be able to show that we have profited from these lessons."[4]

Joshua Lederberg's column had put it more tersely: "We were lucky on this occasion, but it was a near miss. The threat of a major virus epidemic — a global pandemic — hangs over the head of the species at any time."

A new spirit of evangelism began in May 1989 when, moved by the gravity of the threat, the Rockefeller University, the National Institute of Allergy and Infectious Diseases, and the Fogerty International Center cosponsored a conference on emerging infections. The conference focused the alarm of many experts around the world who had been noticing the ominous rise in problems, both in terms of everyday medical practice and in the potential for epidemics, whether viral, bacterial, or protozoal. Concern was expressed about the complacency of the scientific and medical communities, governments, and public, in the face of this escalating threat. Acknowledging the reality of these worries, the Board on Health Sciences Policy of the American Institute of Medicine decided to do something about it.

In February 1991, they convened a committee of nineteen experts, from various related disciplines, which then conducted an eighteen-month study of hard facts. They included virologists, public health experts, immunologists, parasitologists, food toxicologists, and representatives from many other disciplines. In September 1992, in a report coedited by Joshua Lederberg, Robert E. Shope, and Stanley C. Oaks Jr., they issued their cautionary findings. It was this report that warned us "there is nowhere in the world from which we are remote and no one from whom we are disconnected."[5] And complacency now constituted a major threat to health.

In New Orleans, on October 17, 1993, at a presentation to the thirty-first annual meeting of the Infectious Diseases Society of America, Drs. James M. Hughes, Director of the Centers for Disease Control in Atlanta, and Ruth L. Berkelman, the Assistant Director, responded to the expressed concern.[6] They too underlined the seriousness of the message with graphic examples from the preceding year.

In January the E coli germ 0157:H7 had caused the worst ever recorded food-borne outbreak of bloody diarrhea in the states of Washington, Idaho, Nevada, and California. The epidemic had arisen from contaminated ground beef in hamburgers sold through a fast-food restaurant chain. More than 500 cases were identified, 56 of whom had developed kidney failure, and four children had died. In March a new strain of influenza virus had prolonged the influenza season, causing confusion during the hantavirus outbreak and raising fears it might herald the much awaited pandemic. In April the biggest waterborne outbreak of diarrhea in American history took place in Milwaukee. Caused by a cryptosporidium, it had infected approximately 400,000 cases and hospitalized more than 4,000.[7]

Just twenty-five years after the Surgeon General had testified to the U.S. Congress that it was time to close the book on infectious diseases, alarm bells were ringing clangorously to the contrary. In an article with the provocative title "The Conquest of Infectious Diseases: Who Are We Kidding," Berkelman and Hughes spelled out the stark reality: "Infectious diseases are the leading cause of death worldwide."[8] Even in the United States, HIV infection, pneumonia, and influenza rank among the ten leading causes of death. Antibiotics are declining in effectiveness for many hospital- and community-based infections and even old diseases long thought defeated, such as tuberculosis and measles, are emerging once again as a serious problem.

In the prevention of infections, nothing is more important than a vigilant system of surveillance. On April 15, 1994, the eminent authorities Ruth Berkelman, Ralph Bryan, Michael T. Osterholm, James LeDuc, and James M. Hughes outlined the crumbling foundations for infectious disease surveillance in the United States. "During the past decade, state and local support for infectious disease surveillance has diminished as a result of budget restrictions." They went on to illustrate in practical terms how in twelve states there was no longer a person dedicated to food-borne disease surveillance, despite the proliferation of new epidemics. As an illustration of governmental delusion, there was no federal financial support to states for their notifiable disease surveillance system and many state health departments received no federal support at all.[9]

The warnings of Berkelman and her colleagues makes disquieting reading: state health departments with no more than skeleton staffs to conduct surveillance for most infectious diseases, staff freezes on replacement posts, privatization (that blinkered dogma of the accountancy-directed brain), public health departments reluctant to add newly diagnosed infections to their surveillance list because of personnel shortages, with the result that many reportable infections were significantly underreported.

For example, in the 1993 multistate outbreak of food poisoning caused by eating contaminated hamburgers, the outbreak actually began in Nevada, where there were fifty-eight cases of bloody diarrhea and acute kidney failure. The system of passive reporting for this germ in Nevada failed to make the correct diagnosis in a single case. The laboratories did not even include the pathogen in their infection screen so that the realization of the true nature of the epidemic was delayed until the state outbreak fulminated into a major epidemic in three other states, including Washington, where the real diagnosis was made. To realize just how important this type of accurate surveillance is, one only has to consider that when the correct diagnosis was made, no less than 250,000 hamburgers needed to be withdrawn from public sale. Only then was the epidemic terminated.

Early warning from surveillance on infections is vital. In the words of Berkelman and colleagues, that America in 1993 should have witnessed these two severe epidemics "is not surprising." Both emerged as health threats while attention to public health functions, required to detect and control infections, were diminishing. Surveillance is also critical in assessing the effectiveness of control measures once an outbreak is recognized and challenged. Little wonder that these concerned experts began their U.S.-directed article with the terse warning: "Our ability to detect and monitor infectious disease threats to health is in jeopardy."

A similar complacency shackles the efforts in every other developed country, including Britain. Few if any countries have bothered to gather their local experts in this way or to compile a summary document that would expose the frailties of their national systems to outside scrutiny. Yet everywhere in the world, the disquiet in knowledgeable circles is escalating. In a recent article published in the *Lancet*, Haines, Epstein, and McMichael issued their own unequivocal warning. "Our capabilities for health monitoring and rapid response are seriously fragmented, with insufficient coordination and communication, let alone provision for future needs."[10]

A single consideration in such fragmentation and incoordination is dengue fever. Although the disease has been recognized since 1780, recent decades have seen a recrudescence of the plague threat from this particular agent, an arbovirus within the family of *Flaviviridae.* Dengue fever is caused by four different strains of the virus, called serotypes. Infection with one strain does not confer immunity to the others. On the contrary, following infection, with the development of immunity to the causative strain, infection with one of the other three can precipitate the potentially lethal dengue hemorrhagic fever, an illness that mainly afflicts children and which is a growing threat in large areas of the world.[11]

A resurgent pandemic of dengue began in Southeast Asia after the Second World War. By 1975 dengue hemorrhagic fever had become a leading cause of hospitalization and death in children in many of the afflicted countries. By the 1990s it had, during a second epidemic wave, expanded to India, Sri Lanka, the Maldives, and Pakistan. China and Taiwan saw their first epidemic in thirty-five years, at the same time the epidemic swept into Singapore and proliferated into the Pacific, reaching East Africa in the 1980s. Nowhere has it been more dramatic than in Central and South America. The main vector of the virus, the mosquito *Aedes aegypti,* had been brought close to eradication in South America during the Pan American Health Organization efforts against yellow fever in the 1950s and 1960s. But an abandonment in the United States in 1970, followed by a gradual erosion of efforts elsewhere, had allowed the mosquito to return. Over the last two decades this has resulted in major epidemics of dengue, with virus strains rapidly spreading throughout the subcontinent that have resulted in outbreaks of dengue hemorrhagic fever in Venezuela, Colombia, Brazil, Guyana, Suriname, and Puerto Rico.

Today, the virus and its vector mosquitoes extends in a broad swathe across the globe encompassing tropical and subtropical regions. The result is that dengue has supplanted yellow fever as the most important mosquito-borne viral disease affecting humans globally, with tens of millions of cases yearly and a staggering 2.5 billion people now living in the geography of risk. With the arrival into the United States of the newly proliferating tiger mosquito — augmented perhaps by future global warming — the threat of dengue could become much more extensive. No effective dengue vaccine is available, though attenuated candidate vaccines are under trial.

2

The headquarters of the World Health Organization stands at the top of a hill on Avenue Appia in the beautiful Swiss city of Geneva. It is a bright, sunny building of seven long rows of office windows above a graceful lobby with a marbled floor and staircase. An abstract painting, about thirty feet square — a donation from the people of Brazil — covers the lobby wall, and a bronze bust of Dr. Andrija Ŝtampar, who was the president of the WHO's first assembly, gazes down on passersby with an expression redolent of the lofty ideals that inspired its first foundation. The WHO is the organization that monitors the global threat of infections, that initiates and coordinates responses to such a threat. Recently there has been growing disquiet about its capacity to respond to the threat of emerging infections. In an article with the provocative title "WHO in Retreat: Is It Losing Influence?," Fiona Godlee concluded that the World Health Organization was badly in need of leadership if it is to evolve to meet the health challenges of the next century.[12]

James W. LeDuc is the technical officer in charge of arbovirus and hemorrhagic fever viruses. LeDuc stands well over six feet, athletically slim with sparkling blue eyes and a thick head of prematurely white hair. When I asked him, in the light of the aforementioned worries, if we are prepared globally for the threat of emerging infections and particularly the threat of emerging viruses, he was honest in his reply. "No — we are not prepared!"

Today, high-profile diseases, such as AIDS, attract extra-budgetary funding. This means that other diseases, less prominent in the public mind, must be covered by in-house core budgets, which are continually dwindling. Among those not-so-popular diseases, shackled by underfunding, are emerging viruses. Though there are other technical officers in charge of departments — such as AIDS, influenza, and viral diseases linked to mass vaccination programs — in his responsibility with regard to arboviruses and hemorrhagic fever viruses, Jim LeDuc works entirely alone. He has no technical assistants. There is no young doctor or scientist learning from his example. When, as frequently happens, a problem turns up in some part of the world at a time when he is already occupied with a serious threat in a different part of the world, the inquiry meets with his office answering machine.

So just how frequently is Jim LeDuc presented with the emergence of a new infection? If by "new" one implies the conventional grouping of newly emerging diseases together with the reemergence or unforeseen resurgence of an already established one, he is involved with a significant outbreak at least every one or two months. Most are caused by familiar viruses. Outbreaks caused by new viruses dangerous enough to be brought to his attention take place less frequently, but still more than once a year.

Dengue is a growing and constant worry. Japanese encephalitis, though controlled with vaccination in Japan and Korea, is extending its threat in less affluent Asian countries. Recently it broke out for the first time on an island in the Torres Strait off the coast of Australia. The message can become worrisome through repetition, however much the developed world ignores it. Ignore the suffering of poorer countries and the problem comes around to our own backyards. When I called to see Jim LeDuc, he was occupied with a serious outbreak of Crimean-Congo hemorrhagic fever in Kosovo, which is in the former Yugoslavia. At the same time he was attempting to wrap up the investigation of the new Ebola outbreak in Zaire while still planning to travel to India to coordinate a number of important responses, including perhaps reemerging dengue.

If we are to contain a new epidemic, particularly an epidemic caused by an emerging virus, our ability to respond quickly, accurately, and decisively is vital. Delay in diagnosis, with consequent failure even to realize that a new plague threatens, will not only result in a much greater local loss of life but, in the event of an epidemic or pandemic strain, will allow time for the plague to spread widely, making containment much more difficult, even in the most developed setting. Dr. Gordon Smith concluded in 1977, in the opening to a conference dealing with Ebola: "We can conclude that unless 'new diseases' that occur in relatively remote areas cause large or severe epidemics, or affect hospital staffs, or occur in the 'parish' of a virus research institute, they are unlikely to be investigated or their cause discovered." That message, overwhelmingly clear during the African Ebola outbreaks, remains poignantly relevant today.

Never has the world more desperately needed a system of early warning stations distributed about the most likely sources of emergence that would alert us to the first sign of danger. The most likely sources of emergence, which lie within the tropics and subtropics, and in particular the great rain forests and their hinterland savannahs, are far from the home base of the highly organized CDC, but squarely within the territorial concerns of the World Health Organization. But the WHO has no diagnostic laboratories of its own. It relies on the charity and interest of a network of collaborating scientists and their laboratories globally. And there has been an alarming decline both in numbers and in the quality of the diagnostic armamentarium of this fragmented and sadly neglected global network.

The 1956 outbreak of Kyasanur Forest disease was recognized because a high monkey mortality was investigated and identified by the local Virus Research Institute in Poona, India. O'nyong-nyong fever, in spite of the scale of the epidemic, might have been dismissed as dengue without the investigation of the Virus Research Institute in Entebbe, Uganda.

In the 1950s and sixties, many satellite organizations sprung into existence throughout the tropics and subtropics, affiliated to institutes such as the Pasteur, in France, the tropical medicine schools in Britain, Belgium, and Germany, and several American institutions, notably the Rockefeller Foundation. These organizations might perhaps have reflected the old colonial paternalism. Yet the importance of those satellites is illustrated by the fact that between 1951 and 1971, the Rockefeller Foundation's eight satellites discovered no less than sixty new viruses. Since then political changes and the assumption that infectious plagues were on the wane have resulted in many of these being closed down. In consequence, our world today is less prepared in terms of surveillance than it was even in the sixties and seventies.

When he first took up his post at the WHO, Jim LeDuc circulated a questionnaire to these collaborating laboratories. He inquired about the level of technical skill they retained, what staffing, what reagents, for example, they possessed for diagnosing unusual viruses. The replies were ominous. These were reputedly national-level laboratories, yet half to two thirds could not even test for dengue fever and a quarter or less could test at all for hemorrhagic fevers such as Ebola or Marburg. In LeDuc's shocked reflection: "That was the eye-opener for me." What hope was there that such laboratories, underequipped and lacking suitably trained personnel, would recognize or diagnose the emergence of a new plague virus? Such, if the situation is left unaltered, is the forlorn hope of the world.

At the Pasteur Institute, Bernard Le Guenno is working alone. As we saw during the Ivory Coast outbreak of Ebola, even this dedicated scientist working in such high-tech surroundings needed assistance with reagents from Tom Ksiazek at CDC before he could make his diagnosis. At the Institute of Tropical Medicine in Belgium, the virology department of international repute that was formerly directed by Stefan Pattyn is now focused exclusively on AIDS. Tom Ksiazek, in LeDuc's estimation, uncomplainingly does the work of ten men. Jim LeDuc is rightly concerned that many public health agencies have been cut to the bone, some to the extent that frustrated scientists have abandoned their vocations in disillusionment. "We are all facing cutbacks, but there is a line beyond which we cannot go."

Underfunded and distracted by overwork, the dedicated professionals such as Ksiazek and LeDuc nevertheless shoulder the burden for the good of humanity. One of LeDuc's proposals to improve this wholly unsatisfactory situation is to set up a training fellowship with the Pasteur Institute, or perhaps Porton Down, where young scientists could acquire a more advanced expertise before returning to their home bases, whether in developing or developed countries. When proposed to Kenya, the initial response was laudably enthusiastic. But at present the idea is hampered by funding needs and the lack of an organizational structure. LeDuc believes that the funding required, estimated at a modest $5 million a year, would follow, perhaps with assistance from the EEC, if and when the organizational problems were solved. He deserves every support.

3

The governments and public health authorities of developed countries present a brisk confidence in their ability to cope with any health emergency. But what is the justification of such confidence in the face of such revelations? How equipped are we to deal with the emergence of a powerfully aggressive epidemic virus in our community? What would be the impact of its arrival in, say, the United States, Britain, Japan, Germany, Canada, Italy, Spain, or France?

David Simpson is now the Professor of Microbiology at the Royal Victoria Hospital, Belfast. He believes that Britain, for example, is lacking in the necessary infrastructure to deal with a viral outbreak, such as the hantavirus that appeared in America.[13] David Bishop, until recently director of the Institute of Virology at Oxford, in discussing the containment of the Reston outbreak in America, reinforced this opinion. "Had this happened in Europe — in England, for instance — I have my doubts whether the infection could have been contained as effectively as at Reston."[14]

Porton Down is a collection of low brick buildings set in the rolling Wiltshire countryside. Over the entrance of the main building are low relief plaques depicting three of the labors of Hercules. The allegories represent humankind's glorious struggle against the forces of adversity and pestilence. These laboratories have long been independent of their origins as an army chemical and biological warfare establishment. Today, the Centre for Applied Microbiology and Research (CAMR) includes some of the best equipped viral diagnostic and investigation facilities in Europe. Graham Lloyd, a genial and knowledgeable scientist, is the director of the hemorrhagic fever and virus research laboratory, which is equipped to deal with BSL-4 viruses in what is essentially an extended BSL-3 technology, with a remarkable system of interlinked laminar flow cabinets.

Lloyd believes that Britain, in Porton, possesses good investigative facilities. It is true that Britain has wisely retained the isolation facilities at Coppetts Wood, with the added support infrastructure of the Public Health Laboratory service at Colindale, which collates a matrix of infectious diseases consultants and public health authority structures distributed around the country. But what Britain, and every other European country, lacks is the hard focus and coordination found at the CDC, where all of the necessary responses, from diagnostic microbiology, virology, molecular biology, and epidemiology are condensed in a single organizational infrastructure, directed by a single pyramidal executive.

No Western European country has the equivalent of the CDC, whether in its BSL-4 facility, its mass of experience, the motivation and cutting-edge awareness that comes from day-to-day involvement, and its rapid task force mentality, which is geared to investigate and control a potentially explosive epidemic situation. Unfortunately, this leads to a series of potential pitfalls. The biggest problem, as always, is the funding dilemma. How do you convince governments that they need to spend money on insurance against an uncertain threat in the future when they are already overstretched by existing demands? In Lloyd's words, "We have never really revisited the mechanisms for housing patients during such a catastrophe. We have never reappraised whether or not we have a sufficiently sensitive detection mechanism in place."

Surprisingly, no formal mechanism of global communication appears to exist — Porton, for example, relies on word of mouth from the WHO to warn them if a major threat is looming. It is not difficult to predict what is likely to happen in the event of a catastrophe. Direct contact would certainly be made by the Department of Health, as and when they became aware of the crisis, but time would inevitably be lost in the initial incredulity, while people have little knowledge of plague viruses reacted as they did with the AIDS outbreak before searching for appropriate zones of responsibility. "I'm sure they would expect answers, but I'm not sure we could give them all the answers or that we could give them the necessary logistical support in a major outbreak."

There can be no doubt whatsoever that the initial confusion, delays, and misunderstandings at the start of the AIDS epidemic fomented and greatly amplified the resultant disaster. The formal responses, not least the need for an executive of knowledgeable experts rather than accountancy-orientated cynics in positions of control and administration, need to be programmed and maintained in a state of permanent readiness throughout the world. This needs to be implemented now, not in the pandemonium of an emergency.

One important consequence of the communication between governments, departments of health, and medical training authorities that would hopefully arise from this would be better training of medical graduates on the potential of new epidemics, which in turn might improve the anticipation and diagnostic skills in the acute medical arenas. Graham Lloyd lists numerous examples of referring hospitals and doctors who appear entirely ignorant of the global situation with regard to emerging or newly emerged infections. Doctors in such disciplines as general practice, accident and emergency departments, and acute medical wards, while they might readily think of malaria, will otherwise indulge in lengthy and unrevealing diagnostic workups before the thought of an emerging infection will even cross their minds. The results are all too predictable, and illustrative of the very real danger during a new epidemic: dangerous delays, even in sick patients admitted to hospitals who tell their doctors that they have recently traveled abroad.

There appears to have been no serious evaluation of the potential for the various countries now affiliated through the EEC to rationalize the present fragmentation and develop a coordinated system of focus, cooperation, and response. Yet the pooled expertise, educational potential, and laboratory support infrastructure needed for a first-class response is ready and waiting for such an initiative. Intergovernmental structures should consider this potential. A similar focused approach is needed in Asia and Australasia, which could draw upon the existing expertise of Japanese and Australian centers. South Africa, with its exemplary past record, is already equipping itself to provide a service for Africa. Such developments would greatly assist the very difficult global role of the World Health Organization, where the initiatives being suggested by Jim LeDuc and experienced colleagues such as R. H. Henderson merit major support and expansion.

In response to the American disquiet on emerging infections, the WHO called together a meeting in Geneva on April 25–26, 1994, where a large panel of international experts, including Joshua Lederberg, Stephen Morse, Ruth Berkelman, and R. H. Henderson, exchanged information on their own experiences of dealing with emerging and reemerging diseases, discussing ways of responding to the various problems posed by these threats and how the WHO should formulate its coordinating global role. Some remarkable insights emerged.[15]

The forum was honest in acknowledging the difficulties and in confirming Jim LeDuc's fears. Since the 1970s, changes in health care priorities, diminished resources, and the need to focus the limited available manpower and funds on the AIDS epidemic had resulted in a serious and progressive erosion of the global infrastructure surrounding infectious diseases. Notably, there had been a major scaling down in disease-oriented programs, a deterioration of the surveillance efforts, and a serious loss of available technical expertise. Experts in communicable diseases trained before the 1970s who were now either retired or nearing retirement were frequently not replaced or their positions were transferred to some specific support program. All of this had "directly affected the global capacity to recognize and respond to the new, emerging and re-emerging diseases." As LeDuc had discovered, when key reference centers around the world lacked the resources to identify even common pathogens, it was impossible for them to recognize truly new ones.[16] All were agreed that this constituted a dangerous situation.

The experts went on to discuss the various risk scenarios that have been described above. Two in particular now place our modern world in greater threat of a pandemic spread of a new infection, and particularly an emerging virus, than has ever been the case in previous history.

4

Few would deny the claims of the ecologists that the massive increase in human population has put the world's ecology under threat. At present close to 6 billion live in a world where there were no more than 1.5 billion a century ago, and experts are predicting further increases in the next generation to 9 billion or more. The greatest expansions of population are in the poorest countries, where the medical infrastructure and money spent on health is at its lowest. Poverty has not decreased, as some optimists had anticipated, in response to technical innovation. Instead it has increased steadily for half a century and is now burgeoning on a leviathan scale, with tragic consequences in suffering and misery.[17]

Political unrest scourges the Third World on a scale that has rarely been seen before. Political and economic oppression is often the prime cause of large-scale movements of people across countries and continents. History suggests that times when great masses of human population are on the move, particularly when this is associated with war or great social upheaval, are also harbingers of the most dangerous plagues. There are ominous parallels between the present state of human social evolution and the great plagues of history. The bubonic plagues of Justinian and the Black Death took place during periods of major social, climatic, and ecological change. At the time of the Black Death, Europe was overpopulated, racked by hunger, and in the grip of a mini ice age. If McNeill is right, the devastating arrival of smallpox and measles also accompanied the upheavals of interaction between the late Roman, Egyptian, and Middle Eastern societies, and in particular with the coincident rise of teeming metropolises and rapid sea travel. We have already seen the consequences in the Rwandan refugee camps and in Eritrea, as in a wide range of Africa's other sempiternal problems, and during and following the Asian wars. The current rise in tuberculosis in eleven Western European countries is in part a reflection of the global problem, carried into our wealthy backyards by refugees.

Many Third World governments, stressed by conflict and hunger, have been cutting back on their outlays on health and sanitation. Meanwhile large-scale environmental changes, such as deforestation, damming of rivers, or increased irrigation, all known to increase the epidemic threat, are, in Levins's words, "Increasingly encouraged with a sense of economic urgency that thwarts ecological criticism."

We need to remind ourselves of our own origins and what would constitute a rational and reasonable diversity of life on earth. Today's mass of humanity derived from social groups of hunter-gatherers who were sparsely distributed over wide terrains. Epidemics only became possible in human terms with the dawn of civilization and in particular with the evolution of major population centers, epitomized by cities.

Cities are not the invention of our modern industrial revolution. They existed thousands of years ago in ancient Egypt, Babylonia, Greece, and Rome, and in prehistoric America in the civilization of the Maya, Aztec, and Inca. No matter how imperfect, many would see them as the flowering of human society: and how revealing this is about our deep and atavistic need for community existence. Man is a herd animal and epidemics are herd manifestations. It is hardly accidental that the first known epidemics coincided with the rise of city-based civilizations. The old Roman empire, with its multitudinous 50 million population, was decimated by a series of epidemics, one of which, the epidemic of Galen, killed the emperor Marcus Aurelius as well as millions of Roman citizens. McNeill has charted some of the long list of epidemics that ravaged ancient China.

The great towns and cities are the Achilles heel of modern society. Even in our most affluent suburbs lurks the capacity for rapid amplification of a spreading infection. And nowhere is this vulnerability greater than in the sprawling shanties that surround many of the large cities in developing countries, where the health infrastructure is already tenuous.

Little surprise then that Murphy and Nathanson have cited the human population increase as one of the major causes for the increasing emergence of new virus diseases.[18]

Today, in the words of Levins, Awerbuch, et al., "we are living in a time of major climatic, vegetational, demographic, technical, social and political change, and this must also be a time of epidemiological change in which many surprises are likely."[19] Today a potentially disastrous epidemic will circumambulate the world with the speed of a passenger jet.

At the April 1994 World Health Organization meeting, Dr. Robert Steffen summarized the magnitude of this potential. In 1993 alone, approximately 500 million arrivals were recorded in Africa, the Americas, East Asia, and the Pacific, Europe, the Middle East, and South Asia. Of these, approximately 40 million were people traveling between developing and developed countries. At the same meeting, Graham Lloyd illustrated the importation of serious infections across such global boundaries. Porton Down is currently diagnosing 500 to 800 cases of dengue fever arriving in Britain every year, all imported by travelers arriving from the epidemic zones. The newly created International Society of Travel Medicine is currently appealing for the "creation of a new global strategy to detect emerging microbial threats to health."[20]

This huge volume of human traffic through international airline flights so links the cities of the world in a close-knit matrix of vulnerability that we can no longer limit our considerations to a local or even national level. The entire world has become the hunger-gatherer's home village, potentially one vast amplification factory for any such emerging epidemic virus.[21]

5

What follows is not intended to be an exhaustive coverage of the various recommendations that have emerged from authorities such as

Murphy and Nathanson, the Shope-Lederberg report, and the responses respectively of the CDC and WHO. Readers who wish to study those recommendations need to review their recommendations in their original detail.[22]

In essence these authorities make a series of recommendations, which include, as a first vital step, the construction of an ambitious surveillance network that would alert every country and major laboratory in the event of a globally threatening emergence. This would involve a major revitalization of the existing infectious diseases reporting mechanisms, with their preprogrammed algorithms of communication at local and national levels. It would also require a statutory system of rules to replace the present somewhat arbitrary mechanism of informing colleagues. The necessary resources must be quantified and then found, some to be devoted to training a new generation of motivated health care and scientific recruits who will play a key role in speeding up the recognition of future threats and subsequently containing them. Since the threat is likely to arise in developing countries, they will need the assistance, both financial and technical, of more affluent countries.

Improvements in perception and training within the scientific professions will not be enough. Public education needs to be part of an enlightened response to a very real and growing threat. Only through such education are we likely to see changes in human behavior and demographics that are equally vital as part of any program to mitigate the risk. The educational message needs to be transmitted widely to industry, so that modification in technology might lessen the threat, nowhere more essential than in developing countries, with their large-scale changes in land use.

There are two vital, indeed fundamental, prerequisites in all of this. Governments and funding bodies need to be convinced of the global and local dangers so that scarce resources can be allocated to these initiatives. Equally importantly, scientists such as Jim LeDuc cannot be left to work without a good deal more high-grade professional assistance. Quite apart from the day-to-day importance of his work, such rare accumulated knowledge needs to be imparted to a new generation.

In their intelligent approach, based upon understanding of the ecological and social implications of the threat of infection, the Harvard Group are to be commended. Such progressive thinking, particularly in their intercalation of health with social and ecological concerns, deserves major support and funding. As an example of the type of investigation they suggest, there are two pioneering projects currently under way in the rain forests of Brazil and Papua New Guinea, where teams of scientists are looking for less destructive — and microbially less hazardous — ways to accommodate humanity's needs with the protection of the rain forest ecology.[23]

In Levins's words, "These days, scientists no longer predict that the history of human infection will progress steadily towards the total elimination of infectious disease. More likely the pattern will be one of disease turnover. With a new acceptance that infectious disease will always be part of the human experience comes the realization that scientists will have to adopt a new approach to understanding the patterns of disease evolution."[24]

Thankfully, it is not all bad news. Already an interesting exception to the global withering of facilities is the Centre for International Medical Research, in Gabon, which has been achieving considerable success in a range of activities, from molecular biology to ecology and epidemiology. The Pasteur Institute in Dakar has been putting a similar enlightened foot toward the future, and in consequence is attracting a high caliber of scientist, with high-quality training of locally recruited personnel. In New Delhi, in August 1995, an international meeting of nine countries, including Bangladesh, Bhutan, India, Indonesia, and Thailand, met to begin to identify common strategies aimed at tackling emerging and reemerging infectious diseases in Southeast Asia. In America, the inspired Joe McDade, assisted by Stephen Morse and others, has undertaken the editorship of a pioneering journal, *Emerging Infectious Diseases,* which can only foster education and understanding among professional colleagues. But these are isolated instances, and most advanced countries continue to cut back on their support for such Third World initiatives.

Will the world continue to ignore these renewed warnings? Measures to contain plagues are costly and have to compete for funding with the manifold demands on the purse strings of health authorities and governments. If the world fails to respond with sufficient alacrity and concern, what will be the price of that failure?

NINETEEN

Virus X

The Doomsday Scenario

Some may say that AIDS has made us ever vigilant
for new viruses. I wish that were true.
Others have said that we could do little better
than to sit back and wait for the avalanche.
I am afraid that this point of view is much
closer to the reaction of public policy and
the major health establishments of the world. . . .

JOSHUA LEDERBERG[1]

1

The Rockefeller University stands on York Avenue, sandwiched between the skyscraper of New York Hospital and the looming bulk of the Queensboro Bridge. Dr. Joshua Lederberg, the president of the university, is something of a legend among microbiologists and geneticists. He has provided the world with a brilliantly original vision of the microbial contribution to the biosphere, through the global nexus of genetic information and the way it is handled by the complex interactive matrix of life on earth. For almost twenty years he has also been warning the world about the danger of emerging diseases and particularly that from new viral epidemics.

Just how serious a threat do emerging infections pose? I considered the implication of this as, arriving early on York Avenue, I walked around the back of the university to gaze out over the two-hundred-yard expanse of East River. It was a pleasant sunny day in September 1994, and below me the polluted metal-green water flowed almost noiselessly into the crepuscular shadows of the gargantuan ironwork bridge.

It is scary, as Lederberg acknowledges, to view pandemics as a natural evolutionary program.[2] Could such a pandemic virus, or a pandemic resulting from a change in an existing virus, threaten the whole human species? Such questions are almost too appalling to contemplate. Yet they must be faced.

Lederberg is white-haired, heavyset, and speaks with a deep Brooklyn-tinged accent. A Rabelaisian wit colors the parry and thrust of his conversation. His office is situated in the Founder's Hall, connected through a battered and groaning elevator with a reception area where a bronze bust of John D. Rockefeller vies for attention with oil paintings of famous microbiologists, including Simon Flexner, who identified the dysentery germ.

Admitting that the arrivals of Marburg and Lassa shocked him into awareness, Lederberg so began his lonely task of attempting to warn the world to the danger. He made it his business to learn more about the ecology and evolutionary potential of viruses. At the root of the complacency was the fact that public health traditions give names to infections and their causative microbes and then make static entities of them. But germs, and viruses in particular, are not static at all. They are all too capable of taking us by surprise.

We know now that the most dangerous new pandemic is likely to be viral in origin; and the most likely trigger for the emergence of this virus will be human intrusion into a wilderness ecology. I no longer view the aggressive attack as random but designed by genomic intelligence, under the influence of evolutionary laws, to maximally injure and kill the transgressor. In previous history, this aggression would have eradicated the hunter band or primate troop, perhaps even the village they carried it back to, or the equivalent animal colony. We have witnessed what such pandemics do in nature, as in the examples of yellow fever in South American monkeys, or myxomatosis in Australian rabbits. The fact that such a virus might annihilate all or most of a species, and potentially a new host — the attacking strain dying out itself in an evolutionary dead end — does not signify the same Pyrrhic failure at all in symbiotic thinking. In the symbiotic hypothesis, the sacrifice was in keeping with the plan, reflecting an overall gain for the survival of the original virus-host symbiotic partnership.

There are two junctional zones that facilitate such a perilous scenario. The first is the original ecological intrusion, where people or animals first encounter the virus-host cycle in nature. The second junctional zone lies in a very different geography, the genomic landscape of the transgressor species. At this time the virus is maximally primed for aggression. Yet still the vast majority of viral attacks fail at this stage, because the transgressor, though immunologically naive to the virus, is sufficiently hostile in its subtle barriers and chemistry to preclude a successful viral penetration. However, in a tiny percentage of such contacts, successful invasion takes place. The chance of this happening increases greatly if the symbiotic host is very close genetically to the intruder species, for example, chimpanzees to people. In time there will be a gradual evolution, from primed aggression, driven by symbiosis, toward the behavioral amelioration of coevolution, driven by natural selection. The ideal, almost universal, outcome, provided the coevolution is ultimately successful, will be a new symbiotic relationship.

This is how familiar viruses such as smallpox and measles have evolved with people. The danger period, the species threat from a lethal pandemic, must therefore arise very early: at the beginning of this developing relationship, when aggressive symbiosis is still the predominant mechanism. But as the elegant mathematics of May and Anderson and others have demonstrated, natural selection may at this stage also decide that the most successful viral progeny within the evolving swarm are the most aggressive. It is startling to consider that both natural laws might, at the same time, be programming the virus for maximum aggression.

How might the human race appear to such an aggressively emerging virus? That teeming, globally intrusive species, with its transcontinental air travel, massively congested cities, sexual promiscuity, and in the less affluent regions — where the virus is most likely to first emerge — a vulnerable lack of hygiene with regard to food and water supplies and hospitality to biting insects!

The virus is best seen, in John Holland's excellent analogy, as a swarm of competing mutations, with each individual strain subjected to furious forces of natural selection for the strain, or strains, most likely to amplify and evolve in the new ecological habitat.[3] With such a promising new opportunity in the invaded species, natural selection must eventually come to dominate viral behavior. In time the dynamics of infection will select for a more resistant human population. Such a coevolution takes rather longer in "human" time — too long, given the ease of spread within the global village. A rapidly lethal and quickly spreading virus simply would not have time to switch from aggression to coevolution. And there lies the danger.

Joshua Lederberg's prediction can now be seen to be an altogether logical one. Pandemics are inevitable. Our incredibly rapid human evolution, our overwhelming global needs, the advances of our complex industrial society, all have moved the natural goalposts. The advance of society, the very science of change, has greatly augmented the potential for the emergence of a pandemic strain. It is hardly surprising that Avrion Mitchison, scientific director of Deutsches Rheuma Forschungszentrum in Berlin, asks the question: "Will we survive?"[4]

We have invaded every biome on earth and we continue to destroy other species so very rapidly that one eminent scientist foresees the day when no life exists on earth apart from the human monoculture and the small volume of species useful to it. An increasing multitude of disturbed viral-host symbiotic cycles are provoked into self-protective counterattacks. This is a dangerous situation. And we have seen in the previous chapter how ill-prepared the world is to cope with it.

It begs the most frightening question of all: could such a pandemic virus cause the extinction of the human species?

2

Joshua Lederberg is unequivocal in warning us that the survival of the human species is not a preordained evolutionary program. One day our sun will explode into a red giant, devouring the earth in its violent swell. Long before this, human extinction may well result from some other cosmic, geological, or environmental catastrophe — perhaps sooner still from an Armageddon of our own making. For sadly, the species threat from a pandemic virus will derive as much from human behavior as will a nuclear Armageddon.

In virological circles such a doomsday virus is often referred to as "the Andromeda strain," after the bestselling thriller by Michael Crichton. But this is a misnomer. The Andromeda strain was not a virus at all, but a crystalline entity that came to earth on a meteorite from outer space. It was devoid of DNA, RNA, or even protein, and imbibed carbon and oxygen from its environment, replicating as the perfect nanomachine. If ever an extinction strain does threaten the human species, it is more likely to be a virus, and it will emerge from the diversity of life on earth. For this reason, I have called it Virus X, with X the logical derivative of extinction. What would be the likely properties of such a virus?

To cause our extinction, Virus X would need to take two steps. First it would have to kill everybody, or almost everybody, it infected. The qualifier is needed because the end of human civilization might not require the death of all of its members. Has any virus, or any other infectious agent, ever caused such a lethality? The answer, unfortunately, is yes — though the emergence of such catastrophic lethality is rare. Rabies was uniformly lethal in humans for at least four thousand years of history until Louis Pasteur discovered the first vaccine treatment. A member of the genus Lyssaviruses, from the Greek *lyssa,* which means "frenzy," rabies is one of over one hundred members of the family of *Rhabdoviridae,* which infect an incredible range of life, from plants to reptiles, fish, crustaceans, and mammals. In the opinion of Hervé

Bourhy, an expert at the Pasteur Institute, the human rabies virus lives in a symbiotic cycle with bats, from which it is capable of infecting a wide variety of mammals, particularly foxes, coyotes, jackals, and rodents.[5] Could anybody conceive a more sinister expression of aggressive symbiosis? The virus is programmed to infect the centers in the animal brain that induce uncontrollable rage, while also replicating in the salivary glands to best spread the contagion through the provoked frenzy of biting.

We should be very thankful that the manner of spread precludes the virus from ever causing a human pandemic. HIV-1 still appears almost uniformly lethal, though occasional reports of survivors are now appearing, and the Zairian Ebola virus was a close rival, with 90 percent fatality to the people it infected. However, with HIV-1 and Ebola, the worry of a species threat is far greater than with rabies.

Could such lethal agents ever take the second step, and become sufficiently contagious to infect all or virtually all of the human species? The reassuring fact is that the vast majority of emerging viruses, including those with such huge lethality, fail in practice to become pandemic. We know some of the explanations. An infection directly contracted from an animal or biting insect will never pose such a problem because the numbers infected will be limited by the extent of contact. This was the case, for example, with the *Sin nombre* hantavirus. Food- or water-borne epidemics, though they might infect large numbers of people, can be interrupted by appropriate recognition and introduced measures of hygiene. Even sexually transmitted disease, such as AIDS, can be controlled by mechanical prophylaxis and a reduction in promiscuity; and blood-borne infections by control of contaminated supply, syringes, and needles. As far as we are aware — and one always has to qualify extrapolation based on past experience with caution — the only route of contagion likely to prove universally threatening to humanity would be person-to-person spread by the respiratory route.

The potential for respiratory spread of a plague microbe is unique. Each adult inhales about 10,000 liters of air each day, and we cannot avoid inhaling one another's expired discharges. Given a few minutes in a crowded room, the infected individual will, by coughing or sneezing, transmit the microbe to many of the others present. This was seen most tragically and historically in the switch from bubonic to pneumonic plague during the Black Death. Notable among present-day viruses that spread with very high levels of contagion by the aerosol route are the rhinoviruses and corona viruses that cause common colds. We are all familiar with the rapid spread of the influenza virus, which has long proved its capability to cause pandemics. But to threaten our species, or to provoke a near enough catastrophe to destroy human civilization, a virus would need to combine the infectivity of influenza with the lethality of HIV-1 or Ebola Zaire.

Cedric A. Mims is an eminent British virologist who has spent much of his working life studying the factors that might make such a grim scenario possible. I traveled to London to discuss this with him in the cloistered setting of the Royal Society of Medicine.

In his opinion, although the most dangerous emerging infections of recent years have for the most part been viruses, for example, Ebola and hantavirus, "one striking characteristic shared by all these infections is that none of them is effectively transmitted from person to person."[6] Yet acute respiratory infections are a frequent visitor to the evolutionary hothouse provided by the modern human ecology of overcrowded urban centers. And these agents are likely to remain a step ahead of traditional vaccines or antimicrobial drugs.

The main requirement for respiratory transmission is multiplication within the lining of the respiratory passages, which results in large enough quantities of infectious viruses being thrown out by the act of sneezing or coughing. Many such viruses do infect the respiratory tree, and they are transmitted from direct multiplication there, as, for example, colds and nonpandemic flu. Most of these cause no more than a localized irritation, spreading across the surface epithelium and being shed without invading more deeply. What additional properties would enable a more lethal virus to acquire aerosol transmission? In Mims's opinion, it is the viruses that become blood-borne infections that are most likely to carry both the transmissibility and lethality for serious pandemic. And a number of viruses have already demonstrated this potential.

Measles was once the equivalent of a BSL-4 category virus that spread by the aerosol route. The precise tracks of its terror are now lost in the mists of time, yet its first emergence must have been horrifyingly contagious and lethal. Smallpox is also transmitted by aerosol. Yet even smallpox, though it had been ameliorating by coevolution in people over millenia, was still causing 2 million deaths annually just prior to its eradication. Measles and smallpox spread to vital organs by the bloodstream and then recirculate to localize in the lungs, where they are amplified in the lining before being coughed or sneezed out to infect others. This has a fundamental relevance to the Virus X scenario.

In realizing that most emerging microbes derive from animal sources, Mims has concluded that infections recently derived from such a source experience a key block when it comes to secondary spread from person to person. The precise nature of this block is very important. Here science has little hard factual information on the operative mechanisms. The transmission of infectious diseases is a woefully neglected subject. What we do know is that, once evolved, the respiratory means of spread is very difficult to interrupt. The speed of transmission and the numbers infected make isolation impracticable: the standard surgical face masks, though psychologically reassuring, would not trap a virus. But Virus X would also face a series of very difficult obstacles.

As the virus successfully invades the body through the lungs, it must now gain entry into the blood, there to survive furious attack by every force and device of the immune system. This is probably the site of major block for many newly emerging viruses. If the virus survives this attack it begins the next phase of invasion, targeting certain vital organs, where its main cycles of replication — and its lethality — will focus. Here again, science is ignorant of how and why viruses choose one particular target organ over another. The localization of mumps in salivary glands, polio in the anterior horn cells of the spinal cord, or rubella in the fetus remains a mystery. One possibility is the presence of chemical attractors on the lining of blood vessels, called vascular addressins.[7] From a pandemic perspective, there is a further obstacle, when the virus, greatly amplified in its target organs, now reinvades the bloodstream and, through further unknown mechanisms, localizes for a final time in the lungs. Here it must find its way back to the cells lining the alveoli, bronchi, or upper air passages, where, like measles or smallpox, it multiplies before being explosively propelled into the surrounding air by coughing and sneezing.

The importance of each step in this concatenation of genesis is illustrated by the *Sin nombre* epidemic. It was the failure of the hantavirus to cross successfully from the lining of the capillaries to the adjacent air sacs — a distance of mere microns — that prevented a lethal pandemic from originating in the United States in 1993.

Although viral stability becomes an issue once the mass of infective particles is expelled into the air, in fact even a very delicate virus will often survive the mere seconds that would be needed for reinhalation by another potential victim in a crowded room. Many respiratory viruses are a good deal tougher than this. Measles virus is transmitted by droplet nuclei less than four microns in diameter, which remain suspended in the air for hours after being expelled. Chicken pox can spread over the distances between wards in hospitals. Mims recounts the striking case of an outbreak of foot-and-mouth disease on the Isle of Wight in England, which was caused by the virus being windblown fifty miles across the English Channel from France while still retaining its infectivity.

3

People who doubt the very notion of future pandemics are comforted by two propositions: faith in the capacity of science to cure or to prevent infections and confidence in the human immune defenses.

In Levins's words, such "faith in our technical means of cure and prevention has been naively reductionist." He quotes the example of

DDT. This familiar insecticide kills a mosquito when tested in a bottle. From there we assume that all we have to do is to distribute DDT throughout nature and the mosquito is wiped out. Instead the DDT also kills off the mosquito's natural predators while the mosquito itself evolves a resistance to it. To assume that microbes are static is a cardinal error. Such a faith may be understandable, even plausible, but is thwarted at each step by the action of unknowns. Genomic intelligence, the capacity of a living organism to change, its labyrinthine ramifications with the diversity of life in its local ecology — these are major unknowns. So the future can only be unpredictable.

What then of that second recourse to optimism, the remarkable capacity of the human immune defenses?

Humankind has survived not for the 200,000 years usually attributed to it but for an estimated three and a half billenia of evolutionary ancestry. Throughout such aeons of dramatically changing global climate and ecology, our ancestors must have encroached on multitudinous viral cycles, and must by implication have survived very many serious pandemics. The abundance of retroviral sequences in the human genome is ample testimony to this survival. The extinction scenario must have threatened us many times. We did not survive for three and a half billenia because of antibiotics or vaccines. We survived because of our natural resistance to infections. Our immune system evolved as a survival mechanism in the face of a continuously evolving threat: even today, it is almost certainly evolving more rapidly from generation to generation than ever before.[8]

This continuing battle between immunity and microbial invasion is complex at a molecular level, involving curious entities such as "V genes" and "superantigens." Because of it, our defenses cope admirably with almost every microbial assault. Yet past pandemic microbes must have defeated even these formidable defenses.

Various microbes have evolved ingenious individual strategies to achieve this. The tuberculosis germ, for example, is wrapped in a waxy coat impervious to digestion by our phagocytic cells. Like many of the furiously evolving common germs, it also has a marked capacity to mutate resistance to antibiotic drugs. Other microbes inactivate complement, interfere with interferon, control the chemical mediators of inflammation such as cytokines and chemokines, or, like the trypanosomes that cause sleeping sickness, rapidly alter their surface antigens so as to appear invisible to the recognition arm of the immune defenses. Yet not a single scientist in the world predicted the extraordinary strategy of HIV-1, which destroys the very cells destined to defend against it. The Ebola virus likewise has the tactic of suppressing the immune responses, though in some as yet mysterious ways.

Viruses evolve and modify such strategies using a variety of genetic mechanisms, from point mutations to genetic reassortment and recombination. And although the formidable array of defenses of the immune system, combined with the requirements of any new strain of virus to function and adapt to the restricted ecology of its host, ensure that the emergence of such dangerous strains are relatively few, the threat they pose is "a real and potential problem without boundaries."[9]

There are worrying indications of recent near misses. Lassa virus can localize in the throat but not in enough quantity to result in aerosol spread. A good deal more alarming is that same potential for Ebola. The studies being carried out at USAMRIID on the Ebola Reston are starkly illuminating.

There was abundant epidemiological evidence within the Hazelton monkey facility for spread of the Ebola virus from room to room by aerosol. The pattern of illness in the monkeys was primarily a respiratory one, with running noses, coughing and sneezing, and a pneumonia at autopsy. Meticulous examination of the autopsy tissues confirmed virus replication, with inclusions and budding virions in a variety of pulmonary structures, including the lining cells of the airspaces themselves, the type II alveolar epithelial cells. Masses of virus were found inside the airways, and it was possible to transmit the virus to other monkeys experimentally through very small numbers inhaled by aerosol.[10]

The Ebola virus is widely regarded as the most dangerous acute virus to emerge in modern history. Though thankfully the Reston strain did not ignite a human epidemic, there was evidence it did cross the species barrier. No demonstration could more convincingly evoke the capacity of a newly emerged BSL-4 agent to combine its terrible lethality with respiratory spread.

With AIDS too there is grounds for worry. Because the virus will not infect ordinary laboratory animals, a strain of immunologically deficient mice has been bred to permit study of the dynamics of infection. These mice, genetically altered to allow grafting with human lymphocytes, allow scientists to study the effects of HIV-1 in human cells grafted into the animals. In 1990 the research workers were shocked to discover that when the two HIV-1 viruses came together inside the infected cells they swapped parts of their genomes, creating new hybrids. In still more alarming instances, where the mice were coincidentally infected with their own retroviruses, the introduced HIV-1 viruses swapped part of their genomes, including the coding for structures such as the envelope proteins, with the mouse "endogenous" retrovirus. This opened up a potential for radically altered hybrids.[11] It also alarmed many eminent retrovirologists, including Howard Temin, who questioned the further use of mice in the investigation of AIDS in this way.[12] Subsequent researches have confirmed that AIDS viruses are spectacularly "recombinogenic." Such juggling of genes between different AIDS viruses infecting the same patients is one of the reasons why an AIDS vaccine is proving so elusive.[13]

Recombination between viruses is now regarded as one of the major pathways of viral evolution.[14] It has worried a good many other scientists, including Professor Paul Sharp of Nottingham University: "I would not want to create a scare, but this ability to create hybrids gives you an evolutionary jump and we don't know what the properties of a hybrid may be."[15] An escalation of concern arose from the finding by Lusso, Gallo, and colleagues that HIV-1 viruses that had recombined with mouse leukemia virus inside transplanted human cells now had the capacity to infect cells from respiratory epithelium. While this new "tissue tropism" is still just a finding in the laboratory — and therefore a long way from reproducing an event in nature — it worries a great many experts, including Lederberg, Gallo, and Kilbourne. It shows all too clearly that the worst-case scenario is possible: and that nobody can really predict the degree of certainty or uncertainty of its possibility. Even so, some scientists regard such thinking as apostasy, consoling themselves that if HIV were capable of infecting the respiratory epithelium it would not withstand drying and so would not be infectious to others through coughing or sneezing. But a new surface envelope, as a result of such recombination, might well alter the viral resistance to drying. And as Mims stresses, a modified virus would not need to survive long in air, particularly if kept moist for a short time in the aerosolized droplet, for transmission to prove successful in a crowded ambience.

Had HIV-1 spread by aerosol from the beginning, we would not even have registered its existence for several years, since symptoms of first infection are mild or nonexistent. This sinister spread would have precluded any effective public health measures. In such a dreadful circumstance, the global village, with its closely woven nexus of great cities, packed with utterly naive human populations, would have become one vast amplification zone. From the cities, an altogether more perilous gyre would have surged and widened, diffusing into every crack and crevice of town and village, to become a single pullulating universal wave. Human immunity would have proved no defense against it. Aerosol-spread HIV-1 in about the year 1980 would have proved the ultimate doomsday singularity, the terrifying arrival of a true Virus X.

4

That same self-deluding complacency lies behind our present refusal to come to terms with such a threat. It goes like this: since no such extinction event has ever wiped out humanity in the past, it is impossible that it will happen in the future.

But extrapolating future trends from past history can only be deceiving. In humbling contradiction is the statistic that 99 percent of all of the species that have ever evolved on earth have suffered extinction. The mean survival of a mammalian species is put at about a million years. While suspected causes vary, in fact the fossil record is insufficient to attribute a cause to most of these with any accuracy.[16] But plague infections have been suspected as a possible contributor, possibly a major one, and caused the extinction of at least one species during recent history. Swayne's hartebeest was wiped out by an epidemic of rinderpest, introduced into Africa in the nineteenth century by Indian bullocks used by Kitchener's army to pull gun carriages.[17]

Past history cannot be a template for the future, for we have altered the earth's ecology on such a scale that a host of unknown major variables have entered the scenario. And, shocking as it might seem, the extinction scenario has actually been tested by the deliberate hand of man. The myxomatosis pandemic in rabbits is now regarded as the classic experiment.

In 1859 rabbits, a totally alien species, were introduced into Australia as a source of food. Lacking natural predators, their population underwent an explosive expansion, with consequent destruction of grassland and farmland. In 1950, in an attempt to reduce their numbers, wild rabbits in the Murray Valley in southeast Australia were infected with the myxoma virus. This virus, of the genus, *Leporipoxvirus,* lived in a symbiotic relationship with the Brazilian rabbit, *Sylvillagus brasiliensis,* which is a denizen of the tropical forest. The consequences for the Australian rabbits were altogether predictable. The myxoma virus is not spread by aerosol, but as an arbovirus by biting mosquitoes. Nevertheless the prevalence of the vector was so high it became as efficient as a true aerosol spread. Viral "traffic" between the two species began slowly, from May to November, in the rabbits burrows. Suddenly, in December, the Australian summer, augmented perhaps by a proliferation of the insect vector during a wet spring, the epidemic exploded. Over the course of just three summer months, 99.8 percent of the rabbit population of the entire southeast, a land area equivalent to the whole of Western Europe, became infected and died.[18]

Following this initial annihilation, the tiny surviving population of rabbits began to coevolve with the virus. So, year by year, the mortality fell until, seven years after the introduction of myxomatosis, the lethality was now just 25 percent. Selective changes took place both for more resistant rabbits and for less lethal strains of virus.

At present the AIDS epidemic, which moves much more slowly because of its mode of transmission, may be showing the earliest signs of a similar evolution, as a handful of survivors are being reported. In time, if left to its natural course, even the lethal HIV-1 will evolve, as more and more people survive, until eventually the new human strains of virus coevolve with their human symbiont. After many centuries, the future progeny of HIV-1 may cause no more illness than it currently causes chimpanzees or the SIVs their host monkeys.

How likely is it that, given the enormous diaspora of viruses in nature, there might already reside one or many species that could, with minimal accommodation, assume the Virus X of our worst nightmares?

The human species is essentially a monoculture — comparable in many respects to the rabbits that overpopulated southeastern Australia. The myxomatosis experience, together with the evidence from many past human pandemics, suggests that a percentage of people will be resistant even to the most virulent extinction strain. That resistance derives from the genetic differences that exist between all except identical twins, spread over all of the races and differing populations of people throughout the world and the widely varying ecologies they inhabit.

Total human extinction is therefore unlikely as a result of a viral pandemic, but a near miss, such as the rabbits experienced with myxomatosis, or even a lesser global lethality, would prove so catastrophic socially and psychologically that we can derive only limited comfort from this.

5

Influenza is a very good illustration of an emerging virus because, although it has long been familiar, indeed is one of the oldest of known viruses, it is still emerging.

The influenza virus is so important it has a family dedicated to itself, the *Orthomyxoviridae.* Each member has a negative sense RNA genome in seven or eight segments, each a gene that codes for a specific protein. When two viruses coincide in one animal host, they can, like the AIDS viruses, reshuffle their genes in the genetic process termed "reassortment." There are three major genera of the virus, labeled A, B, and C. They might look the same, but influenza A has a much more rapid rate of evolution than the others and it is this genus that is associated both with serious epidemics and most grave of all, killer pandemics.[19]

Two proteins on the surface of the virus are visible as spikes on the electron microscopic views. These are the hemagglutinin, or H protein, and the neuraminidase, or N protein. They are the virus's means of attaching to the host cell. The human immune system recognizes this and makes antibodies to them. Influenza A has numerous subtypes of both the H and N proteins: to explain the terminology, a strain labeled H2N28 carries the H2 and N28 antigens on its spikes. It is the continuing emergence of new subtypes that gives the virus its emerging potential.[20]

Just how effective this can be is illustrated by a mini outbreak on a commercial jet airliner in 1979. The plane, carrying fifty-four persons, was delayed on the ground in Alaska for three hours, during which the ventilation system was inoperative and at a time when a single passenger was suffering from flu. Within a few days, 72 percent of the other passengers came down with the same virus.[21]

One way in which these spike proteins can alter their antigenicity is through "point mutations." So the commonest way in which a new strain of virus emerges is through the accumulation of these individual mutations in the H and N proteins. This is called antigenic "drift." The familiar outbreaks of influenza sweeping through our populations each year arise in this way, and so doctors have to re-immunize vulnerable people before winter, because the immunity from the previous year no longer works. But there is another mechanism that can cause rather more dramatic changes in these virus antigens. This antigenic "shift" happens when two very different influenza viruses concomitantly infect a single animal — usually the pig but also possibly a human. At this time a major reassortment of genes takes place. The viruses swap the genetic coding for whole H and N antigens.

But where is the pool of such antigens that makes such reassortment possible?

The natural reservoir of influenza is, curiously enough, the aquatic birds of the world. Wild ducks and other water fowl already harbor all of the fourteen known H antigens. So the potential for very many new pandemic strains of influenza already exists in this natural gene bank. All these different influenza viruses replicate in the digestive tracts of the wild fowl, who excrete them into the surrounding water. When scientists drew samples from the great lakes of Canada during the winter, they found extensive contamination with different populations of influenza virus. Virtually every wild fowl in the world is infected with one or more viruses. Yet, in the words of the expert Robert Webster, "In the wild duck, these influenza viruses cause no disease."[22]

Where in the epidemic strains that emerge just about every year only a percentage of the H and N antigens have changed, in a pandemic strain the entire surface envelope of the virus is new. This means that the virus appears totally foreign to the human immune system.

Nancy Cox is the director of the Influenza Division at the Centers for Disease Control in Atlanta. In her words, "In this sort of instance, with a population lacking immunity and a virulent pandemic strain on the move, we have a very dramatic situation developing." This happened in the winter of 1918, when tens of millions of people died from a great influenza pandemic. Tragically, investigation of a single Third World epicenter during this pandemic, India, suggests that perhaps 20 million people died there alone. There have been two less severe pandemics since then, in 1957 and 1968. The media tends to label these "Chinese flu" because there is a tendency for pandemics to arise in China, where farmers keep common herds of ducks and pigs.

On the wall of her office, Nancy Cox has two maps of the world, decorated with spreading contour lines and various colored pins. Each map bears a year in its legend, 1957 and 1968. Like many other influenza experts around the globe, Nancy is anticipating a new pandemic of the influenza virus. She believes that its past behavior will give her a good idea of what to expect. Experts who have to deal with pandemic strains of influenza spend much of their time tracking the virus in this way. For example, with a glance at her maps, Dr. Cox can gauge at what stage Australia was hard hit in both '57 and '68.

On the lower part of the same wall is a map of China, with six different sites ringed. All these are currently being monitored by observers hoping to pick up the new strains as soon as they arrive. China is not the only possible site of emergence, and there are observers elsewhere too, notably Italy. When the pandemic strain arrives a dramatic race will begin. The aim is to incorporate the antigens of the new virus as quickly as possible into a vaccine. For in the event of a new pandemic, such celerity in response will become a global matter of life and death. With the pandemic strain the world would have no more than months to prepare and distribute sufficient new vaccine in advance of the rate of spread of the pandemic influenza virus.

To give some indication of the size of the public health undertaking, a virulent pandemic strain would infect up to 50 percent of the population of the world. The fastest the pharmaceutical industry can prepare and distribute such a vaccine is about six months. To protect America alone would demand some 250 million doses of vaccine, all the way from preparation to manufacture, distribution, and inoculation. Globally, the task is gargantuan. Meanwhile the virus, one of the most efficiently contagious respiratory viruses ever known, would be fulminating in the amplification zones of our modern cities, fanned by the wind of modern airline travel. And while such affluent countries as America, Japan, Australia, and Western Europe might have some chance of winning the race, there is much less chance in poorer countries, where two thirds of the human population wait virtually without any preparation or protection at all.

Influenza, through history, has shown a definite and consistent pattern of repeated epidemics and pandemics. Those historical records proclaim that another pandemic of influenza is now overdue.

6

What can we really conclude, based upon known fact and likely projection for the future?

We can be certain that there will be many outbreaks of serious infection. Most will behave as we have seen with the hantavirus, Legionnaire's disease, Lyme disease, and the various outbreaks of Ebola. People will die locally but the outbreaks will not evolve into epidemics. Less frequently, but with equal certainty, they will become epidemics, and a small minority of these will extend to pandemics.

The exact timing of such pandemics cannot be predicted. But the consequences are all too tragically obvious: there will be great suffering and a great many deaths. Anybody who has lived through the past forty years will be familiar with the threat of nuclear annihilation. How curious that as this fear recedes, though it has hardly disappeared, it has been replaced by a renewed fear of the threat from humankind's most ancient enemy, disease-causing microbes.

We evolved in harmony with, sometimes in competition with, the spectrum of life on earth, in all of its wonderful and complex variety. When one form of life becomes so dominant, overwhelmingly upsets this equilibrium, changes will inevitably occur to the ecological whole. In our exploitation of all life on earth, in our intrusion into every crevice of the biosphere, we have become a threat to ourselves.

Life, including the humble bacterium and virus, is not predictable. Its unpredictability is sewn into its genetic structure, an adaptation designed for survival. The twenty-first century will see a continuation of the present escalation of emerging viruses and they will bring attendant risks. Can we afford to take such chances? I do not think so, no more than I would advocate a gamble with nuclear weapons. On the contrary, we must do everything in our power to mitigate the danger.

The universe is surely not benign. The night sky, so seemingly peaceful, is permeated by unimaginable scenes of violence, where stars explode into supernovas and galaxies collide. We cannot forsake our future to chance, for the roulette wheel is the red tooth and claw of evolution.

APPENDIX

Emerging Infections

This list is not intended to be comprehensive. It omits major lists of new infections in animals and plants over the same period. It is also instructive to note the overwhelming role of ecological intrusion or other human behavioral factors, notably agriculture, in the disease emergence. In their natural hosts, most if not all of the viruses are clearly symbiotic, causing little or no disease. Further details of all of these will be found elsewhere in the book or in the standard texts such as *Fields Virology.*

EMERGING VIRUSES

1930* ***Rift Valley Fever Virus***
Influenza-like illness causing hundreds of thousands of infections in sheep, goats, and cattle and, to a much lesser extent, hemorrhagic fever in humans. **Africa.** (*Ungulates/mosquito*)

1933 ***Eastern Equine Encephalitis Virus***
Infection of the brain and central nervous system, mainly in children, with between 50 percent and 75 percent mortality. **North America, from Canada to Florida, South America, and the Caribbean.** (*Wild birds/mosquito*)

**The dates refer to the isolation of the causative microbe. Clinical disease will often have predated this, sometimes by great periods of time.*

1936 ***Western Equine Encephalitis Virus***
Less virulent disease, otherwise similar to Eastern Equine encephalitis. **The Americas generally, also related viruses widely in Russia, Europe, Scandinavia, and New Zealand.** (*Birds/mosquito*)

1938 ***Venezuelan Equine Encephalitis Virus***
Massive epidemics among horses, with extension to humans. Illness in humans follows the pattern of encephalitis described above, but probably milder, with a high percentage of subclinical disease. **Tropical and subtropical America.** (*Rodents/mosquito*)

1943 ***California Encephalitis Virus***
Acute infection of the brain and central nervous system, usually in children. This was the forerunner of a number of related viruses, all causing a similar pattern of infection (see La Crosse, Snowshoe hare, etc.). **Western United States and Canada.** (*Rodents, including rabbit/mosquito*)

1950 ***Hantaan Virus***
The cause of Korean hemorrhagic fever (see index). A potentially fatal disease in humans, leading to kidney failure. **Asia and Eastern Europe.** (*Rodents*)

1952 ***Sindbis Virus***
Fever with arthritis. Distributed over vast areas of **Europe, Africa, Asia, and Australia.** (*Birds/mosquito*)

1954 ***Mayaro Virus***
Fever, rash, arthralgia, and arthritis. Alphavirus related to Chikungunya and Sindbis. **South America, particularly Brazil.** (*Animal host uncertain/mosquito*)

1956 ***Chikungunya Virus***
Fever with rash and crippling arthralgia and arthritis. Caused by an alphavirus of the same family as Sindbis and O'nyong-nyong. **Tropical Africa,** where the reservoir is probably primates/mosquito, and **Asia,** where it is spreading from human to human, through mosquito.

1957 ***Kyasanur Forest Virus***
A hemorrhagic fever caused by a flavivirus. **Asia (India).** (*Rodent or bat/tick*)

1958 ***Junin Virus***
Hemorrhagic fever with high lethality. **South America (Argentina).** (*Rodent*)

1959 ***O'nyong-nyong Virus***
Explosive outbreaks of fever and severe joint pain. **African savannah or tropical forest (Kenya).** (*Animal host unknown/mosquito*)

1960 ***La Cross Encephalitis Virus***
The commonest encephalitis of the "California" group above. **Eastern United States.** (*Chipmunk or squirrel/mosquito*). Other viruses from the same group found over the years since isolation include Snowshoe hare. **Canada, Alaska, northern United States.** (*Snowshoe hare/mosquito*) San Angelo. **Western United States.** (*Origins unknown*). Tahyna. **Europe.** (*Domestic animals and rabbits/ mosquito*) Lumbo. **East Africa.** (*Origins unknown*) Inkoo. **Finland, USSR.** (*Origins unknown*)

1960 ***Ross River Fever Virus***
Fever, severe arthritis, and rash. An alphavirus disease, similar to O'nyong-nyong. **Australia and the Pacific.** (*Animal host uncertain, but may be mammal or marsupial/mosquito*)

1961 ***Oropouche Virus***
Major epidemics, with thousands of people infected (see index). **South America, notably Amazon basin.** (*Sloths and monkeys/midge*)

1966 ***Machupo Virus***
Hemorrhagic fever with high lethality (see index). **South America (Bolivia).** (*Rodent*)

1967 ***Marburg Virus***
Hemorrhagic fever with 30 percent lethality (see index). Caused by filovirus. **Germany, but source of virus was East Africa.** (*Animal reservoir unknown, though infection contracted from primate*)

1967 ***Igbo Ora Fever Virus***
Fever, rash, arthritis. Alphavirus related to Chikungunya and O'nyong-nyong. **Africa.** (*Animal host uncertain/mosquito*)

1970 ***Lassa Virus***
Major epidemics of hemorrhagic fever with 10 percent lethality, now endemic (see index). **Africa.** (*Rodent*)

1970s ***Trivittatus Encephalitis Virus***
Frequently found in mosquitoes in the **United States, especially in the Midwest,** and recently isolated for the first time from a human case. (*Animal host unknown/mosquito*)

1973 ***Rotaviruses***
Major pandemics caused by various strains of rotavirus, a genus within the family of *Reoviridae.* Note that the mortality from diarrheal illness in infants in developing countries is staggering — in 1975, the total infections were estimated at 450,000,000 yearly, with 1 to 4 percent fatality (*Fields Virology*). (*Origins unknown, but the very high antibody frequency in children, particularly in developing countries, indicates the virus has long been coevolving with humans, spreading by the fecal-oral route*).

1974 ***Parvovirus B19***
The cause of over 90 percent of the aplastic crises in children with sickle-cell anemia and other hemolytic anemias. Now also associated with fetal infection and some cases of sterility in women. Spread by respiratory route, causing febrile illnesses with rash in children. **Global distribution.** (*Original source unknown but probably coevolving with humans*)

1976 ***Ebola Sudan Virus***
Hemorrhagic fever, caused by Ebola filovirus with 50 percent lethality (see index). **Africa (Sudan).** (*Animal source unknown*)

1976 ***Ebola Zaire Virus***
Hemorrhagic fever, caused by Ebola filovirus with 90 percent lethality (see index). **Africa (Zaire).** (*Animal source unknown*)

1977 ***Seoul Virus***
Hemorrhagic fever with renal syndrome (see index). **Asia and most of Europe.** (*Rodents*)

1980 ***Human T-lymphotrophic Virus (HTLV-1)***
The cause of adult T-cell leukemia and of central nervous system disease (see index). **Asia, notably Japan, Central Africa, the Caribbean, and northeastern South America.** (*Original reservoir probably primates but now coevolving with people*)

1982 ***Human T-lymphotrophic Virus-2 (HTLV-2)***
The cause of hairy cell leukemia. **The Americas,** given the scanty knowledge to date. (*South American primates*)

1983 ***Jamestown Canyon Encephalitis Virus***
A virus of the Melao group, related to the "California" group. Recently 25 percent of young adults in Michigan were found to be seropositive. **North America.** (*White-tailed deer/mosquito*). Substrains of virus include South River. **Northeastern United States, Quebec.** (*Animal and arthropod unknown*). Keystone. **Eastern United States.** (*Animal unknown/mosquito*). Serra do Navio. **Brazil.** (*Animal and arthropod unknown*)

1983 ***Human Immunodeficiency Virus-1 (HIV-1)***
Human acquired immunodeficiency syndrome, or AIDS. **Global.** (*African primate, probably the chimpanzee*)

1985 ***Human Immunodeficiency Virus-2 (HIV-2)***
A milder form of human AIDS. **Mainly West Africa.** (*African primate*)

1986 ***Human Herpesvirus-6 (HHV-6)***
The cause of pandemic rash and fever in children, Rosea subitum. **Global.** (*Origins unknown, coevolving with people*)

1988 ***Hepatitis E Virus***
The cause of epidemic human liver disease, previously classified as Non-A Non-B hepatitis, now discovered to be caused by a waterborne calcivirus. **Tropical zones of Asia, Africa, and South America.** (Origins unknown, now coevolving with people)

1989 ***Hepatitis C Virus***

Transfusion related and sporadic outbreaks of liver disease. **Global.** (*Virus first isolated from chimpanzee, suggesting primate origins*)

1989 ***Ebola Reston Virus***

Outbreak in monkey quarantine facility in the United States (see index). **Asia, notably the Philippines.** (*Animal reservoir unknown*)

1991 ***Guanarito Virus***

The cause of Venezuelan hemorrhagic fever. **South America.** (*Rodent*)

1992 ***Barmah Forest Virus***

A mosquito-borne alphavirus, found only in Australia. Causes polyarthritis in humans (similar to Ross River Virus). First isolated from mosquitoes in 1989. First human cases in arid northern and central region of **Western Australia** 1992.

1993 ***Sin Nombre Hantavirus***

Highly lethal disease, mainly in young people (Hantavirus pulmonary syndrome). **United States, particularly the Southwest.** (*Rodents*)

1994 ***Sabia Virus***

Sporadic cases of hemorrhagic fever. **South America (Brazil).** (*Rodents*)

1994 ***Equine Morbillivirus in Australia***

Outbreak of severe respiratory disease in horses in Brisbane affecting twenty horses, thirteen of which died. Two further outbreaks soon followed. Three human infections with one fatality. Virus found in flying foxes, which are thought to be the natural host.

1994 ***Ebola Gabon***

Outbreak of Ebola fever in gold-panner encampments far northwest of **Gabon.** Virus contracted from two patients butchering a chimpanzee found dead in the rain forest. Total of about forty cases in two attacks. Virus similar to Zairean strain, but significant differences in sequences. Serological evidence had suggested a local Ebola virus since 1982.

1995 ***Human Herpes Virus-8***
The cause of Kaposi's sarcoma in AIDS patients. **Global.** (*Origins unknown*)

1995 ***Ebola Ivory Coast Virus***
Epidemic among chimpanzees that infected one human contact. **Africa.** (*Origins unknown*)

1995/1996 ***Argentinean Hantavirus***
A novel strain of hantavirus (Andes virus) reported from **Argentina,** causing hantavirus pulmonary syndrome in eighteen people, which may spread by direct transmission from person to person.

1996 ***Human Monkeypox Virus***
Epidemic in **Zaire** affecting seventy-one people, with six deaths.

1996 ***Lyssavirus in Australia***
First report of a new endemic virus causing a rabies-like encephalitis in the flying fox (bat) in northern **New South Wales, Australia.** Human implications as yet uncertain.

1996 ***Ebola Gabon***
New outbreak in July (see 1994) in town of Booué. All primary cases again associated with butchering chimpanzees found dead in the rain forest. One patient treated in **South Africa,** where a nosocomial infection was subsequently reported in a nurse. In all, between 1994 and 1996, a total of forty-three deaths.

EMERGING BACTERIA

1975 ***Lyme disease*** (*Borrelia burgdorferi*)

1976 ***Legionnaire's disease*** (*Legionella pneumophilia*)

1976 ***Acute and chronic diarrhea*** (*Cryptosporidium parvum*)

1977 ***Bowel infection*** (*Campylobacteri jejuni*)

1978 ***Toxic shock syndrome*** (*Staphylococcus aureus*)

1982 ***Hemorrhagic colitis; Hemolytic Uremic syndrome*** (*Escherichia coli 0157:H7*)

1983 ***Cat scratch disease*** (*Afipia felis*)

1983 ***Peptic ulcer disease*** (*Helicobacter pylori*)

1985 ***Persistent diarrhea*** (*Enterocytozoon bieneusi*)

1986 ***Persistent diarrhea*** (*Cyclospora cayatanensis*)

1991 ***Atypical babesiosis*** (new species of *babesia*)

1992 ***7th pandemic of cholera*** (*Vibrio cholerae 0139*)

1992 ***Cat scratch disease*** (*Bartonella henselae*)

1994 ***Flesh-eating bug*** (*Beta hemolytic streptococcus*)

EMERGING PROTOZOA

1991 ***Conjunctivitis, disseminated disease*** (*Encephalitozoon hellem*)

1993 ***Disseminated disease*** (*Encephalitozoon cuniculi*)

REFERENCES

CHAPTER ONE

1. I have described these discoveries in Ryan, F. (1993). *The Forgotten Plague.* Little, Brown and Co., Boston.

2. Stewart, Gen. W. H. (1967). A mandate for state action. Presentation to the Association of State and Territorial Health Officers, Washington, D.C., December 4.

3. McNeill, W. H. (1977). *Plagues and Peoples.* Basil Blackwell, Oxford.

4. Dubos, R. (1959). *Mirage of Health.* George Allen & Unwin, London.

5. Lederberg, J. (1993). Viruses and humankind: intracellular symbiosis and evolutionary competition. Chapter 1 in *Emerging Viruses,* ed. Stephen S. Morse. Oxford Univ. Press, New York and Oxford.

6. Science Notebook. *Washington Post,* May 22, 1995. Also, *Daily Telegraph,* May 26, 1995.

7. *Daily Telegraph,* September 29, 1994.

8. Grossman, D., and Shulman, S. (1995). The biosphere below. *Earth: The Science of Our Planet* (June).

CHAPTER TWO

1. I have changed the names of patients from the 1993 Four Corners outbreak to protect them and their families.

CHAPTER THREE

1. Fenner, F., Henderson, D. A., et al. (1988). *Smallpox and Its Eradication.* World Health Organization. Also Denevan, W. M. (1992). *The Native Populations of the Americas in 1492.* Univ. of Wisconsin Press, Madison. There has been some recent controversy on the speed and manner of smallpox spread in early America. See Pringle, H. (1996). The plague that never was. *New Sci;* July 20, 1996.

CHAPTER FOUR

1. Medawar, P. (1994). Viruses. *National Geographic* (July).

2. A more accurate scientific definition is included in the dictionary addenda.

3. I am grateful to George Lomonossoff at the John Innes Centre for providing me with copies of the original papers.

4. Murphy, F. A., and Kingsbury, D. W. (1990). Virus taxonomy. Chapter 2 in *Fields Virology,* 2nd ed. Raven Press, New York.

5. George Lomonossoff interview. See also, Lomonossoff, G. P., and Wilson, M. A. Structure and in vitro assembly of Tobacco Mosaic virus. Chapter 2 in *Molecular Plant Virology,* Vol. 1, ed. J. W. Davies. CRC Press, Boca Raton, Florida

6. See 3 above.

7. *Fields Virology,* op. cit.

8. Ibid.

9. Gubler, D. J., and Clark, G. G. (1995). Dengue/Dengue haemorrhagic fever: the emergence of a global health problem. *Emerging Infect Dis;* 1 (2): 55–57.

10. Murphy, F. A., and Nathanson, N. (1994). The emergence of new virus diseases: an overview. *Seminars in Virology;* 5: 87–102.

11. *Emerging Viruses,* op. cit.

12. Shope interview.

13. Johnson, K. (1993). Emerging viruses in context: an overview of viral hemorrhagic fevers. Chapter 5 in *Emerging Viruses,* op. cit.

14. See the appendix to the present volume.

15. Lee, H. W., and Lee, P. -W. (1976). Korean haemorrhagic fever: demonstration of causative antigen and antibodies. *Korean J Intern Med;* 19: 371–83. Also, Lee, H. W., Lee P. -W., and Johnson, K. M. (1978). Isolation of the aetiologic agent of Korean haemorrhagic fever. *J Infect Dis;* 137: 298–308.

16. McKee, K. T., Jr., LeDuc, J. W., and Peters, C. J. (1991). Hantaviruses. Chapter 22 in *Textbook of Human Virology,* 2nd ed., ed. Robert B. Beishe.

17. Yanigihara, R. (1990). Hantavirus infection in the United States: epizootiology and epidemiology. *Rev Inf Dis;* 12: 449–56. Also, Lee, P. -W., Amyx, H. L., et al. (1982). New haemorrhagic fever with renal syndrome-related virus in indigenous wild rodents in United States. *Lancet;* 2: 1405.

CHAPTER FIVE

Note: Elizabeth W. Etheridge has written an excellent history of the CDC, *Sentinel for Health: A History of the Centers for Disease Control.* University of California Press, Berkeley, 1992.

1. Judson, H. F. (1979). *The Eighth Day of Creation.* Jonathan Cape, London.

2. LeDuc, J. W., Childs, J. E., and Glass, G. E. (1992). The hantaviruses, etiologic agents of hemorrhagic fever with renal syndrome: a possible cause of hypertension and chronic renal disease in the United States. *Ann Rev Publ Health;* 13: 79–98. Childs, J. E., Frusic, M., et al. (1988). Evidence of human infection with a rat-associated hantavirus in Baltimore, Maryland. *Am J Epidemiol;* 127: 875–78. See also, Childs, J. E., Korch, G. W., et al. (1985). Geographical distribution and age-related prevalence of Hantaan-like virus in rat populations of Baltimore, Maryland, USA. *Am J Trop Med Hyg;* 34: 385–87. Childs, J. E., Korch, G. W., et al. (1987). Epizoology of hantavirus infections in Baltimore: isolation of a virus from Norway rats and characteristics of infected rat populations. *Am J Epidemiol;* 126: 55–68.

3. See 1 above. See also, Watson, J. D. (1968). *The Double Helix.* Weidenfield and Nicholson, London.

4. For an interesting discussion of Mullis's discovery, see Baskin, Y. (1990). DNA unlimited. *Discover* (July): 77–79.

5. Nichol, S. T., Spiropoulou, C. F., et al. (1993). Genetic identification of a hantavirus associated with an outbreak of acute respiratory illness. *Science;* 262: 914–17. See also, Marshall, E. (1993). Virology without a virus. *Science;* 262: 834. And, Marshall, E. (1993). Hantavirus outbreak yields to PCR. *Science;* 262: 832–36.

CHAPTER SIX

1. Wilson, E. O. (1993). *The Diversity of Life.* Penguin, New York.

2. Professor Verghese is the Chief of the Academic Health Center in the School of Medicine at Texas Tech University in El Paso. He has written a delightful book, *My Own Country: A Doctor's Story of a Town and Its People in the Age of AIDS* (Random House, New York, 1995), which describes, with humor and poignancy, the personal pain and the triumphs of AIDS in his patients.

3. Parmenter interview. See also, Stone, R. (1993). The mouse–piñon nut connection. *Science;* 262: 833.

4. Childs, J. E., Ksiazek, T. G., et al. (1994). Serologic and genetic identification of *Peromyscus maniculatus* as the primary rodent reservoir for a new hantavirus in the southwestern United States. *J Infect Dis;* 169: 1271–80.

5. The story of the development of this test, by Jenison and Hjelle, is fascinating in itself. See, Jenison, S., Yamada, T., et al. (1994). Characterization of human antibody responses to Four Corners hantavirus infections among patients with hantavirus pulmonary syndrome. *J of Virology;* 68: 3000–3006.

6. Nichol interview. Also, Nichol, S. T., Spiropoulou, C. F., et al. (1993). Genetic identification of a hantavirus associated with an outbreak of acute respiratory illness. *Science;* 262: 914–17. Spiropoulou, C. F., Morzunov, S., et al. (1994). Genome structure and variability of a virus causing hantavirus pulmonary syndrome. *Virology;* 200: 715–23. Hjelle, B., Jenison, S., et al. (1994). A novel hantavirus associated with an outbreak of fatal respiratory disease in the southwestern United States: evolutionary relationships to known hantaviruses. *J of Virology;* 68: 592–96. Hjelle, B., and Jenison, S. (1994). Probable hantavirus pulmonary syndrome that occurred in New Mexico in 1975. *Ann Intern Med;* 120: 813.

7. Zaki, S. R., Greer, P. W., et al. (1995). Hantavirus pulmonary syndrome: pathogenesis of an emerging infectious disease. *Am J Path;* 146: 552–79.

8. This research is still continuing, though limited by the rarity of new cases these days.

9. Centers for Disease Control and Prevention (1993). Update: hantavirus disease — United States. *MMWR;* 42: 612–14.

10. There was a linked concatenation of these. See Centers for Disease Control and Prevention (1993). Outbreak of acute illness — southwestern United States. *MMWR;* 42: 421–24. Ibid. (1993). Update:

Hantavirus-associated illness — North Dakota. *MMWR;* 42: 612–14. Ibid. (1993). Outbreak of acute illness — southwestern United States. *MMWR;* 42: 421–24. Ibid. (1993). Update: Outbreak of hantavirus infection — southwestern United States. *MMWR;* 42: 441–43. Ibid. (1993). Update: Outbreak of hantavirus infection — southwestern United States. *MMWR;* 42: 477–79. Ibid. (1993). Update: Outbreak of hantavirus infection — southwestern United States. *MMWR;* 42: 495–96. Ibid. (1993). Update: Outbreak of hantavirus infection — southwestern United States. *MMWR;* 42: 517–19. Ibid. (1993). Update: Outbreak of hantavirus infection — southwestern United States. *MMWR;* 42: 570–72. Ibid. (1993). Update: Outbreak of hantavirus infection — southwestern United States. *MMWR;* 42: 612–14. Ibid. (1993).

11. Schmaljohn, A. L., Dexin, L., et al. (1995). Isolation and initial characterisation of a newfound hantavirus from California. *Virology;* 206: 963–72. Elliott, L. H., Ksiazek, T. G., et al. (1994). Isolation of the causative agent of hantavirus pulmonary syndrome. *Am J Trop Med Hyg;* 51: 102–108. See also, Altman, L. (1993). Virus that caused deaths among Navajos is isolated. *New York Times,* November 21.

12. Chapman's definitive paper is not yet published, but she presented her data at the fall 1994 ICAC conference. See Chapman, L. E., Mertz, G., et al. (1994). Open label intravenous ribavirin for hantavirus pulmonary syndrome. See also, Chapman, L. E., and Khabbaz, R. F. (1994). Etiology and epidemiology of the Four Corners hantavirus outbreak. *Infectious Agents and Disease;* 3: 234–44.

13. Further studies are taking place at NIH.

14. Lederberg, J., Shope, R. E., and Oaks, S. C., eds. (1992). *Emerging Infections: Microbial Threats to Health in the United States.* National Academy Press, Washington, D.C.

Note on Chapter 6: In addition to the works listed above, the following papers might be helpful to colleagues interested in the hantavirus epidemic: Centers for Disease Control and Prevention (1993). Hantavirus infection — southwestern United States: interim recommendations for risk reduction. *MMWR;* 42. Duchin, J. S., Koster, F. T., et al. (1994). Hantavirus pulmonary syndrome: a clinical description of 17 patients with a newly recognized disease. *New Eng J Med;* 330: 949–55. See also, Editorial (1994). A new hantavirus infection in N. America. *New Eng J Med;* 330: 1004–1005. Levy, H., and Simpson, S. Q. (1994). Hantavirus pulmonary syndrome. *Am J Respir Crit Care Med;* 149: 1710–13. Hughes, J. M., Peters, C. J., et al. (1993). Hantavirus pulmonary syndrome: an emerging infectious disease. *Science;* 262: 850–51.

CHAPTER SEVEN

1. *The Shorter Oxford English Dictionary,* rev. ed., 1980. Clarendon Press, Oxford.

2. Epstein, P. R. (1993). Algal blooms in the spread and persistence of cholera. *Biosystems;* 31: 209–221. Also, Epstein, P. R. (1992). Commentary: pestilence and poverty — historical transitions and the great pandemics. *Am J Prev Med;* 8: 263–65.

3. News: cholera attacks former Soviet Union. *Br Med J,* October 1, 1994; 309: 827–28.

4. WHO (March 1, 1991). Cholera: the epidemic in Peru — part I. *Weekly Epidemiol Rec;* 66 (9): 61–63. WHO (March 8, 1991). Cholera: the epidemic in Peru — part II. *Weekly Epidemiol Rec;* 66 (10): 65–70. Also — Levine, M. M. (1991). South America: the return of cholera. *Lancet;* 388: 45–46. CDC (1991); Cholera — Peru. *MMWR;* 40: 108–110.

5. Colwell, R. R., Caper, J., et al. (1977). *Vibrio cholerae, Vibrio parahaemolyticus* and other vibrios: occurrence and distribution in Chesapeake Bay. *Science;* 198: 394–96. Colwell, R. R., Brayton, P. R., et al. (1985). Viable but non-culturable *Vibrio cholerae* and related pathogens in the environment: implications for release of genetically engineered microorganisms. *Biotechnology;* 3: 817–20. Colwell, R. R. (1991). Non-cultivable *Vibrio cholerae* O1 in environmental waters, zooplankton and edible crustacea; implications for understanding the epidemiological behavior of cholera. Presentation to the American Society of Tropical Medicine and Hygiene, December, Boston.

6. Tauxe, R. V. (1995). Epidemic cholera in the new world: translating field epidemiology into new prevention strategies. *Emerging Infect Dis;* 1 (4): 141–46.

7. See 2 and 3 above. Also, report by Anna Reid, *Daily Telegraph,* September 30, 1994.

8. Altman, L. K. (1991). "Catastrophic" cholera is sweeping Africa. *New York Times,* July 23.

9. *Time,* August 1, 1994. See also, *Newsweek,* August 1, 1994; and News (1994). Doctors battle to contain cholera in Rwandan camps. *Br Med J;* 309: 289.

10. World Health Organization, personal communication. See also, van Damme, W. (1995). Do refugees belong in camps? Experiences from Goma and Guinea. *Lancet;* 346: 360–62.

11. Cholera attacks former Soviet Union. *Br Med J;* 309: 827–28.

12. See 2 above.

13. Rich hoard vital drugs as plague panic grips India. *Sunday Times,* September 25, 1994. Also, News: Troops battle to contain India's outbreak of plague. *Br Med J;* 309: 827.

14. Ziegler, P. (1969). *The Black Death.* Penguin, New York.

15. Highfield, R. (1994). Era of antibiotics fades as diseases stage comeback. *Daily Telegraph,* September 30.

16. Mudur, G. (1995). India's pneumonic plague outbreak continues to baffle. *Br Med J;* 311: 706. Campbell, G. L., and Hughes, J. M. (1995). Plague in India: a new warning from an old nemesis. *Ann Intern Med;* 122: 511–53.

17. Ryan, F. (1993). *The Forgotten Plague.* Little, Brown and Co., Boston. Also, Murray, J. F. (1991). An emerging global programme against tuberculosis: agenda for research, including the impact of HIV infection.

Bull Int Union Tuberc Lung Dis; 66: 199–201. Murray, J. F., Styblo, K., and Rouillon, A. (1990). Tuberculosis in developing countries: burden, intervention and cost. *Bull Int Union Tuberc Lung Dis;* 65: 2–20. And, D. A. Mitchison, personal communication.

18. I have changed the names of patients from this outbreak.

19. Ryan, S., Rogers, L., and Driscoll, M. (1994). The bug that ate into our imagination and sent us all mad. *Sunday Times;* May 29; 4: 6.

20. For this and the details of Lucien Bouchard, see *Sunday Times,* July 30, 1995.

21. Teletext News, September 17, 1995. See also, Invasive group A streptococcal infections in Gloucestershire. CDR *Weekly* (1994); 4 (21).

22. Connor, S. Enter the super-bug. *Independent on Sunday,* May 8, 1994.

23. Louis Rogers. *Sunday Times,* December 17, 1995.

24. Booster Dose for School Leavers. CDR *Weekly* (1994); 4 (17).

25. See 3 above.

26. Ryan, F. (1993), op. cit.

27. Fisher, J. (1994). *The Plague Makers.* Simon and Schuster, New York. See also, Toner, M. (1992). *When Bugs Fight Back.* A special reprint of this Pulitzer-winning series of reports has been produced by the *Atlanta Journal.*

28. Berkelman, R. L. and Hughes, J. M. (1993). The conquest of infectious diseases: who are we kidding? *Ann Int Med;* 119: 426–28.

29. Lederberg, J. (1987). Genetic recombination in bacteria: a discovery account. *Ann Rev Genet;* 31: 23–46.

30. Avery, O. T., MacLeod, C. M., and McCarty, M. (1944). Studies on the chemical nature of the substance inducing transformation of pneumococcal types. *J Exp Med;* 79: 137–58.

31. Epstein, P. R. (1992). Commentary: Pestilence and poverty — historical transitions and the great pandemics. *Am J Prev Med;* 8: 263–65. See also, Ampel, N. M. (1991). Plagues — what's past is present: thoughts on the origin and history of new infectious diseases. *Rev of Inf Dis;* 13: 658–65.

CHAPTER EIGHT

Note: Important sources of information on Ebola include the limited circulation report "Viral Haemorrhagic Fever — Sudan 1976" that was drawn up as a preliminary report to the WHO by the team in enforced quarantine at the end of the epidemic. Also, Ebola Hemorrhagic Fever in Sudan, 1976. *Bulletin of the World Health Organization* (1978); 56 (2): 271–293. Ebola Hemorrhagic Fever in Zaire, 1976. *Bulletin of the World Health Organization* (1978); 56 (2): 247–70. See also Pattyn, S. R., ed. (1978). *Ebola Virus Haemorrhagic Fever.* Amsterdam and New York.

1. Cenydd Jones drew my attention to this classical quote.

2. From a photograph provided by Don Francis.

3. I am indebted to Dr. William Close for many such intimate details of Yambuku. He has described the events within the mission, barely fictionalized, in his articulate and poignantly touching book, *Ebola: A Documentary Novel of its First Explosion in Zaire by a Doctor Who Was There.* Ivy Books, New York.

4. A copy of the report is to be found in the papers of the late Pierre Sureau, and his diary gives full credit to the African doctor.

5. Ruppol interview.

6. From the photograph in the collection of Bill Close. Note also this is a different Mabolo from the index case.

7. See 5 above.

8. Jones interview.

9. Smith interview.

CHAPTER NINE

1. Eykmans interview.

2. Martini, G. A. and Siegert, R., eds. (1971). *Marburg Virus Disease.* Springer Verlag, Berlin and New York.

3. Simpson interview.

4. See 2 above.

5. Murphy interview.

6. WHO internal meeting memorandum. Viral haemorrhagic fever outbreaks in Zaire and Sudan. Plan of Action (October 18, 1976). Confidential.

7. From Pierre Sureau's wonderfully modest and accurate diary, which is kept by his successor Bernard Le Guenno at the Hemorrhagic Fever Laboratory at the Pasteur Institute.

8. Ibid.

9. Isaacson interview. See also, Isaacson, M., Ruppol, J. F., et al. (1978). Containment and surveillance of a hospital outbreak of Ebola virus disease in Kinshasa, Zaire, 1976. *Ebola Virus Haemorrhagic Fever,* op. cit. See also, ibid., Isaacson, M., Sureau, P., et al. Clinical aspects of ebola virus disease at the Ngaliema Hospital, Kinshasa, Zaire, 1976.

10. From the film *The Plague Monkeys.*

CHAPTER TEN

1. *Preliminary Report: Viral Haemorrhagic Fever — Sudan 1976.* Report compiled by the team prior to leaving the Sudan; copy kindly provided to me by David Smith.

2. Interviews with McCormick, Simpson, and many others.

3. Isaacson interview. See also, Isaacson, M., Ruppol, J. F., et al. (1978). Containment and surveillance of a hospital outbreak of Ebola virus disease in Kinshasa, Zaire, 1976. *Ebola Virus Haemorrhagic Fever,* op. cit. Ibid. Isaacson, M., Sureau, P., et al. (1978). Clinical aspects of ebola virus disease at the Ngaliema Hospital, Kinshasa, Zaire, 1976.

4. WHO/Sudanese investigative team for haemorrhagic fever in the Sudan (Western Equatoria). Report by Dr. Brès following his visit to the Sudan, October 18–November 3, 1976. This is a confidential report.

5. Ibid.

6. WHO. Draft objectives and plan of action of outbreak of haemorrhagic fever in Maridi area. (As modified after discussions in the Ministry of Health, Khartoum, October 20, 1976). Confidential.

7. See 1 above. Also, Ebola haemorrhagic fever in Sudan, 1976: report of a WHO/international study team (1978). *Bull WHO;* 56 (2): 247–70.

8. Arata, A. A., and Johnson, B. (1977). Approaches towards studies on potential reservoirs of viral haemorrhagic fever in southern Sudan. In *Ebola Virus Haemorrhagic Fever.* op. cit.

9. Ebola haemorrhagic fever in Zaire, 1976. Report of an International Commission (1978). *Bull WHO;* 56 (2): 271–93.

10. Rapport du Doctor Simon van Nieuwenhove. Surveillance du Nord-Zaire et Sud-Soudan à la recherche de cas de fièvre africaine (November 17, 1976).

11. *Ebola Virus Haemorrhagic Fever,* op. cit.

12. Ibid. Also, Simpson and Highton interviews.

13. Emond, R. T. D., Evans, B., et al. (1977). A case of Ebola virus infection. *Br Med J* (August); 541–44.

CHAPTER ELEVEN

1. McCormick interview.

2. Cox, N. J., McCormick, J. B., et al. (1983). Evidence for two subtypes of Ebola virus based on oligonucleotide mapping of RNA. *J Infect Dis;* 147: 272–75.

3. From Pierre Sureau's diary of the outbreak.

4. Smith, D. H., Isaacson, M., et al. (1982). Marburg virus disease in Kenya. *Lancet* (April 10); 816–20.

5. Soviet labs breeding new killer germs? *Kayhan,* January 31, 1978. Also, Doubts at NATO of germ war allegations. *Kayhan,* February 1, 1978. Germ warfare has also been reviewed several times by James Adams, the Washington correspondent for the London *Sunday Times:* The red death: the untold story of Russia's secret biological weapons, March 27, 1994; Russia helps Iran's bio-warfare, August 27, 1995; Gadaffi lures South Africa's top germ warfare scientist, February 26, 1995.

6. Centers for Disease Control (1990). Update: Ebola-related filovirus infection in nonhuman primates and interim guidelines for handling nonhuman primates during transit and quarantine. *MMWR;* 39: 22–24, 29.

7. Peters, C. J., Johnson, E. D. (1993). Filoviruses. Chapter 15 in *Emerging Viruses,* op. cit.

8. Geisbert, T. W., Jahrling, P. B. (1992). Association of Ebola-related Reston virus particles and antigen with tissue lesions of monkeys imported to the United States. *J Comp Path;* 106: 137–52.

9. See 7 above.

10. Le Guenno, B., Formentyl, P., et al. (1995). Isolation and partial characterisation of a new strain of Ebola virus. *Lancet;* 345: 1271–74.

11. *Daily Telegraph,* May 17, 1995.

CHAPTER TWELVE

1. Desmond, A., and Moore, J. (1992). *Darwin.* Penguin Books, New York.

2. Traub, E. (1935). A filterable virus recovered from white mice. *Science;* 81: 298–99.

3. McCormick, J. (1985). Arenaviruses. Chapter 44 in *Fields Virology,* op. cit.

4. Peters, C. J. (1994). Arena virus infections. Chapter 49 in *Handbook of Neurovirology,* eds. McKendall and Stroop. See also, Barton, L. L., Peters, C. J., and Ksiazek, T. G. (1995). Lymphocytic choriomeningitis virus: an unrecognised teratogenic pathogen. *Emerging Infect Dis;* 1 (4): 152–53.

5. Webb interview.

6. Johnson, K. M., Kuns, M. L., et al. (1966). Isolation of Machupo virus from wild rodent *Calomys callosus. Am J Trop Med Hyg;* 15 (1): 103–106.

7. See 3 above.

8. There is a good review of this extremely complex immunological phenomenon in Childs, J. E., and Peters, C. J. (1993). Ecology and epidemiology of arenaviruses and their hosts. Chapter 19 in *The Arenaviridae,* ed. Maria S. Salvato. Plenum Press, New York.

9. Ibid. Also, Mims, C. A. (1966). Immunofluorescence of the carrier state and mechanism of vertical transmission in lymphocytic choriomeningitis infection in mice. *J Pathol Bacteriol;* 91: 395.

10. *Emerging Viruses,* op. cit. Also, Morse, S. S., ed. (1994). *The Evolutionary Biology of Viruses.* Raven Press, New York.

11. See 10 above. Also, Morse, S. S. (1991). Emerging viruses: defining the rules for viral traffic. *Perspectives in Biology and Medicine;* 34 (3): 387–409.

12. Wilson, E. O. (1994). *The Diversity of Life.* Penguin Books, New York.

13. May, R. M. (1991). Ecology and evolution of host-virus associations. Chapter 6 in *Emerging Viruses,* op. cit. Also, Anderson, R. M., and May, R. M. (1991). *Infectious Diseases of Humans.* Oxford Science Publications, Oxford.

14. Yates interviews.

15. See 1 above.

16. Ehrlich, P. R., and Raven, P. H. (1964). Butterflies and plants: a study in coevolution? *Evolution;* 18: 586–608. Janzen, D. H. (1990). When is it coevolution? *Evolution;* 34: 611–12. Rennie, J. (1992). Trends in parasitology: living together. *Sci Amer,* January; 104–113. See also 10 above.

17. See 13 above.

18. Van der Groen, G., Johnson, K. M., et al. (1978). Results of Ebola antibody surveys in various population groups. In *Ebola Virus Haemorrhagic Fever,* op. cit.

19. Johnson, E. D., Gonzales, J. P., and Georges, A. (1993). Haemorrhagic fever virus activity in equatorial Africa: distribution and prevalence of

filovirus reactive antibody in the Central African Republic. *Trans Soc of Trop Med Hyg;* 87: 530–35.

20. Fisher-Hoch, S. P., Brammer, T. L., et al. (1992). Pathogenic potential of filoviruses: role of geographic origin of primate host and virus strain. *J Infect Dis;* 166: 753–63. Also, McCormick interview.

21. Stansfield, S. K., Scribner, C. L., et al. (1982). Antibody to Ebola virus in guinea pigs: Tandala, Zaire. *J Infect Dis;* 146: 483–86.

22. Francis interview.

23. Peters, C. J., Sanchez, A., et al. (1994). Filoviruses as emerging pathogens. *Seminars in Virology;* 5: 147–54. See also, Regnery, R. L., Johnson, K. M., et al. (1981). Marburg and Ebola viruses: possible members of a new group of negative strand viruses. In *The Replication of Negative Strand Viruses,* eds. D. H. L. Bishop and R. W. Compans. Elsevier Press, Amsterdam and New York. And, Sanchez, A., and Kiley, M. P. (1987). Identification and analysis of Ebola virus messenger RNA. *Virology;* 157: 414–20.

24. Sanchez, A., Ksiazek, T. G., et al. (1995). Reemergence of Ebola virus in Africa. *Emerging Infect Dis;* 1 (3): 96–97.

CHAPTER THIRTEEN

1. Lederberg, J. (1988). Medical science, infectious disease and the unity of humankind. *JAMA:* 260; 684–85.

2. There is a good introduction to genetics and its applications in Nossal, G. J. V., and Coppel, R. L. (1989). *Reshaping Life,* 2nd ed., Cambridge Univ. Press. See also, McClintock, B. (1950). The origin and behavior of mutable loci in maize. *Proc Natl Acad Sci;* 36: 344–55.

3. Judland, H. F. (1985). *The Eighth Day of Creation.* Penguin Books, New York.

4. Lederberg, J. (1987). Genetic recombination in bacteria: a discovery account. *Ann Rev Genet;* 21: 23–46.

5. A good summary appears in Chapter 8 of Laurie Garrett's excellent book, *The Coming Plague.* Farrar, Straus and Giroux, New York, 1994.

6. Private correspondence with Mitsuaki Yoshida.

7. Takatsuki, K., Uchiyama, T., et al. (1977). Topics in haematology. *Excerpta Medica,* 73–77.

8. Polesz, B. J., Ruscetti, F. W., et al. (1980). Detection and isolation of type-C retrovirus particles from fresh and cultured lymphocytes of a patient with cutaneous T-cell lymphoma. *Proc Natl Acad Sci;* 77: 7415–19.

9. In a private correspondence, Mitsuaki Yoshida has outlined this discovery in compelling detail for me. These were published in Hinuma, et al. (1981), *Proc Natl Acad Sci* and Isao Miyoshi (1981), *Nature.*

10. Chen, J. S. Y., McLaughlin, J., et al. (1983). Molecular characterisation of genome of a novel human T-cell leukaemia virus. *Nature;* 305: 502–505.

11. Montagnier interview. See also, Montagnier, L. (1994). *Des virus et des hommes.* Editions Odile Jacob, Paris.

CHAPTER FOURTEEN

1. Shilts, R. (1988). *And the Band Played On.* Penguin Books, New York.

2. Groopman, J. E., and Gottleib, M. S. (1983). AIDS: the widening gyre. *Nature;* 303: 575–76.

3. Kornfield, H., et al. (1982). *New Eng J Med;* 307: 729. Also Goldsmith, J. C., et al. (1983). *Ann Int Med;* 98: 294.

4. *Emerging Viruses,* op. cit.

5. Maddox, J. (1983). No need to panic about AIDS. Editorial, *Nature;* 302: 749–50.

6. Montagnier, L. (1994). *Des virus et des hommes.* Editions Odile Jacob, Paris.

7. These sociological aspects have been covered in many articles and books, in particular *And the Band Played On,* op. cit. See various of the Panos Dossier publications, including, *AIDS and the Third World,* Panos Institute, Budapest/London/Paris/Washington. And Fineberg, H. V. (1988). The social dimensions of AIDS. *Sci Amer;* October: 106–112.

8. *Fields Virology,* op. cit.

9. Redfield, R. R., and Burke, D. S. (1988). HIV infection: the clinical picture. *Sci Amer;* October: 70–78.

10. Haseltine, W. A., and Wong-Staal, F. (1988). The molecular biology of the AIDS virus. *Sci Amer;* October: 34–42.

11. See 9 above.

12. *AIDS and the Third World,* op. cit.

13. Barin, F., Denis, F., et al. (1985). Serological evidence for virus related to simian T-lymphotropic retrovirus III in residents of West Africa. *Lancet;* December 21–28: 1387–89. Saxinger, W. C., Levine, P. H., et al. (1985). Evidence for exposure to HTL-III in Uganda before 1973. *Science;* 227: 1036–38. Biggar, R. J., Melbye, M., et al. (1985). Seroepidemiology of HTLV-III antibodies in a remote population of eastern Zaire. *Brit Med J;* 290: 808–10. De Cock, K. M., and McCormick, J. B. (1988). HIV infection in Zaire. *New Eng J Med;* 319: 309. Quinn, T. C., Mann, J. M., et al. (1986). AIDS in Africa: an epidemiological paradigm. *Science;* 234: 955–63.

14. Mertens, T. E., Belsey, E., et al. (1995). Global estimates and epidemiology of HIV-1 infections and AIDS: further heterogeneity in spread and impact. AIDS 1995; 9 (suppl. A): 5259–72. See also, Piot, P., Plum-

mer, F. A., et al. (1988). AIDS: an international perspective. *Science;* 239: 573–79.

CHAPTER FIFTEEN

1. Essex, M., and Kanki, P. J. (1988). The origins of the AIDS virus. *Sci Amer;* October 1988: 64–71.

2. Daniels, M. D., Letvin, N. L., et al. (1985). Isolation of T-cell tropic HTLV-III-like retrovirus from macaques. *Science;* 228: 1201–1204.

3. Essex interviews; also, Essex, M. (1989). The biology of human immunodeficiency viruses and related agents of monkeys. In *AIDS in Our Lives.* PPI Publishing, Dayton. See also 1 above.

4. Marx, P. A., Lerche, Li Y., et al. (1991). Isolation of a simian immunodeficiency virus related to human immunodeficiency virus type 2 from a West African pet sooty mangabey. *J Vir;* 65: 4480–85. See also, Gao, F., Yue, L., et al. (1992). Human infection by genetically diverse SIV(sm)-related HIV-2 in West Africa. *Nature;* 358: 495–99.

5. Essex, M. (1994). Simian immunodeficiency virus in people. *New Eng J Med;* 330: 209–10.

6. Khabbaz, R. F., Heneine, W., et al. (1994). Brief Report: Infection of a laboratory worker with simian immunodeficiency virus. *New Eng J Med;* 330: 172–77.

7. Essex interviews.

8. Essex interviews. Also, Marlink, R. G., Ridard, D., et al. (1988). Clinical, haematologic, and immunologic cross-sectional evaluation of individuals exposed to human immunodeficiency virus type 2 (HIV-2). *AIDS Res. Hum Retroviruses;* 4; 137–48. Marlink, R. G., Kanki, P., et al. (1994). Reduced rate of disease development after HIV-2 infection as compared to HIV-1. *Science;* 265: 1587–90.

9. Holland, J. (1993). Population error, quasispecies populations, and extreme evolution rates of RNA viruses. Chapter 19 in *Emerging Viruses*, op. cit.

10. Huet, T., Cheynier, R., et al. (1990). Genetic organisation of a chimpanzee lentivirus related to HIV-1. *Nature;* 345: 356–59.

11. McClure, M. (1990). Where did the AIDS virus come from? *New Scientist;* June 30: 54–57. Desrosiers, R. C. (1990). A finger on the missing link. *Nature;* 345: 288–89.

12. Myers, G., MacInnes, K., et al. (1993). Phylogenetic moments in the AIDS epidemic. Chapter 12 in *Emerging Viruses*, op. cit.

13. Nzilami, N., De Cock, K. M., et al. (1988). The prevalence of infection with human immunodeficiency virus over a 10-year period in rural Zaire. *New Eng J Med;* 318: 276–79. Also, De Cock, K. M., and McCormick, J. B. (1988). HIV infection in Zaire. *New Eng J Med;* 319: 309. Also, McCormick statement on BBC "Horizon" documentary on emerging viruses.

14. Travers, K., Mboup, S., et al. (1995). Natural protection against HIV-1 infection provided by HIV-2. *Science;* 268: 1612–15. There is a wide literature on AIDS vaccines; see Essex, M. (October 1994). Confronting the AIDS vaccine challenge. *Technology Review,* 23–29. Matthews, T. J., and Bolognesi, Dani P. (1988). AIDS vaccines. *Sci Amer;* October: 98–105. Editorial (December 22–29, 1990). AIDS vaccine: hope and despair. *Lancet;* 1545–46.

CHAPTER SIXTEEN

1. Gould, S. J. (1992). *Bully for Brontosaurus.* Penguin Books, New York.

2. Dennett, D. C. (1995). Darwin's dangerous idea. *The Sciences;* May/June: 34–40.

3. May, R. M. (1991). Chapter 6 in *Emerging Viruses*, op. cit. Also, Anderson, R. M., and May, R. M. (1991). *Infectious Diseases of Humans.* Oxford

Science Publications. Anderson and May papers on mathematics of coevolution.

4. The current renewal of interest in symbiosis in biological circles will be readily grasped from two very interesting books. *Symbiosis as a Source of Evolutionary Innovation,* eds. Margulis and Fester. MIT Press, Cambridge, Mass., and London. And, Sapp, J. (1994). *Evolution by Association.* Oxford Univ. Press.

5. See, Rennie, J. (1992). Living Together. *Sci Amer;* January: 104–13.

6. Sapp, J. (1994). *Evolution by Association.* Oxford Univ. Press, New York and Oxford.

7. There is even a new journal called *Symbiosis,* published by Balaban, Philadelphia and Rehovot.

8. Price, P. W. The web of life: developments over 3.8 billion years of trophic relationships. Chapter 18 in *Symbiosis as a Source of Evolutionary Innovation,* op. cit. Lewis, D. H. Mutualistic symbiosis in the origin and evolution of land plants. Chapter 20, ibid. There are many interesting examples and deductions throughout this book as also examples of the macrocycles of symbiosis in Lovelock, J. E. (1987). *Gaia.* Oxford Univ. Press, New York and Oxford.

9. Reisser, W. (1992). Symbiosis redefined: symbiotic features of virus-host interactions. *Symbiosis;* 14: 82–86.

10. See 6 above.

11. Jeon, K. W. Amoeba and x-bacteria: symbiotic acquisition and possible species change. Chapter 10 in *Symbiosis as a Source of Evolutionary Innovation,* op. cit.

12. Price, P. W. The web of life: developments over 3.8 billion years of trophic relationships. Chapter 18, ibid.

13. See references to Chapter 19.

14. From the excellent BBC television Attenborough series, *Living Together.*

15. Fleckenstein, B., and Desrosiers, R. C. (1982). *Herpesvirus saimiri* and *Herpesvirus ateles. The Herpesviruses,* ed. Bernard Roizman, vol. 1. Plenum, New York and London.

16. Kaplan, J. E. (1988). *Herpesvirus Simiae* B infection in monkey handlers. *J Infect Dis;* 157: 1090.

17. See reference 3, Chapter One.

18. Norrby, E., and Oxman, M. N. (1990). Measles virus. Chapter 37 in *Fields Virology,* op. cit.

19. Hirsch, A. (1983). Handbook of geographic and historical pathology, vol. 1. New Sydenham Society, London; 154–170.

20. Mahy, B. M. J. (1993). Seal plague virus. Chapter 17 in *Emerging Viruses,* op. cit.

21. Lederberg, J. (1993). Viruses and humankind: intracellular symbiosis and evolutionary competition. Chapter 1 in *Emerging Viruses,* op. cit.

22. Simpson interviews.

CHAPTER SEVENTEEN

1. Levins, R. Preparing for uncertainty. Typescript kindly provided for me by Richard Levins prior to its publication in a new journal, *Ecosystem Health.* Note some of the other information in this chapter derives from my interview with Levins. See also, Levins, R., Awerbuch, T., et al. (1994). The emergence of new diseases. *American Scientist;* 82: 52–60.

2. Lovelock, J. E. (1987). *Gaia,* op. cit.

3. Foreword, *Emerging Viruses,* op. cit.

4. In the appendix to the present volume, I have listed some of the emerging viruses in the last half century or so. Even this lengthy list does not pretend to be comprehensive.

5. Lederberg interview. Also, Lederberg, J. Mankind had a near miss from a mystery pandemic. *Washington Post,* September 7, 1968.

6. Introduction, *Ebola Virus Haemorrhagic Fever,* op. cit.

7. See the map at the beginning of the book.

8. Calisher, C. H. and Sever, J. L. (1995). Are North American Bunyamwera serogroup viruses aetiologic agents of human congenital defects of the central nervous system? *Emerging Infect Dis;* 1 (4): 147–51.

9. Wilson, E. O. (1994), op. cit.

10. Shope interview.

11. Miller, J. A. (1989). Diseases for our future. Global ecology and emerging viruses. *Bioscience;* 39 (8): 509–17.

12. Aldous, P. (1993). Tropical deforestation: not just a problem in Amazonia. *Science;* 259: 1390. Phillips, O. L., and Gentry, A. H. (1994). Increasing turnover through time in tropical forests. *Science;* 263: 954–58. Pimm, S. L., and Sugden, A. M. (1994). Tropical diversity and global change. *Science;* 263: 933–34.

13. Lovejoy, T. E. (1993). Global change and epidemiology: nasty synergies. Chapter 25 in *Emerging Viruses,* op. cit.

14. See 11 above.

15. Wilesmith, J. W. (1994). Bovine spongiform encephalopathy: epidemi ological factors associated with the emergence of an important new animal pathogen in Great Britain. *Seminars in Virology;* April: 179–87. See also, Pruisiner, C. B. (1995). The prion diseases. *Sci Amer;* January: 30–37.

16. See 4 above.

17. Bedford, I. D., Briddon, R. W., et al. (1994). Geminivirus transmission and biological characterisation of *Bemisia tabaci* (Gennadius) biotypes from different geographic regions. *Ann Appl Biol;* 125: 311–25. Bedford, I. D., Pinner, M., and Markham, P. G. (1994). *Bemisia tabaci* — potential infestation, phytotoxicity and virus transmission within European Agriculture. Brighton crop protection conference, 1994. The rise of the superbug. *Grower,* April 8, 1993. More than just another whitefly. *Horticulture Week;* August 6, 1993.

18. Porta, O., Spall, V. E., et al. (1994). Development of cowpea mosaic virus as a high yielding system for the presentation of foreign peptides. *Virology;* 220: 949–55. Usha, R., Rohll, J. B., et al. (1993). Expression of an animal virus antigenic site on the surface of a plant virus particle. *Virology;* 197: 366–74.

19. Maskell, K., Mintzer, I. M., and Callander, B. A. (1993). Basic science of climate change. *Lancet;* 342 (October 23): 1027–31. Kalkstein, L. S. (1993). Direct impacts in cities. *Lancet;* 342: 1397–99. Parry, M. L., and Rosenzweig, C. (1993). Food supply and risk of hunger. *Lancet;* 342: 1345–57. Dobson, A., Carper, R. (1993). Biodiversity. *Lancet;* 342: 1096–99.

20. Lloyd, S. A. (1993). Stratospheric ozone depletion. *Lancet;* 342 (November 6): 1156–58.

21. Amminikutty, J., and Kripke, M. L. (1993). Ozone depletion and the immune system. *Lancet;* 342: 1159–60.

22. Epstein interview. These bands are illustrated in "Notes for Biodiversity and Human Health Conference," delivered at the Smithsonian in Washington by Epstein. See also, Dobson, A., Carper, R. (1993). Biodiversity. *Lancet;* 342: 1096–99.

23. Freir, J. E. (1993). Eastern equine encephalitis. *Lancet;* 342 (Nov): 1281–82.

24. For a guide to the series, see Haines, A., Epstein, P. R., et al. (1993). Global health watch: monitoring impacts of environmental change. *Lancet*; 342: 1464–69. Editorial. Whose future? Whose World? *Lancet*; 342: 1125–26.

25. See 13 above.

26. Bergh, G., Knut, Y. B., et al. (1989). High abundance of viruses found in aquatic environments. *Nature*; 340: 467–68.

27. Sherr, E. B. (1989). And now, small is beautiful. *Nature*; 340: 429.

28. Epstein, P. R., Ford, T. E., and Colwell, R. R. (1993). Marine ecosystems. *Lancet*; 342: 1216–19.

29. Interview with Christon J. Hurst. See also, Hurst, C. J. (1980). Effects of environmental variables and soil characteristics on virus survival in soil. *Applied and Environmental Microbiol* (December): 1067–79. Hurst, C. J. (1988). Effect of environmental variables on enteric virus survival in surface freshwaters. *Wat Sci Tech*; 20: 473–76. Hurst, C. J., Benton, W., and McClellan, K. A. (1988). Thermal and water source effects upon the stability of enteroviruses in surface freshwaters. *Can J Microbiol*; 35: 474–80.

30. Mahy, B. M. J. (1993). Seal plague virus. Chapter 17 in *Emerging Viruses*, op. cit.

31. Editorial. Berlin and global warming policy. *Nature*; 374: 199–200, 203, 208.

CHAPTER EIGHTEEN

1. McNeill, W. H. (1976). *Plagues and Peoples*. Basil Blackwell, Oxford.

2. *Washington Post*, September 7, 1968.

3. Report of an International Commission (1978). Ebola haemorrhagic fever in Zaire, 1976. *Bull* WHO; 56 (2): 271–93.

4. Smith, C. E. G. (1978). Introductory remarks. *Ebola Virus Haemorrhagic Fever,* op. cit.

5. Lederberg, J., Shope, R. E., and Oaks, S. C., eds. (1992). *Emerging Infections: microbial threats to health in the United States.* National Academy Press, Washington, D.C.

6. Hughes, J. M. (1994). Emerging infectious diseases. *J Infect Dis;* 170: 263–64.

7. Berkelman, R. L. (1994). Emerging infectious diseases in the United States, 1993. *J Infect Dis;* 170: 272–77.

8. Berkelman, R. M. and Hughes, J. M. (1993). The conquest of infectious diseases: who are we kidding? *Ann Int Med;* 119: 426–28.

9. Berkelman, R. L., Bryan, R. T., et al. (1994). Infectious disease surveillance: a crumbling foundation. *Science;* 264: 368–70.

10. Haines, A., Epstein, P. R., et al. (1993). Global health watch: monitoring impacts of environmental change. *Lancet;* 342: 1464–69.

11. Gubler, D. J., and Clark, G. G. (1995). Dengue/Dengue haemorrhagic fever: the emergence of a global health problem. *Emerging Infect Dis;* 1 (2): 55–57.

12. Godlee, F. (1994). WHO in retreat: is it losing its influence? *Br Med J;* 309 (December): 1491–95.

13. Simpson, personal communication.

14. Comments made in the television documentary *The Plague Monkeys,* Reunion Films Ltd.

15. World Health Organization. Report of WHO meeting on emerging infectious diseases. CDS/BVI/94.2.

16. Emerging infectious diseases: memorandum from a WHO meeting. *Bull WHO* (1994); 72 (6): 845–50.

17. Kennedy, P. (1993). *Preparing for the Twenty-first Century.* Vintage, New York. Harrison, P. (1992). *The Third Revolution.* I. B. Taurus, London. Kaplan, R. D. (1994). The coming anarchy. *Atlantic Monthly,* February; 44–76.

18. Murphy, F. A., and Nathanson, N. (1994). The emergence of new virus diseases: an overview. *Seminars in Virology;* 5: 87–102.

19. Levins, R., Awerbuch, T., et al. (1994). The emergence of new diseases. *Amer Scientist;* 82: 52–60.

20. See 16 above.

21. Wilson, M. E. (1995). Travel and the emergence of infectious diseases. *Emerging Infect Dis;* 1 (2): 39–46.

22. Many of the references are already detailed above. But in addition I would recommend the following. Addressing emerging infectious disease threats: a prevention strategy for the United States. *MMWR* (April 15, 1994); 43 (RR-5). A large-format booklet has also been issued by the CDC in the same year and under the same title. See also, New and emerging infections: CDC and NIAID views and future plans. ASM addresses funding uncertainties. *ASM News;* 60 (5): May 1994. Report of the NSTC Committee on International Science, Engineering, and Technology (CISET) Working Group on emerging and re-emerging infectious diseases: global microbial threats in the 1990s. The latter was produced by an interagency working group convened in December 1994 under the aegis of President Clinton's National Science and Technology council. Epstein, D. B. (1995). Prevention and control of emerging infectious diseases in the Americas. *Emerging Infect Dis;* 1 (3): 103–105. And, O'Brien, T. F. (1995). WHONET: monitoring antimicrobial resistance. *Emerging Infect Dis;* 1(2): 66.

23. Gibbons, A. (1993). Where are "new" diseases born? *Science;* 261: 680–81.

24. See 19 above.

CHAPTER NINETEEN

1. Lederberg, J. (1993). Viruses and humankind: intracellular symbiosis and evolutionary competition. Chapter 1 in *Emerging Viruses,* op. cit.

2. Lederberg interview.

3. Holland, J. (1993). Replication error, quasispecies populations, and extreme evolution rates of RNA viruses. Chapter 19 in *Emerging Viruses,* op. cit.

4. Mitchison, A. (April 1993). Will we survive? *Sci Amer:* 136–44.

5. Bourhy interview.

6. Mims, C. A. (1991). Special article. The origin of major human infections and the crucial role of person-to-person spread. *Epidemiol Infect;* 106: 423–33.

7. Mims interview.

8. See 4 above. A useful summary, though largely confined to DNA viruses, is Smith, G. L. (1994). Virus strategies for evasion of the host response to infection. *Trends in Microbiology;* 2 (3): 81–88. See also the interesting Katze, M. G. (1995). Regulation of the interferon-induced PKR: can viruses cope? *Trends in Microbiology;* 3 (2): 75–78.

9. Kilbourne, E. D. (1990). New viral diseases: a real and potential problem without boundaries. *JAMA;* 262: 68–70.

10. Geisbert, T. W., Jahrling, P. B., et al. (1992). Association of Ebola-related Reston virus particles and antigen with tissue lesions of monkeys imported to the United States. *J Comp Path;* 106: 137–52.

11. Lusso, P., di Marzo, F., et al. (1990). Expanded HIV-1 cellular tropism by phenotypic mixing with murine endogenous retroviruses. *Science;* 247: 848–52.

12. Marx, J. (1990). Concerns raised about mouse models for AIDS. *Science;* 247: 809.

13. Robertson, D. L., Sharp, P. M., et al. (1995). Recombination in HIV-1. *Nature;* 374: 124–26.

14. Goldbach, R., and de Haan, P. (1994). RNA viral supergroups and the evolution of RNA viruses. Chapter 6 in *The Evolutionary Biology of Viruses,* op. cit.

15. Editorials (1995). Search for HIV vaccine faces setback. *Br Med J* (March 18); 310: 691.

16. Chaloner, W. G., and Hallam, A., ed. (1994). *Evolution and Extinction.* Cambridge Univ. Press.

17. Ibid. Maynard Smith, J. Introduction. The causes of extinction.

18. Fenner, F., and Kerr, P. J. (1994). Evolution of the poxviruses, including the coevolution of virus and host in myxomatosis. Chapter 13 in *The Evolutionary Biology of Viruses,* op. cit.

19. Webster, R. G. (1993). Influenza. Chapter 4 in *Emerging Viruses,* op. cit.

20. Palese, P. Evolution of influenza and RNA viruses. Chapter 21, ibid. Also, Murphy, B. (1993). Factors restraining emergence of new influenza viruses. Chapter 22, ibid.

21. Moser, M. R., Bender, T. R., et al. (1979). An outbreak of influenza aboard a commercial airliner. *Am J Epidemiol* (July); 110; 1–6.

22. See 19 above. Also, Webster, R. G. (1994). While awaiting the next pandemic of influenza A. *Br Med J;* 309 (November): 1179–80.

INDEX